ORIGIN OF LIFE

ENCYCLOPAEDIA OF EVOLUTIONARY BIOLOGY SERIES

ORIGIN OF LIFE

By
Richa Arora

ANMOL PUBLICATIONS PVT. LTD.
NEW DELHI - 110 002 (INDIA)

ANMOL PUBLICATIONS PVT. LTD.
4374/4B, Ansari Road, Daryaganj
New Delhi - 110 002
Phones: 23261597, 23278000
Visit us at: www.anmolpublications.com

Origin of Life

First Edition, 2004

ISBN 81-261-1523-8

PRINTED IN INDIA

Published by J.L. Kumar for Anmol Publications Pvt. Ltd., New Delhi and Printed at Mehra Offset Press, Delhi.

Contents

Preface

All living organisms share a master plan of structural and functional organization because they arose from a common ancestor long ago by the process of evolution. The tremendous work conducted by scientists in this field has increased our knowledge of evolution many folds. In the limited space, it is not possible to go into the details. Hence, a brief and comprehensive account has been produced in the light of recent work and latest information available from various sources.

The present title is an attempt to present an account of what is known or what is postulated about the life, and the subsequent evolution of the main animal groups. Certain basic scientific facts are included, it is true, but the majority of the contents consist of hypothesis which are never likely to be experimentally verified; and the reader should treat the book as a source of facts, ideas and theories on which thought and discussion can be based.

The present title claims to be a complete, thorough and authentic exposition of the topics dealt with. Emphasis has been laid on introducing basic as well as modern concepts. The language of the book is kept simple in lucid style. The subject matter has been interwoven for keeping up interest and curiosity of readers.

Though care has been taken to verify all passages quoted, but it is hardly likely that, in so large a mass of material, all errors shall have been avoided. The author and the publishers would welcome as a favour any suggestions or corrections submitted by interested readers.

Thanks are due to all friends and colleagues whose continuous inspiration have initiated me to bring out the present work.

There can be no claim to originality except in the manners of treatment and much of the information has been obtained from the books and scientific journals available in the different libraries.

—**Author**

Chapter—1
The Origin and Evolution of the Solar System

A great cloud of gas and dust contracted through interstellar space 4.6 billion years ago, far out along one of the curved arms of our spiral galaxy. The cloud collapsed and spun more rapidly, forming a disk. At some stage a body collected at the center of the disk that was so massive, dense and hot that its nuclear fuel ignited and it became a star: the *sun*. At some stage the surrounding dust particles accreted to form planets bound in orbit around the sun and satellites bound in orbit around some of the planets.

So goes—in very broad outline—the nebular hypothesis of the origin of the solar system. Its central idea was proposed more than 300 years ago. It sounds simple enough, and it makes intuitive sense to the layman; indeed, some version of it is accepted by most astronomers today. And yet beyond the broad outlines there is no consensus among students of the origin and evolution of the solar system. We still have no generally accepted theory to explain how the primitive solar *nebula* formed, how and when the sun began to shine and how and when the planets coalesced out of swirling dust.

It was *René Descartes* who first proposed (in 1644) the concept of a primitive solar nebula: a rotating disk of gas and dust out of which the planets and their satellites are made. A century later (in 1745) *Georges Louis Leclerc de Buffon* put forward a second theory: that a massive body (he suggested a *comet*) came close to the sun and ripped out of it the material that constituted the *planets* and their *satellites*. In the two centuries after *Buffon* the many theories that were propounded tended to follow in the tradition of either Descartes's monistic view or *Buffon's* dualistic one; the balance of favour swung back and forth between them. The most significant early monistic theories were those of *Immanuel Kant* and *Pierre Simon de Laplace,* who elaborated on Descartes's original idea by explaining how the cloud of gas and dust, shrinking to form the sun, would have spun faster and faster because of the conservation of angular momentum: a decrease in the radius of a rotating mass must be balanced by an increase in its rotational speed. Laplace suggested that a series of rings were shed, from whose dust the planets and satellites were formed. At the end of the 19th century dissatisfaction with the ability of the nebular hypothesis to explain the accretion of matter into the planets brought dualistic theories back into favour. Today they have been generally abandoned; it seems clear that most of the material that might have been drawn out of the sun by, say, the approach of another star would have fallen back into the sun or dispersed in space before any solid condensates could coagulate into planets.

A major reason for the wide range of early theories of the origin of the solar system was the lack of observational data—of facts to be explained by a theory. The history of the earth's first few hundreds of millions of years is missing from the geological record, which could therefore offer no clues to the environment in which this sample of a planet was born, and the limited capabilities of telescopes restricted the astronomical data. The early theories were devised to explain only a few observations: the spacing of the planetary orbits increased in a regular way (in accordance with what is known as *Bode's law*); planetary orbital motions and spins tended to have the same direction of rotation; the sun accounted for only a small fraction of the total angular momentum of the solar system, even though it accounted for the greatest fraction by far of the total mass of the system. These few facts provided few constraints on theory, and so the theories proliferated.

In just the past three decades the situation has changed dramatically. We have a vast amount of new information that imposes additional and powerful constraints on any theory. The new knowledge stems notably from new research on meteorites and from the data returned to the earth by spacecraft dispatched to other bodies in the solar system.

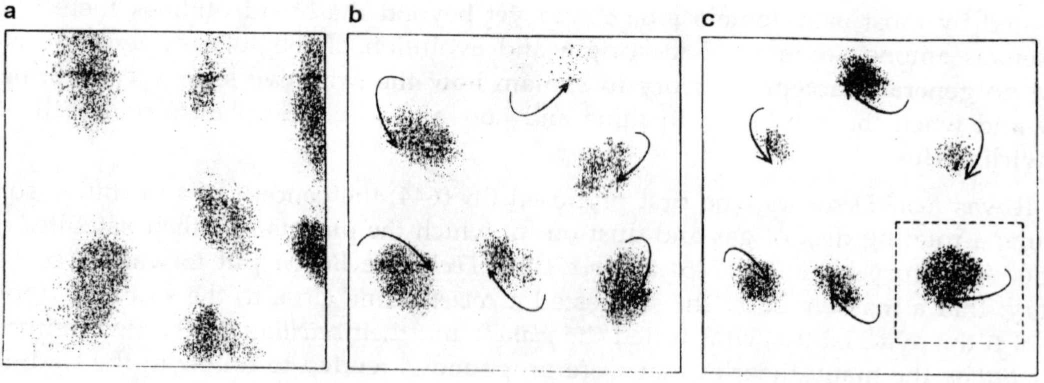

Fig. 1.1: Galaxies formed in the thin, expanding primordial gas (mostly hydrogen, with some helium) when regions of somewhat greater density (a) contracted gravitationally to form protogalaxies (b), rotating because of the net effect of gas eddies within them. The protogalaxies continued to contract gravitationally, and then to rotate faster (c), One of them (*rectangle*) was our own.

The *meteorites* are samples of primitive solar-system material. They are evidently fragments of rather small bodies that have collided and broken up, sending many of their pieces into new orbits that ultimately intersect the earth. They bring to us, trapped in their interior, samples of the gases of the solar nebula. The details of their mineralogy provide clues to the temperatures and pressures in the nebula at the time its individual grains were last exposed to chemical reaction with its gases. From the relative amounts of the products of radioactive decay that remain trapped in the interior of the meteorites we learn how long ago the original elements that gave rise to certain radioactive isotopes were assembled to form the meteorites' parent bodies.

One of the primary scientific goals of the space-probe programme was to advance understanding of the origin of the solar system, and the programme has already borne fruit. Measurements made by spacecraft have refined our knowledge of planetary masses and radii, from which we derive accurate mean densities of the planets and clues to their internal composition. By observing how the gravitational potentials of a planet differ from those of a perfectly uniform sphere we derive constraints on the degree to which the density can vary in different parts of the planet's interior. Determining whether or not a planet has an intrinsic magnetic field tells us something about the planet's internal dynamics. Spacecraft data on the composition of a planetary atmosphere reveal something about the gases that once were incorporated in the planet and about chemical interactions between the atmosphere and the planet's surface. Examining the incredibly detailed images of solid planetary surfaces that have been sent back by spacecraft cameras, we can see how volcanic and other geological processes have operated on other planets. The density of craters tells us about the terminal stages of the planet's accretion and about the numbers of smaller bodies that have wandered through the solar system.

Still other constraints come from the general advances in astrophysics that have marked the past three decades. We now know that our galaxy as a whole is between two and three times older than the solar system; we therefore have good reason to believe that the conditions we see today in the galaxy are not very different from those at the time the solar system was formed. We see regions in our galaxy in which stars have been formed in the recent past and are probably still being formed today; that gives us important information if we believe the sun and the solar nebula formed as parts of the same general process. We have learned much about the birth and death of stars and how elements originate in nuclear reactions within exploding stars and are formed into tiny grains if interstellar dust, and about how those grains concentrate in the dark patches in the sky that blot out the light coming to us from distant stars. Those grains of dust and the interstellar gases that accompany them were the raw material of the solar nebula. Let me now try to weave the many threads of information into a coherent picture of the solar system's formation.

Galaxies form when gas-mostly hydrogen—collapses out of intergalactic space. Many billions of years before the origin of the solar system our galaxy began to take shape in that way. Out of the collapsing gas a first generation of stars was born—stars that still remain spherically distributed around the center of the galaxy, a reminder of its original roughly spherical shape. After those first stars were formed the residual gas, because of its intrinsic angular momentum, settled into the thin disk that is a characteristic feature of all spiral galaxies, and further generations of stars formed from the gas in the disk. The more massive of them evolved quickly, forming heavy elements that were ejected into the interstellar gas. Some of the heavy elements condensed into tiny grains: the interstellar dust. When enough stars had formed in the central plane of the galaxy, an instability developed in their motions that allowed them to cluster together temporarily, forming the spiral arms.

Such arms represent local enhancements of star-population density in the disk; the arms are continuing features that rotate around the center of the galaxy, but the material that constitutes them keeps changing: individual stars spend only about half of their time in one arm before moving on to the next one. Like the stars, the interstellar gas and dust spend about as much time in an arm as they do flowing through the larger spaces between successive arms; the result is that the density of gas and dust is considerably enhanced in a spiral arm.

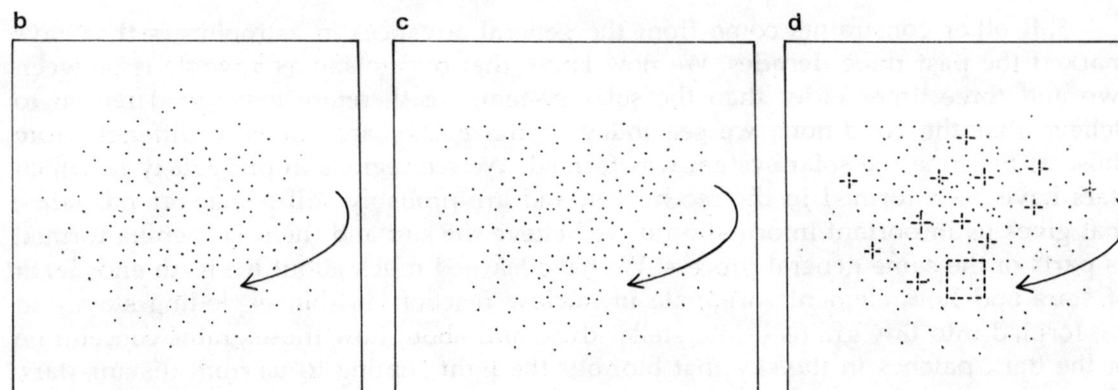

Fig. 1.2: Our galaxy evolved dense lumps of gas contracted within the protogalaxy to form a first generation of stars (a). In time the residual gas settled into a disk in the plane of the galaxy's rotation under the combined influence of gravity and centrifugal force (b). Further generations of stars (*colour*) formed within the disk (c); some stars, evolving rapidly, produced heavier elements through nuclear fusion and ejected them into the disk, where some elements condensed into solid interstellar grains. Instabilities in the motions of gas and stars led to density enhancements that we see as the spiral arms of the galaxy (d). The area in the rectangle is enlarged in the first drawing on the next page.

We know from studying galaxies other than our own that it is in these high-density spiral arms that the new stars of spiral galaxies are formed.

Pressure differences arise within the gas, perhaps as the result of a *supernova explosion;* the gas flows away from regions of higher pressure, but in moving it may tend to pile up somewhere else. Clouds of high-density un-ionized gas accumulate, which typically have a mass of from several hundred to several thousand times the mass of the sun. Gravitational forces tend to pull such a cloud into a more compact configuration. Contraction is opposed, however, by the internal pressure of the gas in the cloud, which tends to make the cloud expand; ordinarily the internal pressure is much stronger than the gravitation and the cloud is in no danger of collapsing. Sometimes, however, a sudden fluctuation in pressure—from a nearby violent event such as a supernova explosion, the formation of a massive star or a large re-arrangement of the interstellar magnetic field—may compress a cloud to a density much higher than normal. Under such conditions, which are quite rare, gravity may win out over internal pressure, so that the cloud begins to collapse to form stars. As the cloud collapses, its interstellar grains shield its interior against the heating effect of radiation from the stars outside. The temperature of the cloud falls, and the internal pressure becomes less effective. The collapsing cloud breaks into fragments and the fragments break into smaller fragments. When one small fragment eventually completes its collapse, it will have formed into a flattened disk, cool at the edges and very hot at the center: a primitive solar nebula.

What was the nature of the solar nebula and how did it evolve? When did the sun form? Why are there planets? How did they take shape? Quite different pictures of the structure of the primitive solar nebula and of its evolution result from different estimates of its size. Such estimates have usually been arrived at by reasoning backward in time from the present masses of the planets. Let me reproduce such an argument.

In a very general way one can divide the materials of the planets into three classes depending on their volatility: *rocky, icy and gaseous.* The major constituents of rocky materials are iron and oxides and silicates of magnesium and other metals, notably aluminum and calcium. All these materials would be in solid form at pressures characteristic of the primitive solar nebula and at temperatures in the range from 1,000 to 1,800 degrees Kelvin (degrees Celsius above absolute zero). The four inner planets and the earth's moon (and at least two of the major satellites of Jupiter) appear to be basically rocky. The rocky solids represent about .44 per cent by mass of the material out of which the sun formed. The present mass of a rocky planet, then, represents about .44 per cent of its share of the primitive solar nebula; the remainder of that share is "missing" because it was too volatile to have been incorporated in the planet.

At a temperature below 160° K. the water in the nebula would be in the form of ice. Ammonia and methane form solids only at a somewhat lower temperature. The ices constitute 1.4 per cent by mass of the material out of which the sun formed. Rock-ice mixtures account for most of the mass of *Uranus* and Neptune and some of the mass of *Saturn* and *Jupiter* (and for the bulk of the mass of most of the satellites of the outer planets and the comets). Arguing as for the inner planets, one can assume that the rock and ice now present in such bodies represent 1.4 plus .44 per cent, or 1.84 per cent, of those bodies' original share of the nebula.

At any temperature likely to have been attained in the primitive solar nebula the very volatile elements-hydrogen and the noble gases such as helium and neon—would

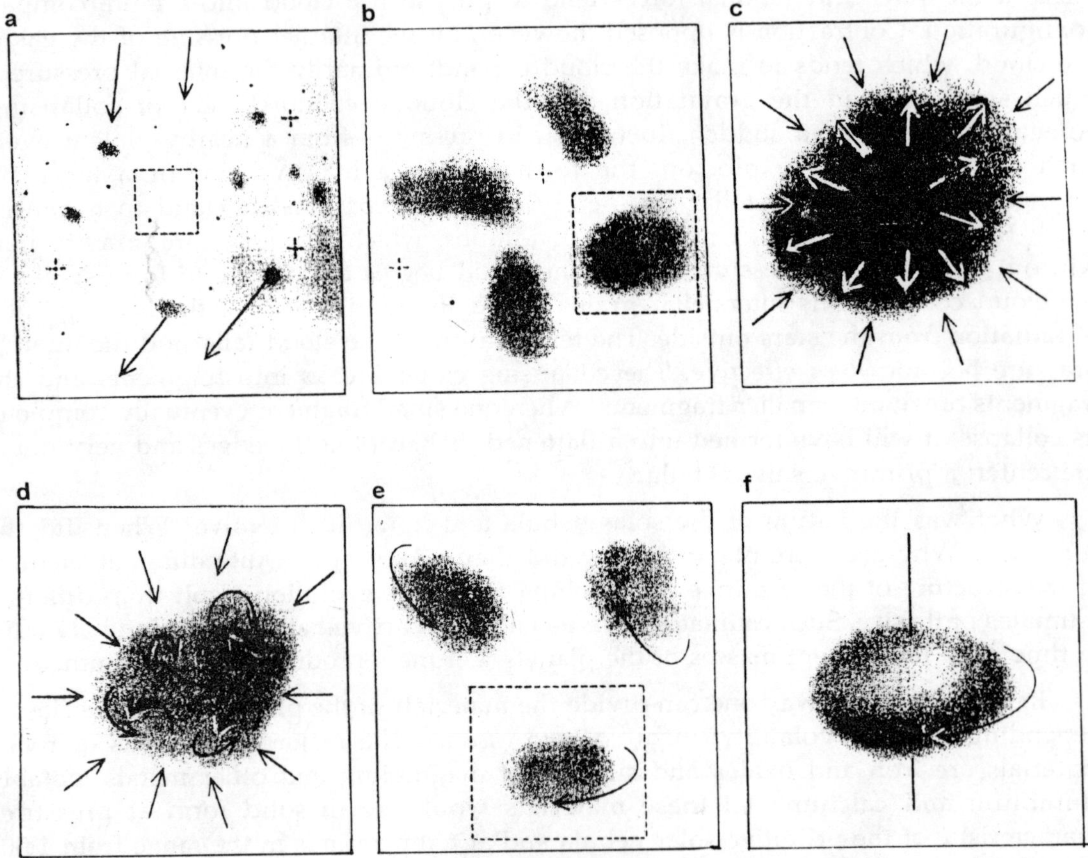

Fig. 1.3: Solar system evolved in a spiral arm about two-thirds of the way out from the center of the galaxy. Stars, gas and dust grains move through the arm, and new stars are born there; massive, short-lived stars outline the arms (a), A supernova explosion or the birth of massive stars creates instabilities that concentrate high-density clouds of gas (b), Gravitational forces contract the cloud, but the cloud's internal pressure opposes contraction (c), If the cloud has enough mass, gravity dominates (d), and the cloud collapses. The collapse generates strong gas eddies (*curved arrows*) and breaks the cloud into fragments (e), each fragment has a net rotation derived from its major eddies. One of these fragments spins faster and its gas settles into a disk that was the primitive solar nebula (f).

remain in the gaseous state. Such gases are incorporated in bodies within the solar system only to the extent that they have been held in planetary atmospheres by gravity and, in the case of hydrogen, held in chemical compounds such as water. (A tiny amount of helium comes from the decay of radioactive elements.) Morris Podolak recently analyzed the structure of the outer planets. He determined that hydrogen and helium constitute about 15 per cent of Uranus' mass, about 25 per cent of Neptune's, about two-thirds of Saturn's and about four-fifths of Jupiter's. In these planets it is necessary to allow for the gaseous components in order to establish the present rock-ice mass.

Thus establishing the rock and the rock-ice masses of the planets and augmenting those masses for the missing constituents that were too volatile to condense it is possible to estimate a minimum mass for the primitive solar nebula: a mass sufficient to account for the formation of the planets. That minimum mass is about 3 per cent of the mass of the sun (Older estimates arrived at a much smaller mass—less than 1 per cent of the sun's—because they did not allow for enough rock and ice in Jupiter and Saturn.)

The 3 per cent figure is definitely a minimum. It assumes that the planets were completely efficient in collecting from the solar nebula all the material that was in condensed form in each planet's orbit in the nebula. For two kinds of solid, however, that collection process might have been quite inefficient. Consider first the tiny unconsolidated grains of interstellar dust, perhaps a micrometer (a thousandth of a millimeter) in diameter, that were not vapourized as the gas-cloud fragment collapsed. The thickness of the nebular disk must have been at least one astronomical unit (the mean distance between the earth and the sun). That dimension is very large compared with the dimensions of any of the planets, which consolidated approximately in the central plane of the disk. Gas-drag effects would prevent large quantities of these small grains from settling through the nebular gas toward the central plane at a significant rate; if much of the gas was instead dissipated inward to form the sun, the grains would have accompanied the gas and could never have become incorporated in the planets.

Larger bodies (centimetres or meters in diameter), on the other hand, would fall rapidly through the gas toward the midplane but might nevertheless not end up in planets. As a result of a difference between the centrifugal forces that act on the solid bodies and those that act on the gas, the solids would rotate around the central spin axis of the nebula more rapidly than the accompanying gas. They would therefore move through the gas with a relative velocity as high as several hundred miles an hour; a head wind of that speed would tend to slow them down so that they would spiral rather quickly through the gas toward the central spin axis and thus be lost to the region of planet formation. For these two reasons the mass of the primitive solar nebula may have been considerably larger than 3 per cent of the sun's mass.

Many of the solar-system theories constructed over the past three decades have involved some version of a *minimum-mass solar nebula*. The concept has a major flaw, however. It assumes that the sun itself was formed directly during the process of collapse and that the primitive solar nebula was marshalled independently around the sun. The trouble is that simple estimates of the amount of angular momentum that must have been contained in the collapsing cloud fragment indicate that it would have been impossible for almost all of the fragment simply to collapse directly to form the sun, leaving a small fringe of nebula to constitute the planets. Such estimates require instead that the nebula's mass be spread out over several tens of astronomical units. The solar nebula itself must have contained substantially more than one solar mass—and probably about two solar massess—of material, with no sun originally present at the central spin axis. Let me first explain the source of the large amount of angular momentum and then show why it indicates that there was not a minimum solar nebula but a massive one.

The strong fluctuations in pressure that led to the rapid compression of the original interstellar cloud and thus brought it to the threshold of gravitational collapse must have stirred the cloud's gases into violent turbulence. Large-scale shearing motions developed—eddies superimposed on eddies. In a wider range of sizes and in many planes and directions. When any one fragment became isolated from such a turbulent cloud, it had a net tendency to spin, derived from the motions of the largest eddies it happened to contain. A fragment's mass, its rate of rotation and its radius combine to endow it with a certain amount of angular momentum, and that momentum must be conserved; as the fragment contracted, it spun faster. The sun turns very slowly, however; in spite of its great mass it accounts for only 2 per cent of the solar system's angular momentum. Most of the original angular momentum of the vast quantities of gas that moved in to form the sun must have been transported outward; a considerable part of the original nebula must therefore have remained at great distances from the sun to take up that angular momentum.

An additional reason for postulating a massive solar nebula is the observation that young stars tend to lose mass at a prodigious rate early in their lifetime; the loss comes as they pass through what is called their *T Tauri stage*, which will be discussed in more detail below:

The combination of the mass that remained in the solar nebula and never became part of the sun and the mass that was once in the sun but was lost in the early sun's T Tauri stage could easily have amounted to as much as one solar mass.

As a result of this kind of reasoning—in effect arguing forward from what is known of the principles of star formation rather than backward from the masses of the present planets—*Milton R. Pine* constructed some numerical models of the massive solar nebula. The models extended out to a radial distance of about 100 astronomical units and contained two solar masses of material. In a typical model the temperature was about 3,000 degrees K. near the spin axis and decreased to a few hundred degrees in the region of planet formation. Such temperatures are considerably higher than the temperatures that characterized the collapse of the original interstellar cloud; they develop in the later stages of compression of the gas, once its density becomes high enough so that its own cooling radiation can no longer escape easily. The escape of this radiation is impeded, however, only during the rapid final stages of the collapse; once the gas stops contracting—once the primitive solar nebula is formed—the radiation can escape relatively quickly, so that in the region of planet formation the nebula will lose most of its heat energy in only a few hundred or a few thousand years.

Planet	*Present Masss (Per cent of Sun's)*	*Augmented Mass (Per cent of Sun's)*
Mercury	.000017	.004
Venus	.000245	.056
Earth	.000304	.07
Mars	.000032	.007
Jupiter	.09547	1.5
Saturn	.02859	.77
Uranus	.00436	.27
Neptune	.00524	.27
Pluto	.00025(?)	.06(?)
Total (Minimum Mass of Solar Nebula)		3.0

Minimum mass of solar nebula is estimated by a adding up the amount of solar material that must have been present (*column at right*) to account for the present mass (*middle column*) of each planet. Solar-nebula mass thus estimated is 3 per cent of mass of sun.

Such a short cooling time (short compared with the time required to form a sun and planets) presents a difficulty for the *massive-nebula model*. As the nebula cools it will flatten into a thinner disk, and thin disks have been shown to be dynamically unstable: they tend to deform into a barlike configuration. (Such a deformation might well be the mechanism by which close pairs of double stars are formed, but that evidently did not happen in the solar system.)

There is another time-scale problem for the *massive-nebula model*. An important process for transporting angular momentum away from the central spin axis so that gas can shrink toward that axis is probably a system of fast meridional currents: gas currents that flow in a plane parallel to the spin axis and at a right angle to the central plane of the nebula. *Pine* estimated that the characteristic time for the outward transport of the angular momentum shed by the inner parts of the primitive solar nebula would be only a few thousand years. *John Stewart* of the Max Planck Institute for Physics and Astrophysics in Munich has shown that gas turbulence must play an important role in a primitive solar nebula and may cause an even more rapid outward transport of angular momentum.

Both the time for cooling and the time for angular momentum transport seem too short compared with the time required for the accretion of the solar nebula. After a fragment separates from the interstellar cloud its central region is likely to be denser, and will collapse more rapidly, than the remainder of the fragment. A small solar nebula will therefore be formed at first when the central region ceases to collapse; that small nebula will grow by accretion of the remainder of the infalling fragment over a period of time—probably between 10,000 and 100,000 years, a lot longer than the cooling and angular—momentum transport times we estimated.

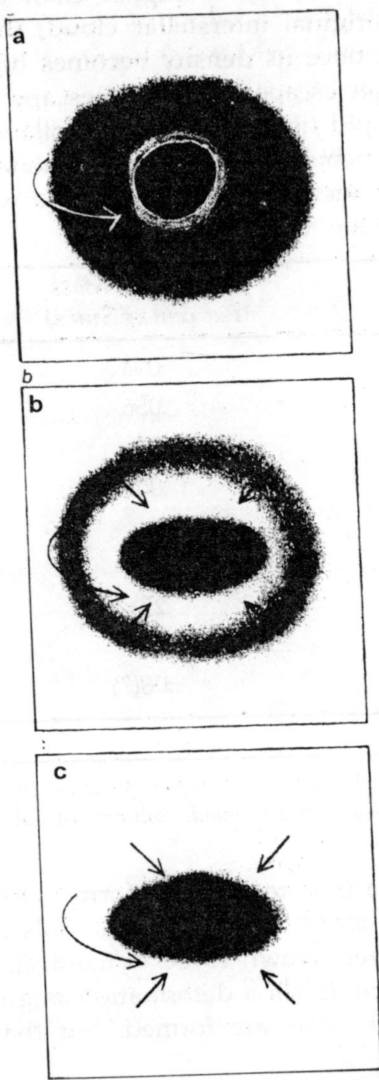

Fig. 1.4: Accretion model of the primitive solar nebula assumes that a central region of the cloud fragment collapses faster than the rest (a). It forms a small solar nebula: a central mass that is not yet the sun, surrounded by a disk of gas and dust grains, with more gas and dust concentrated around the periphery (b). The small nebula then grows by accretion over a long period of time (c).

It is necessary to construct not a single model but an evolutionary sequence of models, beginning with a small solar nebula that grows through accretion over a time interval of perhaps 30,000 years. In this case the time for the redistribution of angular momentum remains short compared with the accretion time, so that much of the mass flows inward to form the sun not at the beginning of the accretion period but throughout that period; the mass of the nebula out in the region of planet formation remains a relatively small fraction of a solar mass throughout the period. As for cooling, the accreting gas is suddenly decelerated when it hits the surface of the solar nebula; the energy of its infall is converted into heat that is radiated from the surface. In the later stages of accretion that process keeps the surface layers in the region of planet formation at a temperature of perhaps a few hundred degrees; the temperature in the interior would be somewhat higher. And meanwhile the steady flow of mass toward the central spin axis diminishes dynamic instabilities within the nebula.

The two pictures of the primitive solar nebula, one derived from the masses of the planets and the other from the principles of star formation, thus seem to be converging to form an intermediate model of the initial solar nebula. In that model somewhat more than one solar mass has collected toward the spin axis but is not yet recognizable as the sun. It is surrounded by a disk of gas and dust amounting to perhaps a tenth of a solar mass. Farther out, beyond the region of planet formation, considerable additional amounts of mass are still falling toward the solar nebula.

The planets were created by the accumulation of interstellar grains and, in the case of the outer planets, the subsequent attraction and adherence of gases. The buildup of solid matter would have begun, in the collapsing gas cloud. Turbulent gas eddies would have accelerated the interstellar grains until they had large enough relative motions to begin to collide with one another. Having been formed out of material in stars and then ejected into interstellar space, where ices and other volatile constituents condensed on their surface, the grains probably had a rather fluffy structure. It would not be surprising if such particles stuck to one another when they collided, forming clumps. As time passed the clumps of grains would collide with one another, sometimes amalgamating into larger clumps and sometimes breaking up into smaller ones. By the time the solar nebula had formed, many clumps were likely to have grown to a diameter measured in millimetres or centimeters.

Clumps of that size could settle through the gas toward the midplane of the nebula in tens or hundreds of years. Since their settling rate would vary with size there would be further collisions, increasing the size of the clumps and accelerating their fall toward the midplane. At that point, however, unless they were somehow able to grow substantially larger they would rapidly be lost to the inner solar nebula as a result of the *gas-drag effect*.

A critical process in planet formation may therefore be a mechanism recently proposed by *Peter Goldreich* of the California Institute of Technology and *William R. Ward* of Harvard University, which would give rise to those larger bodies. They showed that if there is a thin layer of condensed solids at the midplane of the nebula, with very little relative velocity among the particles, then a powerful gravitational instability mechanism will break up the thin sheet into bodies with diameters in the range of the diameters of asteroids: kilometres or tens of kilometres. The instability mechanism gradually operates over larger distances, attracting the asteroid-size bodies into loosely bound gravitating clusters of hundreds or thousands of bodies. The clusters remain unconsolidated because of the large angular momentum contained in their component bodies, which makes them rotate around common gravitating centres. When two clusters approach each other, however, they intermingle; the fluctuating gravitational field in the combined cluster leads to a violent dynamic relaxation of the motions of the bodies, so that many of them coalesce to form cores around which others go into orbit (although some of the bodies would be lost). The clusters interact with one another gravitationally over quite large distances; mutual perturbations gradually build up the velocities of the clusters with respect to one another, leading to further collisions that produce ever larger bodies.

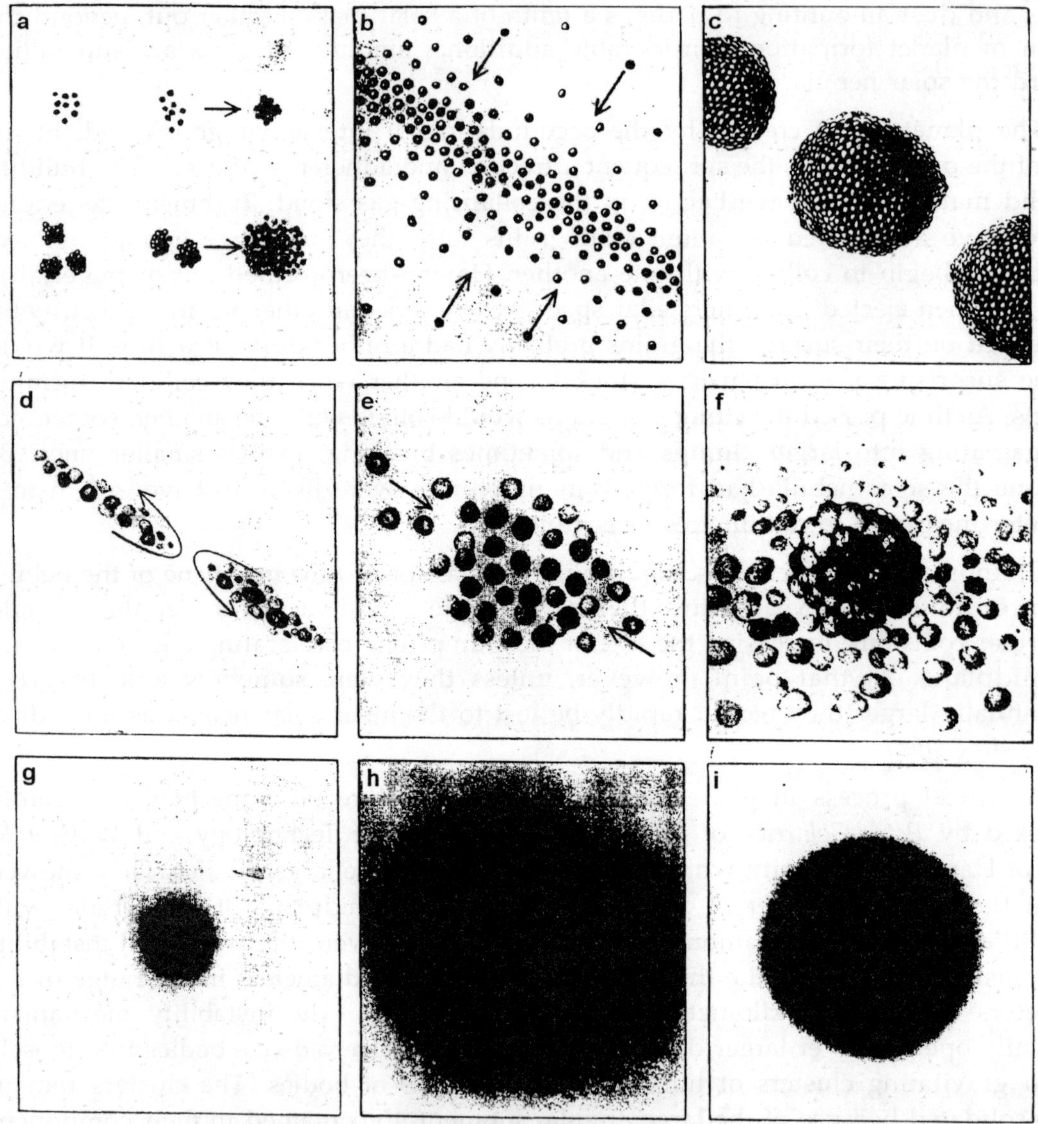

Fig. 1.5: Planets begin to form when interstellar dust grains collide and stick to one another, forming ever larger clumps (a). The clumps fall toward the midplane of the nebula (*b*) and form a diffuse disk there. Gravitational instabilities collect this material into millions of bodies of asteroid size (*c*), which collect into gravitating clusters (*d*). When clusters collide and intermiingle (*e*), their gravitational fields relax, and they coagulate into solid cores, perhaps with some bodies going into orbit around the cores (*f*). Continued accretion and consolidation may create a planet-size body (*g*). If the core gets larger, it may concentrate gas from the nebula gravitationally (*h*). A large enough core may make the gas collapse into a dense shell that constitutes most of the planet's mass (*i*).

The *Goldreich-Ward* instability mechanism would appear to be a powerful first step in the accumulation of planetary bodies. The subsequent stages in the process are still highly speculative and were surely different in different regions of the solar nebula. The interstellar grains whose clumping initiates the accumulation process are those that have not been vapourized by the heat of the nebula; their materials, and therefore the materials of the larger bodies into which they are incorporated, would be different at different distances along the steep temperature gradient: metals, oxides and silicates in the region of the inner planets; similar rocky compounds and water ice farther out; rock, water ice and frozen methane and ammonia still farther out.

In the case of the smaller inner planets the progression to full size may be just a question of successive collisions and amalgamations of rocky bodies. In the case of the outer planets there are other considerations. *Fausto Perri* has recently considered the behaviour of the primitive solar nebula as a large planetary core grows within it. As the mass of the core increases, gas in the solar nebula becomes gravitationally concentrated toward the core; with the continued growth of the core the amount of mass in the gas that is concentrated increases even more rapidly than the mass of the core itself. At some point the core reaches a critical size (which depends on the temperature conditions in the surrounding gas) such that the gas becomes hydrodynamically unstable and collapses onto the planetary core.

The major constituents of *Jupiter* and *Saturn* are the hydrogen and helium in their atmosphere, and we believe it was through this process of concentration and collapse that these planets acquired most of their mass. Hydrogen and helium account for a smaller fraction of the mass of *Uranus* and *Neptune,* probably indicating that their core never grew to the critical size for hydrodynamic collapse; those two planets did, however, grow large enough to retain much of the hydrogen and helium that was gravitationally concentrated toward their core. The inner planets, on the other hand, may be too small ever to have concentrated much of the nebular gas.

When the collapse events took place to form *Jupiter* and *Saturn,* local conservation of angular momentum in the gas would cause it to flatten into a disk around the planetary core. As time went on the two planets would sweep up essentially all the gas in their vicinity within the nebula. One can think of them as forming miniature versions of the primitive solar nebula: a central core of condensed rock and ice taking the place of the sun, with the gaseous disk around the core as the analogue of the solar nebula. Both of these large planets have systems of regular satellites incorporating considerable mass, which probably formed from a gaseous disk by processes quite analogous to the formation of the planets in the solar nebula.

Any theory of the origin and evolution of the solar system must account somehow for the *comets,* its most spectacular but least understood members. *Jan Oort* of the Leiden Observatory suggested some years ago that the comets inhabit an enormous volume of space centred on the sun, starting well beyond the outer planets and extending to a distance of perhaps 100,000 astronomical units. The total mass of the comets in this vast "Oort cloud" is probably equivalent to between one earth mass and 1,000 earth masses, which would account for between 10^{12} and 10^{15} comets. The comets we see are those few whose orbital elements are perturbed by a passing star in just the right way to send them plunging toward the center of the solar system.

A comet is a "*dirty snowball,*" an aggregate of ice and rocky material, in the model first suggested by *Fred L. Whipple* of Harvard University. As a comet approaches the sun, gases are vapourized from it, accompanied by dust particles, to form the characteristic coma and tail. Analysis of the tails shows that the molecules that are vapourized are primarily water but also include exotic organic compounds. The dust, some of which comes to rest high in the earth's atmosphere, consists of fluffy clumps of fine-grained rocky material. A comet, in other words, is apparently an assembly of interstellar grains.

The *comets* must either have been made within the solar nebula and somehow ejected into the *Oort cloud* or else have been made out in the *Oort cloud* itself. *Oort* originally suggested that they were formed near *Jupiter* and perturbed by Jupiter's gravitational field into very large orbits that were subsequently rounded out by stellar perturbations. That would require the formation of a staggeringly large mass of comets, since many more would have been ejected from the solar system than were retained in the *Oort cloud.* Moreover, given the temperatures that must have prevailed near *Jupiter* it seems unlikely that molecules more complex than water would have been in solid form. More complex ices are possible farther out in the nebula, and so *Whipple* and others have suggested that the comets were formed in the neighbourhood of *Uranus* and *Neptune* and sent out into the *Oort cloud* by the gravitational fields of those planets. That proposal, however, meets only one of the objections to the *Oort hypothesis.*

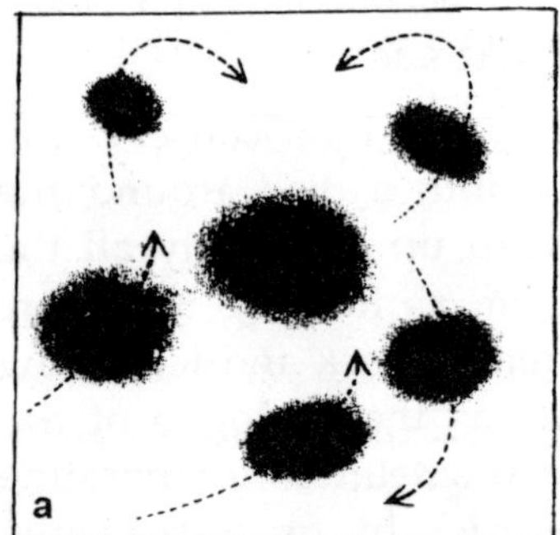

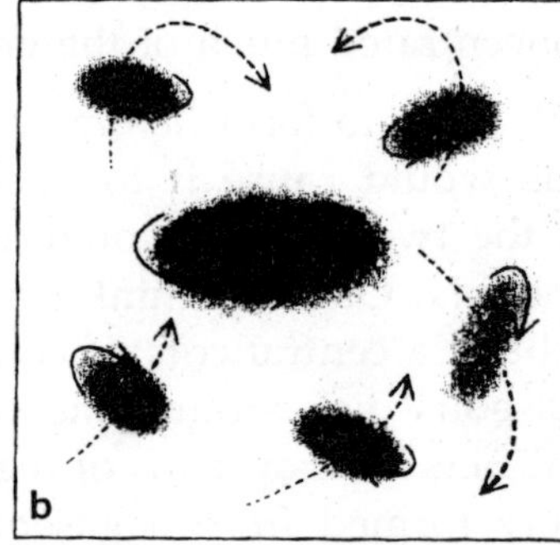

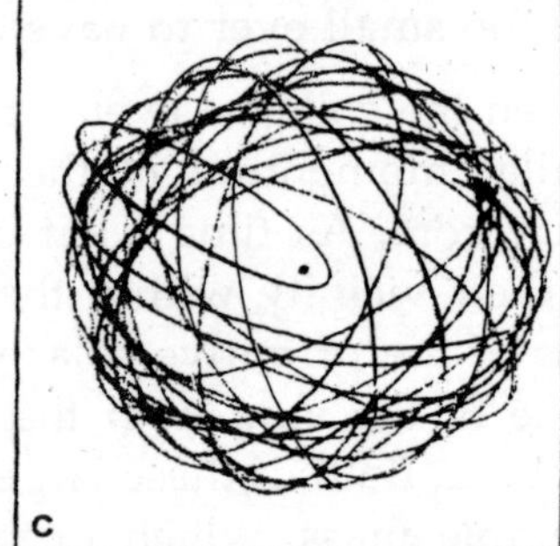

Fig. 1.6: Comets may have formed from small cloud fragments that once were in orbit around the larger fragment that became the solar nebula (a). The small fragments spun down, like the solar one, to form disks in which comets were accumulated much as planets were (b). Eventually starlight could have evaporated the gases of these "cometary nebulas," leaving the comets in enormous orbits around the sun (c). From time to time a comet's orbit is perturbed by a passing star, and the new orbit brings it close to the sun.

The comets were probably formed out in the *Oort cloud* itself. It is true that the collapsing gas of the fragment of cloud that became the primitive solar nebula was never dense enough so far from the center for the interstellar grains to have aggregated into sizable bodies out there. There is another possibility, however. Most of the stars in the galaxy are much less massive than the sun, suggesting that gas-cloud fragmentation sometimes continues at least down to fragments a tenth of a solar mass in size. Fragmentation may have gone further still. Small fragments could have been bound gravitationally to the primitive solar nebula in orbits traversing the region of the *Oort cloud*. Such fragments would form fairly large and very cool disks, ideal places for the comets to form. When the disks were ultimately heated by ultraviolet radiation from external stars, the gases would evaporate away and leave the comets in solar orbits.

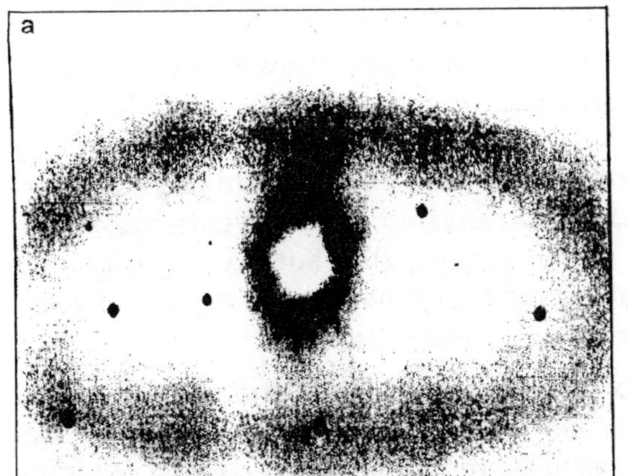

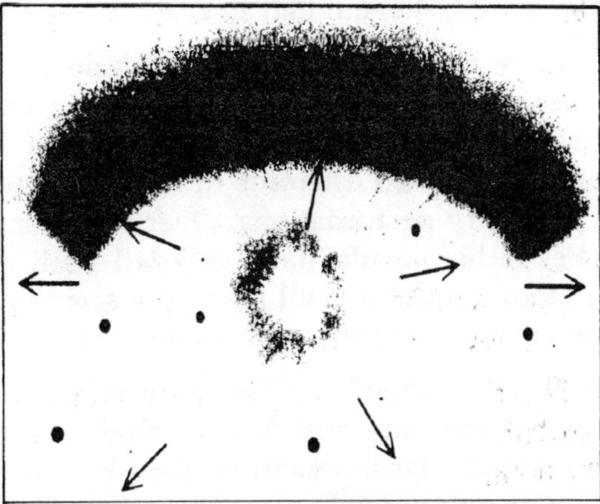

Fig. 1.7: Solar sytem was cleaned up by the "T Tauri wind." When gas contracting toward the center of the nebula reached a sufficient density, the nuclei of bydrogen atoms began to fuse and the sun began to shine (a). During its T Tauri phase the sun lost vast quantities of material. That material constituted an intense solar wind that could have blown away the remaining gas (b).

After the planets had formed, much of the gas of the solar nebula must have remained in orbit around the sun, along with countless small bodies and large amounts of unconsolidated dust. There are only planets and asteroids in orbit now, with very little dust and almost no gas. How was the solar system cleaned up? As mentioned above Young stars characteristically pass through what is called the *T Tauri stage,* when they eject matter at a prodigious rate: as much as one solar mass per million years! There is every reason to believe the sun passed through a similar phase, and the fierce "wind" of that ejected mass undoubtedly dissipated the solar nebula by carrying the residual gas off into space. That early solar wind would have stripped the inner planets of the remains of any primitive atmosphere of hydrogen and helium from the primitive solar nebula; the outer planets must have formed early enough to capture their hydrogen and helium before the solar wind began to blow. Moreover, if the accretion of gas from the original cloud fragment was still continuing, it would have been terminated by the wind, with the infalling gases being ejected back into interstellar space.

What determined when the *T Tauri wind* began? Why did it not blow away the primitive solar nebula long before the sun got so large? The thermonuclear reactions of hydrogen that constitute a stellar furnace are ignited only under extreme conditions of high temperature and high density. *Perri* has recently determined that the temperatures of the primitive solar nebula were so low that compressional heating of the gas could not have ignited the sun at a central density comparable to that of the sun today; the density would have had to be at least 100 times higher than it is now. Only then could the sun have adjusted itself into its present configuration, and only then could the thermonuclear furnace have been ignited and could the intense *T Tauri wind* begin to blow. An enormous amount of mass must have had to be gathered together to achieve such a density. Given the original temperature of the nebula, in other words, the sun had to reach a large mass before it could be a sun.

Once the sun had begun to shine and the *T. Tauri wind* had blown away the gas, the stage was set for the final clean-up of interplanetary space and the completion of planet formation. The orbits of most of the small bodies in the solar system (other than the observed asteroids in their isolated belt) would have been continually modified by planetary perturbations. Over the course of a few hundred million years most such bodies either would have collided with one of the planets (the surfaces of Mercury, the moon and Mars still show the scars of that terminal bombardment) or would have been ejected from the solar system by a major planet, usually Jupiter.

The tiny dust particles were subjected to forces even stronger than gravitational perturbations: the effects of sunlight. The *photons* the particles absorb from the sun carry no angular momentum; the photons the particles radiate, however, carry off some of the angular momentum of the particles' orbital motion. The sunlight therefore acts as a resisting medium for the particles, making them spiral in toward the sun. Larger solids, up to a kilometre in diameter, are perturbed by sunlight in a different way. As such a body rotates the temperature of a section of its surface increases as long as it is on the sunlit side but decreases while it is on the dark side. One hemisphere of the body therefore emits considerably more radiation than the other. That gives rise to a preferential thrust that can perturb the orbit of the body either toward the sun or away from it, depending on the body's direction of rotation. Such bodies will eventually come close to one of the planets, whereupon they will be absorbed by collision or be ejected from the solar system. In these several ways the sunlight could have acted as a broom to sweep away much of the smaller debris left over from the formation of the solar system.

Chapter—2
The Origin of the Earth

It is probable that as soon as man acquired a large brain and the mind that goes with it he began to speculate on how far the earth extended, on what held it up, on the nature of the sun and moon and stars, and on the origin of all these things. He embodied his speculations in religious writings, of which the first chapter of *Genesis* is a poetic and beautiful example. For centuries these writings have been part of our culture, so that many of us do not realize that some of the ancient people had very definite ideas about the earth and the solar system which are quite acceptable today.

Aristarchus of the Aegean island of Samos first suggested that the earth and the other planets moved about the sun—an idea that was rejected by astronomers until *Copernicus* proposed it again 2,000 years later. The Greeks knew the shape and the approximate size of the earth, and the cause of eclipses of the sun. After *Copernicus* the Danish astronomer *Tycho Brahe* watched the motions of the planet *Mars* from his observatory on the Baltic island of Hveen, as a result *Johannes Kepler* was able to show that *Mars* and the earth and the other planets move in ellipses about the sun. Then the great *Isaac Newton* proposed his universal law of gravitation and laws of motion, and from these it was possible to derive an exact description of the entire solar system. This occupied the minds of some of the greatest scientists and mathematicians in the centuries that followed.

Unfortunately it is a far more difficult problem to describe the origin of the solar system than the motion of its parts. The materials that we find in the earth and the sun must originally have been in a rather different condition. An understanding of the process by which these materials were assembled requires the knowledge of many new concepts of science such as the molecular theory of gases, thermodynamics, radioactivity and quantum theory. It is not surprising that little progress was made along these lines until the 20th century.

The Earlier Theories

It is widely assumed by well-informed people that moon came out of the earth, presumably from what is now the Pacific Ocean. This was proposed about 60 years ago by *Sir George Darwin*. The notion was concluded in detail by *F.R. Moulton*, who concluded that it was not possible. In 1917 it was again considered by *Harold Jeffreys*, who thought that his analysis indicated the possibility that the moon had been removed from a completely molten earth by tides. In 1931, however, *Jeffreys* reviewed the subject and concluded that this could not have happened, since then most astronomers have agreed with him.

But although *Moulton* and *Jeffreys* showed the improbability of the origin of the moon from the earth, they proposed theories for the origin of the solar system involving the removal of the earth and the other planets from the sun. Together with *James Jeans* and *T.C. Chamberlin* they proposed that another star passed near or collided with the sun, and that the loose material resulting from this cosmic encounter later coagulated into planets. This idea of the origin of the solar system has been widely held right up to the present.

The evidence gathered by our great telescopes now tells us that most of the stars in the heavens are pairs or triplets or quadruplets. We have determined the masses of multiple stars by means of Newton's laws of motion and his universal law of gravitation, we have also studied the velocities of these stars by significant changes in their spectra and by actually measuring the motions of nearby examples. We find that the two stars of a pair seldom have exactly the same mass and that the ratio of the mass of one star to that of the other varies considerably. *Gerard P.Kuiper* of the University of Chicago concludes that the number of pairs of stars entirely independent of the ratios of their masses; that is, there is very little probability that one ratio of masses would occur more often than another. In fact, it would appear that there is about as much chance of finding a pair of stars in which one has one-thousandth the mass of the other is 999 thousandths as massive as the other.

Of course it would be very difficult to see a double star in which the secondary was only a thousandth as large as the primary, particularly if the second emitted no light. The Sun and Jupiter, the largest of the planets, might be viewed as such a double star: Jupiter weighs about a thousandth as much as the sun, and it shines only by reflected sunlight. Even from the nearest star Jupiter would be invisible. There is much evidence, however, that a double star such as the sun and Jupiter should occur as a regular event in our galaxy, and the same considerations would seem to indicate that there may be as many as a hundred million solar systems within it. Solar systems are almost certainly commonplace, and not the special things that one might expect from the collision of two stars.

The Dust Cloud Hypothesis

Many years ago *E.E. Barnard* of the Yerkes Observatory observed certain black spots in front of the great diffuse nebulae that occur throughout our galaxy. Bart *J. Bok* of Harvard University has investigated these opaque globules of dust and gas, they have about the mass of the sun and about the dimensions of the space between the sun and the nearest star. *Lyman Spitzer, Jr.*, of Princeton University has shown that

Fig. 2.1: Cloud of Dust from which the solar system evolved may have developed this intricate pattern of turbulence, suggested by the German physicist C.F. von Weizsäcker. The dust in each eddy gradually coagulated.

if large masses of dust and gas exist in space, they should be pushed together by the light of neighbouring stars. Eventually, when the dust particles are sufficiently compressed, gravity should collapse the whole mass, and the pressure and temperature in its interior should be enough to start the thermonuclear reaction of a star.

It would seem reasonable to believe that if a star such as the sun resulted from a process of this kind, there might be enough material left over to make a solar system. And if the process was more complex we might even end up with two stars instead of one. Or again we might have triple stars or quadruple stars. Theories along this line are more plausible to us today than the hypothesis that the planets were in some way removed from the sun after its formation had been completed. In my opinion the older hypotheses were unsatisfactory because they attempted to account for the origin of the planets without accounting for the origin of the sun. When we try to specify how the sun was formed, we immediately find ways in which the, material that now comprises the planets may have remained outside of it.

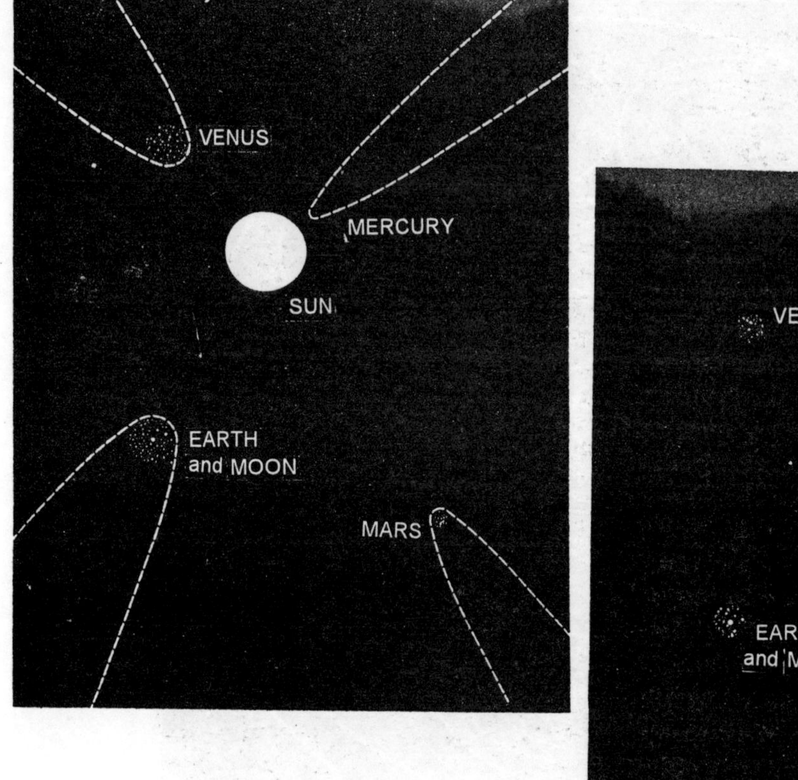

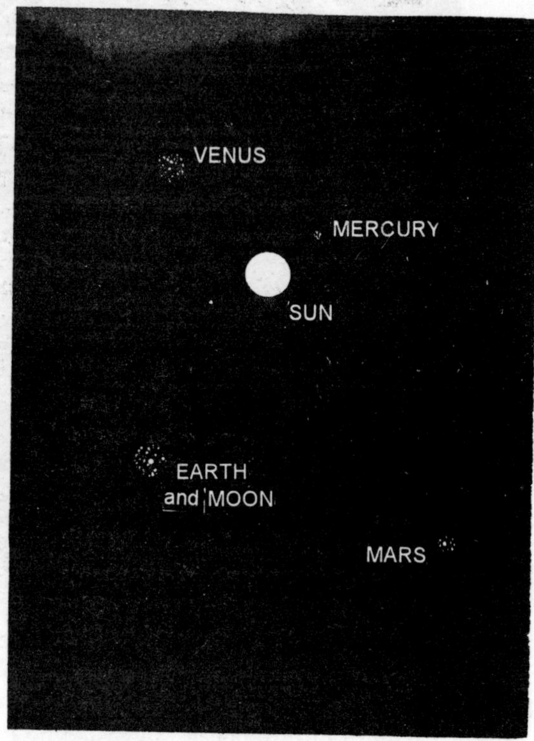

Fig. 2.2: Evolution of the Earth and the planets Mercury, Venus and Mars is depicted in this series of schematic drawings. In the first drawing the primordial dust cloud has coagulated into protoplanets composed of planetesimals. The gases that have coagulated with the planetesimals are driven away (*dotted lines*) by the pressure of light from the sun. In the second drawing the gas has been completely removed from the proto-Planets.

One piece of evidence that must be included in any theory about the origin of the solar system consists in our observation of the angular momentum that resides in the spinning sun and the planets that travel around it. The angular momentum of a planet is equal to its mass times its velocity times its distance from the sun. Jupiter possesses the largest fraction of the angular momentum in the solar system, only about two per cent resides in the sun. Another fact that must be encompassed by any theory is the socalled Titus-Bide law, which points out in a simple mathematical way how the distances of the planets from the sun vary: the inner planets are closer together and the outer ones are farther apart. This is only an approximate law which does not hold very well, and perhaps more emphasis has been put upon it than it deserves.

Some 15 years ago *Henry Norris* Russell of Princeton and *Donald H. Menzel* of Harvard pointed out that there was a very curious relationship between the proportions of the elements in the atmosphere of the earth and the atmospheres of the stars, including the sun. It is particularly noteworthy that neon, the gas that we use in electric signs, is very rare in the atmosphere of the earth but is comparatively abundant in the stars. Russell and Menzel concluded that neon, which forms no chemical compounds, escaped from the earth during a hot early period in its history, together with all of the water and other volatile materials that constituted its atmosphere at that time. The present atmosphere and oceans, they proposed, have been produced by the escape of nitrogen, carbon and water from the interior of the earth. The German physicist *C.F. von Weizsäcker* similarly suggested that the argon of the air has resulted mostly from the decay of radioactive potassium during geologic time, and has escaped from the interior of the earth *F.W. Aston* of Cambridge University also pointed out that the other inert gases, krypton and xenon, were virtually missing from the earth.

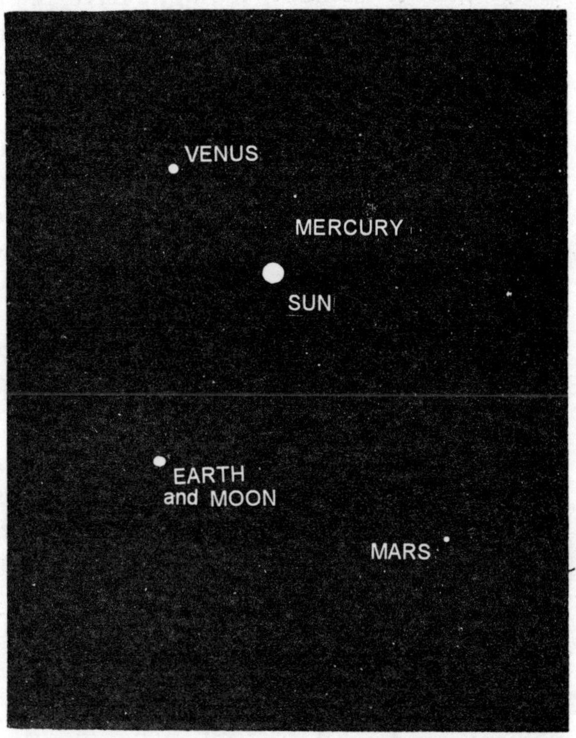

In the third drawing the planetesimals have formed the planets. The relative sizes of the sun and planets and the distances between them have been distorted for purposes of diagrammatic clarity.

The Chemical Approach

My own studies in the origin of the earth started with such thoughts about the loss of volatile chemical elements from the earth's surface. Exactly how did these elements escape from the earth, and when? It was impossible that they were evaporated from a completely formed earth, the evaporation must have occurred at some earlier time in the earth's history. Once the earth was formed its gravitational field was much too strong for volatile gases to escape into space. But if these gases escaped from the earth at an earlier stage, what is the origin of those that we find on the earth today? Water, for example, would have tended to escape with neon, yet now it forms oceans. The answer seems to be that the chemical properties of water are such that it does not enter into volatile combinations at low temperatures. Thus if the earth had been even cooler than it is today, it might have retained some water in its interior that could have emerged later. But meteorites contain graphite and iron carbide, which require high temperatures for their formation. If the earth and the other planets were cool, how did these chemical combinations come about?

Indeed, what was the process by which the earth and other planets were formed? None of us was there at the time, and any suggestions that I may make can hardly be considered as certainly true. The most that can be done is to outline a possible course of events which does not contradict physical laws and observed facts. For the present we cannot deduce by rigorous mathematical methods the exact history that began with a globule of dust. And if we cannot do this, we cannot rigorously include or exclude the various steps that have been proposed to account for the evolution of the planets. However, we may be able to show which steps are probable and which improbable.

Kuiper believes that the original mass of dust and gas became differentiated into one portion that formed the sun and others that eventually became the planets. The precursors of the so-called terrestrial planets—*Mercury, Venus,* the earth and *Mars*—lost their gases. The giant planets *Jupiter* and *Saturn* retained the gases, even most of their exceedingly volatile hydrogen and helium. *Uranus* and *Neptune* lost much of their hydrogen, helium, methane and neon, but retained water and ammonia and less volatile materials. All this checks with the present densities of the planets.

It seems reasonably certain that water and ammonia and hydrocarbons such as methane condensed in solid or liquid form in parts of these protoplanets. The dust must have coagulated in vast snowstorms that extended over regions as great as those between the planets of today. After a time substantial objects consisting of water, ammonia, hydrocarbons and iron or iron oxide were formed. Some of these planetesimals must have been as big as the moon; indeed, the moon may have originated in this way. The accumulation of a body as large as the moon would have generated enough heat to evaporate its volatile substances, but a smaller body would have held them. Most of the smaller bodies doubtless fell into the larger; *Deimos* and *Phobos,* the two tiny moons of *Mars,* may be the survivors of such small bodies.

Massive chunks of iron must also have been formed. On the moon there is a huge plain called Mare Imbrium; it is encircled by mountains gashed by several long grooves. It would seem that the whole formation was created by the fall of a body perhaps 60 miles in diameter; this has been suggested by *Robert S. Dietz* of the U.S. Naval Electronics Laboratory, and by *Ralph B. Baldwin, the author of a book entitled The Face of the Moon*. The grooves must gave been cut by fragments of some very strong material, presumably an alloy of iron and nickel, that were embedded in this body. Of course large objects of iron still float through interplanetary space; occasionally one of them crashes into the earth as a meteorite.

How were such metallic objects made from the fine material of the primordial dust cloud? In addition to dust the planetesimals contained large amounts of gas, mostly hydrogen. The compression of the gases in a contracting planetesimal generated high temperatures that melted silicates, the compounds that today form much of the earth's rocky crust. The same high temperatures, in the presence of hydrogen, reduced iron oxide to iron. The molten iron sank through the silicates and accumulated in large pools.

It now seems that the meteorites were once part of a minor planet that travelled around the sun between the orbits of *Mars* and *Jupiter.* The pools of iron that formed in this body may have been a few yards thick. In the case of the object that was responsible for Mare Imbrium and its surrounding grooves, the depth of the pools must have been several miles. If the temperature of such a planetesimal had been high enough, its silicates would have evaporated, leaving it rich in metallic iron. The object must eventually have cooled off, for otherwise its nickel-iron fragments could scarcely have been hard enough to plow 50-mile grooves on the surface of the moon.

It was at this stage that the planetesimals lost their gases; *Kuiper* believes that they were probably driven off by the pressure of light from the sun. This left the iron-rich bodies that are today the earth and the other planets. The whole process bequeathed a few meaningful fossils of the modern solar system: the meteorites and the surface of the moon, and perhaps the moons of *Mars.*

The Moment of Inertia

Recently we have redetermined the density of the various planets and the moon. The densities of some calculated at low pressures, are as follows: *Mercury,* 5; *Venus,* 4.4, the *earth* 4.4, *Mars* 3.96, and the *moon,* 3.31. The variation is most plausibly explained by a difference in the iron content of these bodies. And this in turn is most plausibly explained by a difference in the amount of silicate that had evaporated from them. Obviously a planet that had lost much of its silicate would have proportionately more iron than one that had lost less.

It is assumed by practically everyone that the earth was completely molten when it was formed, and that iron sank to the center of the earth at that time. This idea, like the conception of an earth torn out of the sun, and a moon torn out of the earth, almost has the validity of folklore. Was the earth really liquid in the beginning? *N.L. Bowen* and other geologists at the Rancho Santa Fe Conference of the National Academy of Sciences in January, 1950, did not think so. They argued that if the earth had been liquid we should expect to find less iron and more silica in its outer parts.

There is other evidence. Mars, which should resemble the earth in some respects, contains about 30 per cent of iron and nickel by weight, and yet we have learned by astronomical means that the chemical composition of Mars is nearly uniform throughout. If this is the case, Mars could never have been molten. The scars on the face of the moon indicate that at the terminal stages of its formation metallic nickel-iron was falling in its surface. The same nickel-iron must have fallen on the earth, but there it would have been vapourized by the energy of its fall into a much larger the energy of its fall into a much larger body. Even so, if the earth had not been molten at the time, some of the nickel-iron might still be found in its outer mantle.

If there is iron in the mantle of the earth, it may be sifting toward the center of the earth; and if it is moving toward the center of the earth, it will change the moment of inertia of the earth. The moment of inertia may be defined as the sum of the mass at each point in the earth multiplied by the square of the distance to the axis of rotation, and added up for the whole earth. If iron were flowing toward the interior of the earth, this quantity should decrease. It is a requirement of mechanics that if the moment of inertia of a rotating body decreases, its speed of rotation must increase. Finally if the speed of the earth's rotation is increasing, our days should slowly be getting shorter.

Now we know that our unit of time is changing; but it is getting longer, not shorter. That is, the earth is not speeding up but is slowing down. Very precise astronomical measurements, some of them dating back to the observation of eclipses 2,500 years ago, indicate that the day is increasing in length by about one- or two-thousandths of a second per day per century. It has been thought that the lengthening of the day was due to the friction of the tides caused by the sun and the moon. But if we attempt to predict changes in the apparent position of the moon on the basis of this effect alone, we find that our calculations do not agree with the observations at all. If on the other hand we assume that iron is sinking to the core of the earth, the changing moment of inertia would also influence the length of the day. Indeed, calculations made in the basis of both the tides and the changing moment of inertia do agree with observations.

In order to make the calculations agree we must postulate a flow of 50,000 tons of iron from the mantle to the core of the earth every second. Staggering though this flow may seem, it would take 500 million years to form the metallic core of the earth. Some calculations indicate that it may have taken as long as two billion years. If this reasoning is correct, the earth was made initially with some iron in its exterior parts, and it could not have been completely molten.

To complicate matters *Walter H. Munk* and *Roger Revelle* of the Scrips Institution of Oceanography have shown that the moment of inertia of the earth is probably decreasing because water is slowly being transferred from the oceans to the ice caps of Greenland and Antarctica, and that this process can account for the lengthening of the day without assuming that iron is moving to the center of the earth, at least not so rapidly as has been calculated. In view of the argument of *Munk* and *Revelle* we really have no evidence for the flow of iron to the center of the earth. However, we have little evidence to the contrary. New observations are needed.

The Last Stages

Let us briefly retell what the course of events may have been. A vast cloud of dust and gas in an empty region of our galaxy was compressed by starlight. Later gravitational forces accelerated the accumulation process. In some way which is not yet clear the sun was formed, and produced light and heat much as it does today. Around the sun wheeled a cloud of dust and gas which broke up into turbulent eddies and formed protoplanets, one for each of the planets and probably one for each of the larger asteroids between *Mars* and *Jupiter.* At this stage in the process the accumulation of large planetesimals took place through the condensation of water and ammonia. Among these was a rather large planetesimal which made up the main body of the moon, there was also a larger one that eventually formed the earth. The temperature of the planetesimals at first was low, but later rose high enough to melt iron. In the low-temperature stage water accumulated in these objects, and at the high-temperature stage carbon was captured as graphite and iron carbide. Now the gases escaped, and the planetesimals combined by collision.

So, perhaps, the earth was formed!

But what has happened since then? Many things. Of course, among them the evolution of the earth's atmosphere. At the time of its completion as a solid body, the earth very likely had an atmosphere of water vapour, nitrogen, methane, some hydrogen and small amounts of other gases. *J.H.J. Poole* of the University of Dublin has made the fundamental suggestion that the escape of hydrogen from the earth led to its oxidizing atmosphere. The hydrogen of methane (CH_4) and ammonia (NH_3) might slowly have escaped, leaving nitrogen, carbon, dioxide, water and free oxygen. But many other molecules containing hydrogen carbon, nitrogen and oxygen must have appeared before free oxygen. Finally life evolved, and photosynthesis, that basic process by which plants convert carbon dioxide and water into foodstuffs and oxygen. Then began the development of the oxidizing atmosphere as we know it today. And the physical and chemical evolution of the earth and its atmosphere is continuing even now.

The Earth

Apollo astronauts have said that the earth, with its blue water and white clouds, was by far the most inviting object they could see in the sky when they were on the moon. Their bias is understandable. They knew from intimate observation what this planet is like and could translate the sight of clouds, oceans and continents into everyday experience—of, say, a sea breeze blowing surf onto a sunny beach.

Probably the thing people like most about the earth, even if they have not put the thought into words, is its pattern of constant movement. On the earth stillness is remarkable for its rarity. Motion extends from the constant shifting of grains in a sand dune and the movement of bacteria and all other forms of life to the ponderous motions of the entire earth as it vibrates during and after an earthquake.

This planet is active. Indeed, it has been active for 4.6 billion years, and it shows no signs of calming down. The earth's atmosphere, oceans, thin crust and deep interior have been in motion since they were formed. Life has been an integral part of the surface for at least four-fifths of the planet's history.

As a consequence of its steady activity over this long span of time the earth has evolved through a series of quite different stages, maintaining during the entire time a state of dynamic equilibrium. The balance involves an exchange of matter and energy between the interior, the surface the atmosphere and the oceans. It also involves sharing the radiation of the sun with the other members of the solar system. The study of geology, aided by work in geochemistry, geophysics and paleontology, has shown how the earth's surface skin has evolved. That knowledge, coupled with a firm theory about the constitution of the earth's interior and certain hypotheses on how the interior moves, provides a means for constructing a theory of how the planet evolved.

The article by *A.G.W. Cameron* describes the origin of the earth and the other planets through the condensation of particular regions of the solar nebula. The original composition of the nebula and its later composition are deduced from the composition of the earth's rocks, of rocks brought back from the moon, of meteorites and of the atmospheres of the earth, Mars, Venus and Jupiter.

The theory of the earth's growth most favoured by people who have studied the subject infers a gradual condensation and accretion of a solid planet as it swept up

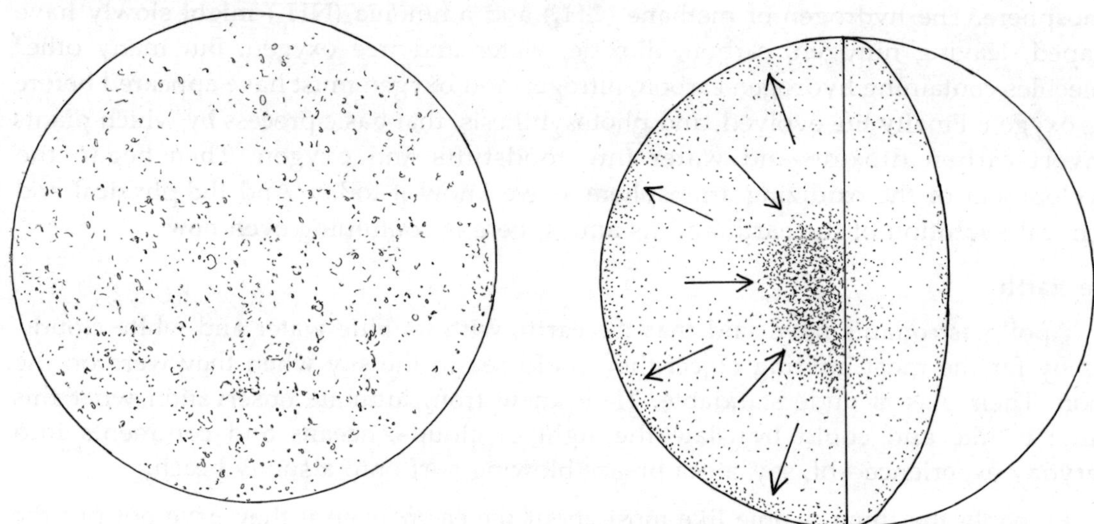

Fig. 2.3: Four Stages in the early evolution of the earth are shown in diagrams, beginning (*for left*) with the sphere that developed by accretion in the first few million years following the condensation of the solar nebula. The young planet is pockmarked by the infall of million of planetesimals; its accretion atmosphere is rich in hydrogen and noble gases. Later (*left*), when some tens of millions of years have passed, a combination of gravitational compression, radioactive decay and impact heating produces melting and differentiation; heavy core and mantle materials sink inward and light crustal materials float outward. The solar wind sweeps away the accretion atmosphere; it is replaced by a primordial atmosphere rich in methane, ammonia and water. The Archaean era (*right*), between 3.7

enormous quantities of small particles from the nebular disk that gave rise to the present solar system. As the planet grew it began to heat up as a result of the combined effect of the gravitational infall of its mass, the impact of meteorites and the heat from the radioactive decay of uranium, thorium and potassium. (Although postassium is not normally regarded as being radioactive, .01 per cent of the element on the earth is the radioactive isotope potassium 40). Eventually the interior became molten. The consequence of the melting was what has been called the iron catastrophe, involving a vast reorganization of the entire body of the planet. Molten drops of iron and associated elements sank to the center of the earth and there formed a molten core that remains largely liquid today.

As the heavy metals sank to the core the lighter "slag" floated to the top—to the outer layers that are now termed the upper mantle and the crust. Accompanying the rise of the lighter elements, such as aluminum and silicon and two of the alkali metals, sodium and potassium, were the radioactive heavy elements uranium and thorium. The explanation for the rise of these heavy elements lies in the way atoms of uranium and thorium form crystalline compounds. The size and chemical affinities of the atoms

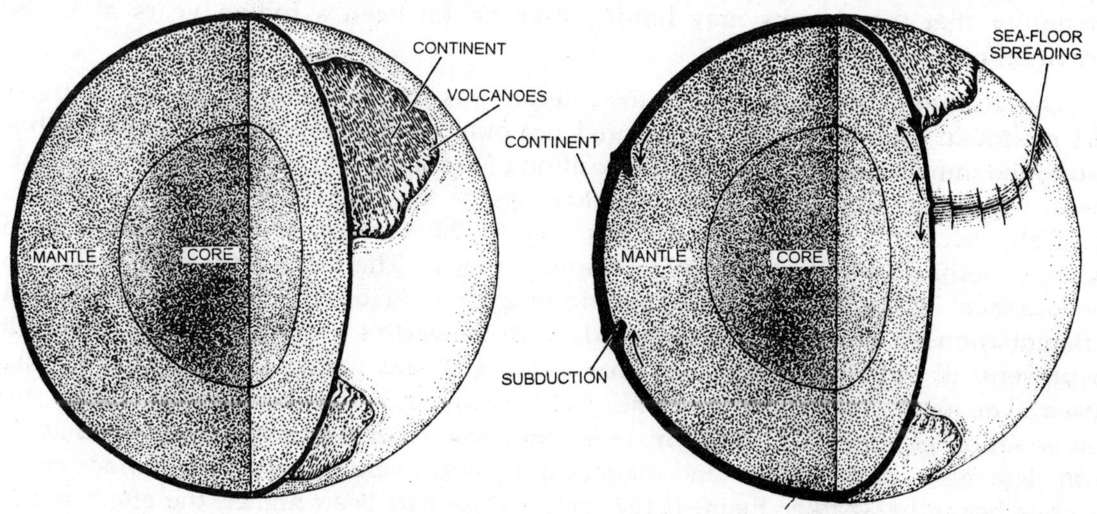

and 2.2 billion years ago, is marked by cooling. As atmospheric water and gases condense, the oceans appear; the earliest continents arise and volcanic activity is intense. Chemical reactions between water, gases and crustal materials give rise to sediments and solutions. The Proterozoic era (*far right*) follows the Archaean era. It harbours a few scant traces of plant and animal life. It ends some 600 million years ago as the Paleozoic era, with its rich fossil record, begins. During that era's 1.5-billion-year span crustal cooling and thickening continue. Crustal plates begin migrating as such mechanisms as sea floor spreading and subduction become established. Long before the end of the Proterozoic era both the interior and the surface of the planet have assumed their modern character.

prevent them from being accommodated in the dense, tight crystal structures that are stable at the high pressures of the deep interior of the earth. Therefore the uranium and thorium atoms were "squeezed out" and forced to migrate upward to the region of the upper mantle and the crust, where they could fit easily into the more open crystalline structures of the silicates and oxides found in crystal rocks.

As the earth was differentiating into a core, a mantle and a crust, the material at the top was also splitting into different fractions. The lower parts of the crust are composed of basalt and gabbro, dark rocks containing calcium, magnesium and iron-rich compounds, mainly silicates. They were derived by partial melting and differentiation of the denser materials of the upper mantle. The basalt and the gabbro were themselves differentiated by fractional crystallization and partial melting and so, as lighter fluids, were pushed up through the crust. In the upper layers of the crust and at the surface they solidified to form the lighter igneous rocks such as granite, which are enriched in silicon, aluminum and potassium.

The question of how much of this "sweating out" process was completed at the early stage remains unresolved. Some geologists argue that a large amount, perhaps the bulk, of the granitic crust had already formed by this stage. Others cite the possibility that the process may hardly have begun even a billion years after the formation of the earth.

One result of the heating up of the interior was the inception of volcanic activity and mountain building. They contributed not only to the shaping of the surface but also to the immense change in the composition of the interior. During that time various gases, which had been locked in the materials of the planet when those materials originally accreted, began to find their way to the surface. They included carbon dioxide, methane, water and gases containing sulphur. The gases must have leaked to the surface in tremendous volume during the period of reorganization and differentiation. At the surface they stayed, since the earth's gravity was strong enough to prevent all but the lightest elements (hydrogen and helium) from escaping into space. The temperature at the time must have been low enough to allow the condensation of water. Dissolved in that water, the other gases combined chemically with elements such as calcium and magnesium, which were leached from surface rocks as rains began to weather them. If the temperature had been higher, the effect of the dense atmosphere with its large content of carbon dioxide would have been to institute the kind of "greenhouse effect" that seems to have arisen on Venus, producing that planet's hot, cloudy atmosphere.

As the surface of the earth cooled and the oceans formed from the condensation of water, the processes of erosion by wind and water began to operate in much the same way that they do today. Liquid water became the dominant mode for transporting and redistributing the debris of eroding mountains. The river systems of the surface are the visible traces of the networks that carry eroded material to the oceans, where much of it accumulated as aprons of sediment along the continental shelves and continental rises. The rest of the sediment is spread as thin layers over the ocean deeps by slow settling and the motions of turbidity currents.

A number of thoughtful geochemists and geophysicists have speculated on a somewhat different chain if events leading to the early accretion of the earth from the condensing solar nebula. According to these views, the earth and the other planets are the products of a gradual condensation of the solar nebula during which certain of the heavy elements, mainly iron, crytallized first, while the lighter fractions of the nebula were still in gaseous form. In what process the core of the accreting planet would have been iron-rich in the first place, with successively lighter fractions, roughly corresponding to the order of their crystallization from a gas, being accreted on the outside as the planet grew.

Whatever the mechanism of accretion, the story of the earth's later evolution (after the first billion years) is largely told by the record contained in the rocks of the crust. What they reveal is best told in terms of a geologic "clock" that began to turn precambrian times. The oldest rocks now known are a series of métamorphosed sedimentary and volcanic rocks, which from their content of radioactive elements can be given an age of about 3.7 billion years. They are much older than most of the very old rocks from the interval of time known to geologists as the *Archaean*. The rocks of that time are roughly defined as being older than 2.2 to 2.8 billion years. (The age of the boundary with younger eras varies in different parts of the earth's ancient rock terrains.) Much of the rock record is fragmentary, but it is tangible, and one no longer has to rely solely on plausible theory.

The Archaean rocks appear to be some what different from the rocks of all succeeding eras in the sense that certain rock types are abundant almost to the exclusion of many other types commonly found later. Archaean rock series tend to be dominated by basalts and andesites, which are volcanic rocks rich in iron and magnesium, deficient in sodium and potassium and relatively low in silica. The sandstones and shales of Archaean time were derived by the weathering and reworking of those volcanic rocks. Large bodies of granite—rocks richer in alkalis and silica—are absent. Such a skewed composition with respect to later rocks suggests that the sweating out of granitic rocks by fractional crystallization and partial melting of less silicic rocks was not as advanced as it became later.

The Archaean rocks also suggest that the tectonic style of the time, that is, the mountain-making activity by which the surface was shaped, differed from the pattern of today. The present theory of plate tectonics visualizes large plates of the lithosphere (which included the crust and part of the upper mantle) moving laterally over the asthenosphere (a hot, plastic and perhaps partly molten layer of the mantle). The driving force is movement in the mantle, although the precise nature of that movement is uncertain. The geologic activity of earthquakes, volcanoes and mountain building is concentrated along the plate boundaries.

Granted that the Archaean rocks are widely dispersed and offer only a few bits of information, the study of the oldest terrains of Archaean areas in Canada and areas of similar age in Africa and Scandinavia does not suggest mountain building along the boundaries of large plates. Many geologists suspect that the Archaean was a time of very thin lithospheric crust, extensive volcanism and some jostling movement of many small, thin "platelets", with "sutures", or crumpled deformational belts, welding them together.

Although the Archaean era differed markedly from the present in tectonic style and in the average composition of its volcanic rocks, it was the same as the present in all essential processes of erosion and sedimentation on the surface. All the earmarks of weathering, mechanical breakup of rocks, transportation by rivers and sedimentation in regions where the crust gradually subsided and allowed great thicknesses of sediment to accumulate are found in Archaean sediments, as was pointed out more than 30 years ago by *Franci J. Pettijohn* of Johns Hopkins University, who was studying early Precambrian sedimentary rocks in the region of *Lake Superior.* Looking at those sandstones, shales and conglomerates, it is difficult to see any significant difference between them and more recent ones, all being the hardened equivalents of the gravels, sands and muds of today.

The erosion and chemical decay of rocks today are profoundly affected by the presence of land plants. It is known, however, that the higher (vascular) land plants did not evolve until two billion years after Archaean time, that is, in the middle of the Paleozoic era. Perhaps before the plants evolved, lower forms of life existed on the land, as they surely did in the sea.

Evidence of algal life late in the Precambrian era was obtained some years ago when the paleobotanity, *Elso S. Barghoorn* of Harvard University, working with the late Stanley A. Tyler, a sedimentologist at the University of Wisonsin, discovered microscopic remains of algal organisms in the Gunflint chert, a dense sedimentary rock made of silica. The *Gunflint chert* has been dated by its content of radioactive elements and their decay products to an age of about two billion years. Since then other organic structures that look like the remains of organisms have been found in even older rocks. The oldest of them, aged about 3.4 billion years, is the Fig Tree chert of Swaziland in Africa.

This kind of search for evidence of ancient life is a painstaking, laborious process. Thousands of rock specimens have to be sawed into ultrathin slices and then polished so that they can be studied under the light microscope and the electron microscope. Although organic carbon had been found in old rocks long before the discovery of the *Gunflint* and *Fig Tree cherts,* one could always hypothesize a variety of ingenious chemical mechanisms to account for it. The more recent evidence of distinctive forms of cellular life in ancient times is difficult to refute.

How life began on the earth is another story. It is the story of plausible chemical mechanisms that can be deduced from certain assumption about the early chemical environment of the surface. One begins by inferring an early Archaean atmosphere (which had been built up by the escape of gas from the interior) that was dominated by water, methane and ammonia. Free oxygen is a product of life and not an antecedent to it. The atmosphere may also have contained appreciable quantities of carbon dioxide.

The existence and character of this atmosphere are related to the fact that the earth is smaller than Jupiter and larger than the moon. Jupiter was able to hold its hydrogen, which was by far the most abundant element in the solar nebula. The moon could not hold any of its gas.

In the earth's envelope of air and below it, in the surface waters of the sea and in large lakes, ultraviolet radiation from the sun was intense. The Surface was not screened from the ultraviolet by a layer of ozone, as it is now, for want of the oxygen (O_2) from which ozone (O_3) is derived. The high energy of the ultraviolet radiation promoted the synthesis of a variety of organic compounds, for example amino acids. Perhaps many of these compounds were already here, since it is now known that a number of simple organic compounds are present in interstellar space.

The synthesis of transitory organic compounds, however, is not the same as making life. The next steps had to be the growth of large molecules and, before long, the growth of the nucleic acids that would eventually provide the genetic mechanism of reproduction so that cells could divide and give rise to new cells like themselves.

One cannot be sure of the range of chemical environments that will support life. (The uncertainty may be diminished by the U.S. spacecraft scheduled to land on Mars next year.) All that is known now is that the earth supports life and that its life depends on the continuous existence of liquid water. At present the earth is the only planet known to satisfy that condition. The earth's continuous record of life for at least the past 3.5 billion years shows that liquid water has been available during all that time.

Once life evolved, it began to exert an important effect on the surface of the earth and the gaseous envelope surrounding it. In the Bitter Springs formation of central Australia, which is a little less than a billion years old, paleobotanists have found cellular algae showing many of the geometric characteristics of the blue-green algae that today, like all other photosynthetic plants, evolve oxygen as a waste product. By the end of the Proterozoic era, which lies between Archaean time and the beginning of the Paleozoic era, there must have been enough oxygen in the atmosphere to support the evolution of higher organisms. They were the metazoans—animal organisms having many cells with differentiated characteristics. All these organisms appear to need at least small quantities of free oxygen for their biochemical processes.

Oxygen is not the only atmospheric gas that comes from life. Methane, for example, is present in minute quantities. Its source seems to be primarily the methane-producing bacteria that yield the abundant "marsh gas" over swamps. The atmosphere also contains other gases that are the distinctive product of life rather than of simpler nonbiological chemical reactions.

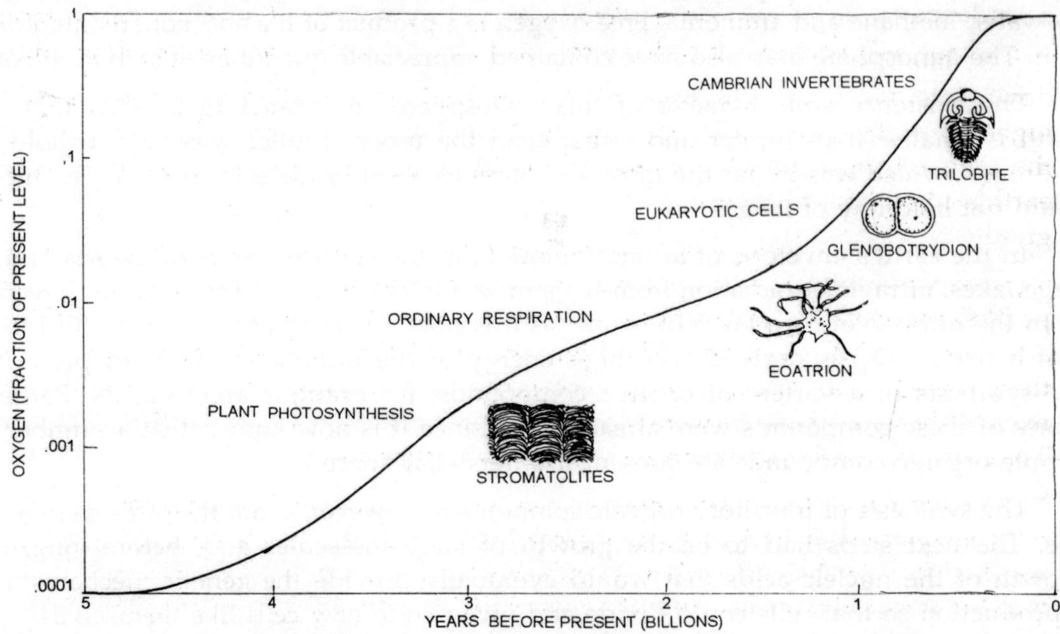

Fig. 2.4: Appearance of Oxygen in significant quantities in the earth's atmosphere, a late development, is an event that remains a subject of controversy. One hypothesis is shown in this semilogarithmic graph. The abscissa intervals are billions of years before present; increases in the oxygen supply, from trace amounts to the present quantity about 20 per cent of the atmosphere), are indicated on the ordinate. The process must have been gradual and must also have been related to an increase in the numbers of photosynthetic plants. The oxygen level may have risen to 10 per cent of its present value a billion years ago but no evidence of oxygen-dependent animal life in any abundance appears until the time of a steep increase in the oxygen supply at the end of the Proterozoic era.

The *Proterozoic era* was a time when the world was populated by bacteria, algae and other primitive single-cell organisms, probably on land as well as in the sea. Their influence on surface processes is seen in the Proterozoic rocks. It is most distinctive in stromatolites: rock formations consisting of the limy secretions of mats of filamentous algae and the sediment trapped by them. Stromatolites are known today in places such as the *Bahamas* and *Bermuda,* where limestone is being laid down in tidal flats. Other evidence of Proterozoic life is found in the existence of a few coal beds formed from masses of carbonized algal remains.

If an observer had looked down on the earth from an artificial satellite in Proterozoic time, he would have described the surface in much the same way that an observer similarly situated would describe it today. Only a sensor that could determine the chemical composition of the atmosphere would reveal any differences. The evidence for the similarities is in the Proterozoic rocks, which are of the same types and abundances as the rocks from all later ages.

By late Proterozoic time the earth-moon system, after early instabilities, had settled down into much the same system we see today. Tides would have been somewhat higher than they are now, but they would not have been grossly different. At about the time the moon became a cold planet the long heating and differentiation of the earth's upper mantle and crust resulted in the extensive intrusion of great bodies of granite rock and in patterns of mountain belts that suggest a plate-tectonic origin.

Another kind of evidence from both Proterozoic rocks and more recent ones reveals periodic reversals of the earth's magnetic poles during much of the earth's history. As a heated rock cools it is magnetized in the direction of the earth's magnetic field, and the pattern is frozen in when the rock solidifies. Similarly, certain sediments that contain magnetic particles will record the direction of the field at the time they were deposited. The causes of reversals lie in instabilities in the fluid motion of the core, which is the driving force that creates the earth's magnetic field.

The same kind of paleomagnetic evidence reveals what has been called polar wandering, although it is not that the North and South poles have moved but rather that the surface features of the earth have shifted in relation to the poles. The conclusion is reinforced by paleoclimatic evidence, that is, the geologic record of ancient climates, such as the occurrence of coal beds in polar regions and glacial deposits near the Equator. From this kind of information it appears that a major continent was at the South Pole in Proterozoic time and that continental drift was already established as a major process affecting paleogeography.

The rocks also record from this time a major glacial epoch, the first for which there is firm evidence. The evidence is insufficient to reveal the details of that ice age—whether it was of the same extent as the recent (*Pleistocene*) ice ages and whether, like them, it had many episodes of glacial advance and retreat. One can only assume that the mechanisms postulated for the Pleistocene ice ages are general ones that are set in motion once a continental land mass lies at one of the poles and restricts the ability of the ocean and the atmosphere to distribute heat evenly around the globe. To an external observer at that time the earth would have looked a little like Mars, except that there were still oceans at the Equator. One of the interesting questions about the earth's glacial epochs is why the earth remained poised at a temperature distribution sufficiently low to give rise to large polar ice caps but not to a complete freeze-over.

Just as human history merges with pre-history, so the most recent 570 million years of the earth's history (starting with the *Paleozoic era*) connect with the nine-tenths of earlier evolution that were long thought to be a mystery. For more than a century the past 570 million years have been regarded as the geologically "known" period; it is therefore often called the *Phanerozoic* from the Greek *phaneros,* "to reveal." Although early geologists recognized that some Precambrian terrains were mappable by ordinary geologic methods, it was the absence of fossils having recognizable affinities with forms of the present that made it unknowable. The strategraphic time scale, a marvelously detailed and precise clock, depends on the rapid evolutionary changes in higher forms of life that are recorded in the fossilized remains of corals, mollusks and thousands of other kinds of metazoan life.

Students of the earth's history never tire of marvelling at the extraordinary speed of the coming of the metazoans. For between three and four billion years, almost its entire history, the earth was populated by single-cell life. Within at most a few hundred million years there—after a fantastic diversity of invertebrate organisms appeared. All the major phyla of the animal kingdom became established quickly, and the vascular plants and the vertebrates soon followed.

Was all of this an accident, a favourable conjunction of continents, seas and environmental niches? Or was it the inevitable consequence of the buildup of oxygen in the earth's atmosphere by photosynthesizing algae? The best guess now is that it was the evolution of the atmosphere to near its present level of oxygen that stimulated biological invention. One such invention was animal shell, which served as armour to protect soft bodies from predators and as a base of attachment for muscles. Shells provide the basis for our understanding of the subsequent course of evolution of both the planet and its inhabitants. A paleobiological record based solely on the soft parts of organisms would provide only the dimmest outlines of the past.

The shells are more than markers of history: they influenced important changes in the dynamics of the earth's exterior. The oceans became populated with organisms that secreted calcium carbonate, calcium phosphate and silica in enormous quantities. Their remains were deposited as sediment, ultimately to become limestone, chert and phosphatic limestone or phosphate rock (a major source of agricultural fertilizer).

The more precise knowledge afforded by Paleozoic records enables geologists to trace the effects of continental drift. In particular it is possible to map more confidently the shape of the early Atlantic Ocean that lay between the European-African land mass and the American before the supercontinent of Pangaea was assembled at the close of the Paleozoic era. The assembly of Pangaea was one of the rare, special events of the later history of the earth, one of the important perturbations of the otherwise more or less evenly ordered evolution of the planet.

One of the major consequences of the assembly of *Pangaea* was the extinction of hundreds of species of invertebrates and the beginning of a wholesale change in the kinds and relative populations of the different animal and plant species. Most of the expanse of shallow shell surrounding each continent disappeared as the continents collided, leaving only one narrow perimeter around the supercontinent. The shelves had harboured the most productive biological populations of the Paleozoic world. The geographic constriction and the concurrent climatic extremes, including the glaciation of parts of what are now Africa, Australia and South America, were enough to decimate many species. The survivors went on to find the new stocks of the post-Paleozoic world.

Pangaea rifted apart in the *Triassic period* (the earliest part of the *Mesozoic era*), and with that event and the following opening of the Atlantic Ocean and the drifting of the continents to their present position the story of the earth's physical evolution is largely told. The oldest parts of the ocean floor that are now preserved came into being at this time, and so began a decipherable history of the world's oceans. It is read from the magnetic "stripes" and the fracture zones of the sea floor formed at mid-ocean ridges and rifting zones.

The new forms of life that evolved early in the Mesozoic era give the appearance of the modern world. Flowering plants appeared, and the lands became covered with the colours of the flowers and foliage of deciduous trees, the grasses and a great number of shrubs and flowers. In the sea new photosynthesizing algae, the diatoms, evolved; they are single-cell creatures secreting thin shells of silica. The diatoms became responsible for much of the primary photosynthetic production of organic matter in the sea.

At about the same time the calcareous foraminifera appeared. They are single-cell animals that live off the plants at the surface of the sea. Their shells of calcium carbonate, raining steadily to the bottom of the oceans, became the source of a new kind of deep-sea sediment, the foraminiferal oozes. The remains of these forminifera became part of another detective story: the deduction of ancient sea temperatures, and thus of world climates, from the isotopic composition and external form of the shells. Both the shape of a shell and the relative proportions in it of the normal oxygen atom (oxygen 16) and the rare heavy isotope (oxygen 18) reflect the temperature of the water in which the animal lived. The temperature of the oceans as measured in this way has revealed an important climatic change.

Over most of the past 50 million years (during much of the *Cenozoic era*) the earth was cooling. This culminated in the past few million years in repeated glaciations. The more recent ones have been witnessed by and have affected the evolution of a new species: man. Already advanced on his course of evolution, man in his primitive cultures was displaced as the glaciers covered much of northern Europe, Asia and North America. In the short 10,000 years since the glaciers retreated to their present ice-cap size (probably a temporary retreat) man became the species that spread and occupied almost every environment of the surface of the planet. As he did so he became the latest of the biological populations to profoundly affect the course of the earth's history. He is only now becoming aware that some of his activities may alter the thin envelope of the atmosphere and the oceans and the fresh waters that make his existence possible.

Chapter—3

Primordial Oceans

Sponges, though useful in the bath and existing as some 5,000 distinct species that range in size from the microscopic to the monstrous, are essentially very simple organisms, little more than water-filtering system made up of one or more chambers through which water flows, carrying nutrients direct to every single cell in the sponge. And since each cell feeds individually, even the largest sponge consumes only microscopic particles.

Structurally, the sponges stand somewhere between a colony of independent cells and true multicellular organisms. They consist of relatively few cell types, there are no distinct tissues or organs, yet the individual cells comprising the organisms do seem to possess some sort of mutual recognition and purpose that enables them to get together and build a sponge. In this respect, sponges are a wonderful example of the single cell as the basis of higher life-forms.

Along the east coast of America, from Nova Scotia all the way down to South Carolina, the elusive oyster beds betray their presence in summer and autumn with occasional patches of bright tomato red. The colour fades away in the winter, returning the following summer. The cause of the colour is a small sponge, known to science as *Microciona prolifer,* which starts out as a thin encrustation on the shells of oysters and eventually grows into a branching, bush-like structure up to 20 centimetres high. The annual colour change is associated with reproduction.

In the summer of 1907, the biologist *H.V. Wilson* was working at Beaufort Harbour, North Carolina, on his research into the generation and regeneration of tissue. Among several experiments he conducted that summer, one is of compelling interest.

Wilson collected a bright red living *Microciona prolifera* sponge and cut it up into tiny pieces. Then he put the pieces in a muslin bag which he suspended in a dish of filtered sea water. He shook the bag about and squeezed it gently with a pair of forceps. Clouds of red cells passed though the muslin and into the water. Soon the bag was all but empty and the red sponge had been reduced to a fine sediment on the bottom of the dish.

Wilson watched closely. Within an hour the tiny component cells of the sponge began together in clumps. After a while thin protrusions grew from the clumps and reached through the water, like arms, when they touched another arm, or another clump of cells, the two groups fused into one larger clump, and so the process went on until eventually all the cells were fused together in a single crusted mass on the bottom of the dish. At this point the most amazing thing happened. The cells began to sort out their individual roles in the formation of a sponge and moved to take up position accordingly. As the days went by small open chambers appeared in the crusted surface, canals formed between them and the branching vent tubes began to grow, slowly at first and then rapidly, as though the cells had regained confidence in their collective existence. Within a week the sponge that Wilson had reduced to a fine and red sediment was completely rebuilt.

Later, *Wilson* tried the same experiment on a member of the jellyfish family that has a more complex structure than the sponge, with two distinct layers of tissue. Here too the animal that *Wilson* cut apart and squeezed through his muslin bag reformed itself from the disassociated cells.

Sponges represent the lowest form of multicellular development in animals; indeed, for a long time they were not considered to be wholly animal at all and were classified as zoophytes—'animal plants'. They are also among the most ancient; fossil evidence traces their existence back to the last part of the Precambrian, 700 million years ago. Thus sponges bridge the gap between the world of the microscopic single cell and the world of multicellular higher organisms, but here their contribution to the story of life on Earth ends. Sponges have remained essentially the same ever since; having found their niche, no further evolutionary development occurred.

The mainline story of life, leading from the single cell to the first multicellular organisms, and on to the first backbone, the first fish, the first land animals, the first mammals and so on to man, continues with the single cells that gathered together to form quite different creatures—the soft-bodied jellyfishes and flatworms. Wilson's experiments provided solid enough support for the contention that jellyfishes and their kind must have played an important role in the early evolution of higher multicellular organisms, but the contention was based on theory only and tangible fossil evidence proving their existence at that stage in the story of life was not forthcoming. This was hardly surprising, given the soft-bodied nature of the living animal. After all, who would expect to find a fossil jellyfish? And since no one expected to find any, no one looked for them. Discovery, when and if it came, would yet again be a matter of chance.

By the 1870s the era of geographic exploration in North America was long past; the broad details of the land, lakes, rivers and mountains were known and mapped from coast to cost, but the finer details still awaited definition. Fortune-seekers wanted a hint of where the next gold strike might be; industrialists urged geologists to seek out workable coal and iron deposits; academics wanted to find new areas of research in their subjects. So when surveyors mapping a route through the Rocky Mountains for the proposed Canadian Pacific Railway sent back word of fantastic mountain peaks, glaciers, lakes and forests in the previously little-known region of the high Burgese Pass, they caught the attention of many interested parties. Among them was George Dawson of the Canadian Geological Survey, who proceeded to make a preliminary study of the area in the early 1880s. He was followed in 1886 by another geologist who described sanctions, shale and limestone deposits thousands of metres thick, dissected and laid bare by uplifting and erosion. Meanwhile the mention of glaciers attracted the attention of *George Vaux* from Philadelphia, who initiated the first glaciological studies of the region and was assisted in the mountains by his sister, Mary, Mary Vaux married *Charles Dolittle Walcott* (1850-1927), a paleontologist who was to become world famous for his discoveries in the Burgess Pass region.

In a series of expeditions between 1910 and 1917 *Walcott* defined a sequence of nearly 4,000 metres of sedimentary rock laid down during the Cambrian period between 570 and 500 million years ago. Over 40,000 fossils were collected from the sequence, among which Walcott described some 150 species belonging to 119 genera. All this can be found in a couple of thousand pages of scientific publications, but the most interesting, and possibly the most significant, aspects of Walcott's work concerned a thin band, about one metre thick, of fossilbearing shale that he discovered on the south-western flank of Mount Wapta, 2,440 metres above sea level.

Charles D. Walcott was a self-made man in the best American tradition. From relatively humble beginnings, with little formal education, he scaled the heights of science and society in the United States. He was born and grew up in Oneida County, New York State, and liked to recount how his scientific career began there one day in the 1860s when he was a boy riding on the back of a wagon going from Trenton Falls to Trenton. The road was still rough in those days, and as the horse was heaving the wagon out of a dry river crossing, a wheel rode up over a loose block of sandstone and split it clean open. As the fragments shot away from the wheel, young *Charles Walcott* looked round and caught a glimpse of what had been trapped in the sandstone for hundreds of millions of years—a fossil trilobite, beautifully preserved. He jumped down to pick it up, and his passion for collecting fossils began right there.

Soon after, *Walcott* took a job in a Trenton handware store, and spent most of his spare time collecting and reading, especially about the Cambrian rock systems in Europe. His ambition was to study the Cambrian systems in America, then still undefined, but first he wanted to study geology and paleontology at Harvard university. It took him until he was 23 to save up enough money to pay for his college course, but once he began his studies he soon made his first significant contribution to science with some distinctive work on the trilobites he had collected around Trenton.

Trilobites—three-lobed animals—were named by *Johna Walch* in 1771. Walch said they were molluscs, but over the years their affinities had been disputed. The problem was that only the hard back armour of the creatures had ever been found. In 1808 a

French savant, Pierre Latreille, offered a simple and precise solution to the problem; if it could be shown that the trilobites had no legs then they must have been molluscs; if they had legs then they were woodlice, he said.

Walcott was the first to find trilobites with legs and settled the matter definitely with a note he published on them in 1876. Subsequently he published notes on the eggs of the trilobite, on the injury one had sustained to its eye while moulting, and on the animal's capacity to roll up for protection—just like some modern woodlice. Walcott became an acknowledged authority on the subject, and his expertise soon earned him a position with the United States Geological Survey.

Nearly 30 years later, in 1907, *Walcott* became Secretary of the Smithsonian Institution in Washington DC. Now at last he had time to explore further afield the Cambrian sequences that had always been his first interest. In the autumn of 1909 he followed the trail that his brother-in-law, *George Vaux* had blazed through the mountains flanking the Burgess Pass in the Canadian Rockies. Along the track that traverses the western slope of the ridge between Mount Wapta and Mount Field his horse stumbled and *Walcott* decided to take a short break. Sitting there on the narrow trail, he idly picked up a small block of dark shade that had tumbled from a cliff some distance above. He took his hammer and split the block, as geologists are apt to do as a matter of course, and to his considerable astonishment found within it the delicate fossil impressions of several soft-bodied marine animals. Quite by chance he had discovered what no one had thought would ever be found: the fossil evidence of creatures that bridged the gap between single-cell organisms and the main thread of life's story.

Thereafter the *Burgess Pass* and its fossil-bearing shales became an annual pilgrimage for *Walcott* and his family. From 1910 to 1917 they spent several months of each year there. Science was the reason for the expeditions, of course, but there is also an element of the idyllic in Walcott's accounts of their journeys along old Indian trails through the moutnains and upland meadows. High up in the Rockies they occasionally encountered snow even in July, and there was always some in September, but while on the move their tents were simply canvas sheets stretched over an open frame of stripped spruce poles. From the base came in the forest just below the ridge to the east of Mount Burgess there was a beautiful view of the Emerald Glacier. They kept another tent up at the quarry—no trees or shelter of any kind up there—and frequently stayed there overnight. The best fossils they found came from the bottom metre of the deposit and they had to use explosives to get at them.

The Burgess Shale deposit is more than 2,000 metres above sea level today; 550 million years ago it was a mudbank, 100 metres beneath the sea. The Earth's relentless workings have moved seabed to mountain top and in the process turned fine silt and mud to rock. Within the layers of this rock are preserved exquisite fossil images of life from a time when the Sun shone harshly from a sky unmellowed by atmosphere even at dawn and dusk.

To preserve the soft body tissue of the animals that lived all that time ago required very special conditions indeed. The Burgess Shale community occupied ledges on a steep undersea cliff. From time to time the mud would slip off the ledge and the entire community would be carried away into deep, cold, oxygenless water where the animals did not decay, but instead were slowly flattened in their complete and original form as the mud gradually compacted under the weight of yet more sediment falling from above. Twenty-five centimetres probably compressed and solidified to rather less than one centimetre of rock. The result of many underwater landslides was one metre of shares containing the water-thin images of life in the seas over 500 million years ago.

Buried in the Burgess Shale are creatures that bear no resemblance to anything ever seen anywhere else on Earth. One, a segmented swimmer that Walcott named *Opabinia,* has five compound eyes on short stalks on its head, and a long flexible trunk with spiked pincers at the tip which it used to catch its prey. Another, aptly named *Hallucigenia,* has seven pairs of stilt-like legs, and seven tentaches on its back. In one slab of rock, 15 of these creatures were found clustered around a large worm, as though gathered for a feast.

It is barely possible to imagine those muddy ledges 100 metres below the surface of the sea, 550 million years ago. There is life, but is it recognizable among these oddities that have no common names, nor even common existence except as specimens in museum cabinets? There are jellyfish, and there are sponges like the ones that Wilson cut up and squeezed through his muslin bag, and rising from the mud, quivering across the ledge, then burrowing into the mud again is a slim silvery creature, like a narrow leaf, about eight centimetres long. The front end has a small coronet of tentacles around an opening through which it sucks in water. There is nothing that could be called a head, only a small, light-sensitive spot that might evolve to be an eye; there are no fins or limbs... but there is something familiar about it. Like no other creature in those early seas, it undulates, sending a series of rhythmic waves down the length of its body; the waves push the water backwards and move the animal forwards. To do this it must have a series of muscles attached to a firm internal structure. This animal has a rudimentary backbone.

Walcott called this creature *Pikaia.* About 30 specimens are known in all, and the rudimentary backbone, which is called the notochord, is evident in most, while in some the transverse bands of muscle that powered the creature can be made out too. Across vast expanses of time and space, *Pikaia* stands at the very beginning of the vertebrate progression that ultimately produced mankind.

In the early 1800s a vertable revolution in thinking about the formation and age of the Earth was taking place. The science of geology was being founded. Among several important developments it had been recognized that the layers of rock found in cliffs and gorges could be identified by the distinctive fossils they contained, and could therefore be correlated from place to place. Thus the red sandstone of *Devon* were shown to be identical to those in Scotland—the layers contained similar fossils and therefore must have been laid down at roughly the same time. Now it was possible to divide the history of the Earth into a sequence of geological periods and to give an indication of their relative ages.

Until the 1830s the oldest known period was the *Devonian,* named after Devon, the region in which its characteristic red sandstones had first been studied. Then two good friends, *Adam Sedgwick,* former student of theology, and *Roderick Murchison,* former army officer, entered the field of geological investigation, and defining the Earth's most ancient rocks became something of a contest.

In 1835 *Murchison* found a formation of fossil-bearing rock in Wales that lay beneath the *Devonian* and was therefore much older. He called it the *Silurian* after the Silures, an ancient tribe which had resisted the Roman invasion of Wales.

During the course of the following year, Sedgwick found another formation lying, he claimed, beneath Murchison's Silurian and therefore even older. He called it the *Cambrian,* from Cambria, the ancient name of Wales itself.

Thereafter the two men travelled extensively through Britain and Europe, establishing wherever they could the boderline between the Silurian and the Cambrian. Increasingly, their interpretations tended to overlap. Often, Sedgwick's Cambrian would be Murchison's Silurian, and vice versa. Finally they became irreconcilable enemies, Murchison contending that there was no boundary at all between the formations; the *Cambrian* was merely a local aspect of his Lower Silurian, he claimed.

Sedgwick resisted vigorously but to little effect, especially once Murchison became Director of the British Geological Survey and pushed through a ruling that the name Cambrian should be deleted from all official maps and publications.

The antogonism persisted throughout their lives, and only after the two men had died, did it become clear that Sedgwick had been right all along. The Cambrian is a highly important period of the Earth's history. The Burgess shales are from the Cambrian period. The *Silurian* and *Murchison* defined it, on the other hand, is no longer regarded as a complete unitary period. Geologists now generally refer to Muchison's Lower Silurian as the Ordovician (after another ancient British tribe) and to his Upper Silurian as simply the Silurian.

In late *Ordovician,* and in *Silurian* times, about 400 to 450 million years ago, the land mass that is now the British Isles lay much close to the equator than it does today; the same was true of North America. Tropical reefs fringed a seashore where Wenlock Edge is situated in Shropshire, England, and where the Niagara Falls make such a spectacular show in the United States. There were shallow seas with warm clear waters, ideal conditions that promoted the prolific growth of corals, algae and shellfish. Among the corals there were crinoids, looking like delicate sea lilies but actually animals with fronded heads waving from long stalks, and there were all sorts of trilobites, some so commonly found as fossils that they have been given local names. There *'Dudley locust'* is often found rolled up, just like a woodlouse. The strawberry-headed' trilobite has eyes on stalks, and the 'bald-headed' trilobite is well known for its large compound eyes, remarkably well preserged and easily seen with a magnifying glass.

The shallow Silurian seas were host to a great variety of organisms, each ensconced in its particular niche, but here and there something entirely new was nuzzling its way across the sea floor and into the story of life. They were small creatures, never more than ten centimetres long, with a wiggling tail and a disproportionately large head. Usually only the heavy plate-like bones of the head were found preserved in the fossil beds, so at first they were thought to have been water beetles, or molluscs, a new trilobite perhaps, or even a kind of tortoise. That they might represent some kind of ancestral fish was hardly considered, but in fact these oddities from the *Silurian period*, the first-known expression of the vertebrate form since its appearance in *Pikaia*, opened one of the most important chapters in the story of life: the Age of Fishes.

At the close of the *Silurian period* 395 million years ago convulsions shook the Earth again. Volcanoes and earthquakes raised to the height of mountain regions that had formerly lain under the sea. This was the beginning of the *Devonian period*. The weathering and wearing down of the newly raised land-forms produced huge quantities of pebbles, sand and mud, which were deposited in inland basins, and in vast coastal deltas. These deposits compacted and solidified to become the Old Red Sandstone formations that are most notably developed in Scotland, in the Welsh borderland and in Devon.

In the early nineteenth century these deposits were exposed to the hammers, chisels and prying fingers of inquisitive men, and everywhere they found fossils of fish. No wonder that the young Swiss expert on fossils fish, *Louis Agassiz* (1807-73), was invited to Britain in 1833 to examine the thousands of specimens in the museums and private collections.

Among the British fossil fishes were a number of the puzzling, disproportionately large armoured heads. A few still had the tail attacked—'like a crab in front and a mermaid behind', commented one zoologist of the day. One specimen in particular, beautifully preserved in a small slab of Old Red Sandstone from Lower Devonian strata exposed at Glamis in Scotland, had a heavily muscled tail, fringed with fins, extending from the large, armoured, box-like head. It was about the size of a minnow, the ribs were visible, and even the individual vertebrae of its backbone could be made out. *Agassiz* named the specimen *Cephalaspis lyelli*, in honour of *Charles Lyell* (a founding father of geology), and he was so certain of its piscine affinities that he classified it and all the other puzzling armoured heads as fish. Subsequently they were acknowledged to be a distinct group of primitive fish and given the name ostracoderm, meaning bony shield.

The strong flexible backbone makes the ostracoderms the first known true vertebrates, but they were very primitive fish indeed. Their tail could propel them forward, but they had no fins capable of lifting them from the sea floor and therefore they could not swim in true, fish-like fashion. The heavy heads weighted the ostracoderms down and kept them nuzzling along the bottom, where they scooped in mud and sand through a simple circular mouth, filtered out edible scraps and expelled the remainder through slits on either side of their heads. They did not have jaws.

One hundred million years stretch between the rudimentary backbone of *Pikaia* in the Burgess Shale and the firm backbone of the ostracoderms in the *Devonian*, and absolutely nothing is known of what happened in between. No fossils have been found with evidence of any intermediate stages. Like so much of the story of life on Earth, it is as though the progression is a series of giant leaps from landing to landing up a stairway that has no steps; like mankind's journey to the moon, perhaps. If you take away all the words and pictures recording that giant step, what hard evidence is left? Just a few footprints.

Nevertheless, having arrived, the ostracoderms flourished. For 60 million years, from 410 to 350 million years ago, these primitive fish exploited their aquatic niche. They multiplied, and diversified into a variety of different species; then, almost as suddenly as they had entered the story of life, they disappeared from it, leaving only their fossil remains in the rocks as evidence of their existence, and today's parasitic lamprey and scavenging hagfish as their sole descendants and inheritors of their once revolutionary lifestyle.

The most probable reason for the sudden disappearance of the jawless primitive fish was the development from among their kind of a creature with jaws that acquired the habit of eating them. Sixty million years in the same ecological niche does two things to a species; firstly, creatures become set in their ways and less capable of adapting to changes in environmental circumstance; secondly, the continuous exploitation of one niche sets a premium on the discovery of another. At some stage of the ostracoderms' existence, the reshuffling of the genetic code that occurs with every generation threw up variations in the structure of the head that bestowed survival advantages upon the individuals possessing them. Progressively, the interplay of genetic variation and survival advantage produced a fish with jaws. And just as the jawless fishes had multiplied and diversified dramatically, so the advent of the jaw set off an explosion in the number and variety of jawed fishes—largely at the expense of their jawless antecedents, who became a prime source of food.

Again, there is no fossil evidence of the developmental stages that came between the jawless and the jawed fish, but that is not altogether surprising. Only a minute fraction of the creatures alive at any one time die in the very special conditions that will lead to their preservation as fossils, thus *Silurian* and *Devonian* fossil fish are plentiful only because there were millions and millions of them alive during those millions of years. By the very nature of things, the progressive changes in head structure that bestowed survival advantage and eventually produced the jaw would only have been present in a very small number of individuals to begin with, and the evidence of their existence is likely to turn up in the fossil record only when there are enough of them to be numerically significant. Given that the fossil record compresses millions of years into a few metres of rock, the gradual increase of the living creatures to significant numbers over a very long period of time is inevitably reduced to an apparently sudden arrival of their fossil remains in the sequence.

In the north west of Australia, near a place that the *Aborigines* call *Gogo* some strange cliffs rise over 300 metres high from the dry ranch lands of a cattle station. The cliffs are composed of ancient coral. Today they face an arid land dotted with dense masses of porcupine grass and the odd stunted mulga tree; 400 million years ago they were reefs along the borders of lagoons. The sea was here then, and rivers flowed into it from the land behind the reefs. When the rains were heavy inland the rivers were laden with mud which periodically stifled the lagoons with sediment, killing fish and trapping their corpses in the muddy sands that settled on the bottom of the lagoons.

Sometime later the lagoons filled entirely, the sea retreated, the Earth convulsed and in a period of worldwide continental upheaval, Australia rose above the waters. Now rain and rivers began to erode the sandstone that had filled the lagoons. The process continued for millions of years, eventually leaving the ancient coral reefs—much harder rock—standing alone.

The natrual events that created the *Gogo* landscape are easily summarized, but defining the nature and origin of its curious features was a much more difficult task. Months of field survey and laboratory analysis were involved, but in the end there was an unexpected bonus. Strewn across the land at the base of the ancient coal cliffs, on what was once the ocean floor, geologists found among the dunes and scrubby bush a profusion of curious nodules. Someone took a hammer, broke one open and found a beautifully preserved fossil fish inside. Subsequently the team took large number of these nodules back to the laboratory and soaked them for several months in baths of acetic acid. Gradually the exterior stone crumbled and fell away, revealing a startling array of complete fossil fish.

Apparently, the fish that died in the lagoons nearly 400 million years ago had acted as catalysts when the muddy sea floor turned to rock; as they fossilized the mud and sand around them became particularly hard, resisting erosion while the rest of the sandstone deposit crumbled away. But the interesting thing was that these were fossils of placoderms, the world's earliest true fish, the jawed killers that replaced the jawless ostracoderms in the story of life. The placoderms (or 'plate-skinned' fish) gained their ascendency during the early part of the *Devonian period*. Their fossil remains have been found in many places, but the Gogo collection is unique in its variety and detail.

The *placoderms* represent the arrival of the vertebrate killer in the story of life on Earth. There were many different species and they came in all sizes: some were gigantic—up to nearly ten metres long. Like their jawless antecedents, most were armoured in some way, with heavy scales attached to bony plates in the skin, and they all had well-developed lateral fins, usually in two pairs: pectoral fins just behind the head and pelvic fins at the rear, as in modern fish. Although some species hugged the sea floor, the placoderms were accomplished swimmers. Guided by sight, propelled by muscles pulling on a strong internal skeleton, steered by paired fins, armed with jaws and a formidable array of teeth, the entire ocean was their niche, and any creature their prey. Confronted with such predators, virtually all the jawless fish species soon dwindled to extinction.

One could speculate long and hard on the evolution of the jaw, the tooth, the gills, the eye and the fin, but the significance here is that after nearly three billion years of life on Earth, this combination of features in the placoderms set the stage for the most spectacular and prolific radiation of forms. This is that point from which sprang all the diverse vertebrates: the amphibians, the reptiles, the birds and the mammals, not to mention the fishes themselves. Today more than 21,000 different species of fish are known, which is more than all other vertebrate species combined.

While the placoderms were ruling the oceans, the water's edge still marked the limit for life on Earth. An oxygen-rich atmosphere now enveloped the planet; rain fell and scoured its way to the seas; lakes and swamps filled land depressions—there was no shortage of aquatic environments and so far nothing had ventured into the terrestrial environments. The move was inevitable, however, and the plants were the first to break the barrier, beginning with algal patches at the water's edge. Then reed-like plants evolved and spread further from the water. Bushes and trees evolved from the reeds and by the end of the *Devonian period*, 345 million years ago, forests covered a good deal of the land surface—very strange looking forests, with trees standing high on exposed branching roots, their stems as scaly as a reptile's skin. Insects evolved from among the aquatic invertebrates and followed the vegetation ashore, but for a long time the vertebrates remained on the other side of the water's edge.

One might imagine that among the vertebrates flourishing in the oceans some species must have perceived the advantages awaiting any that managed to find their way ashore. It seems hardly credible that such a wealth of vegetation, standing ready to feed the adventurous, would not have attracted the vertebrates ashore. But two important factors rendered such a move improbable. Firstly, the aquatic vertebrates did not have the physical equipment—legs, teeth, digestive organs and so forth—that would have enabled them to reach and eat land plants, and secondly, it is extremely unlikely that they possessed any mental faculty remotely resembling a sense of perception. Their lives were entirely directed by the instinctive responses that had evolved over many generations to suit their immediate environment. They were programmed to live under a particular set of circumstances and, for the most part, could not survive in any other—much less perceive and take advantage of alternative opportunities. So the line of aquatic vertebrates from which the land-dwelling species evolved could not have been drawn from the water by the attractions of the terrestrial environment. On the contrary, they were stranded on the shore and managed to survive there when the climate changed and their aquatic environment shrank dramatically. Many aquatic species became extinct when this happened, but some possessed certain physical characteristics enabling them to adapt to the changing environment, and these species eventually became the ancestors of the terrestrial vertebrates.

As the Devonian drew to a close the Earth's climate became more extreme. Violent rains alternated with intense droughts. Oceans receded. Vast expanses of the continents turned to desert, and as drought tightened its grip, rivers ceased to flow; strings of ponds that marked the watercourses turned steadily to mud and finally dried rock hard. Around the world lakes and rivers dried out completely, trapping millions of fish (and thus creating many rich fossil deposits) as their aquatic environments turned terrestrial.

Numerous species of fish must have migrated from salt to fresh waters as the *Age of Fishes* progressed, but those doing as drought became commonplace in the *late Devonian* must often have found themselves in water appreciably warmer than the sea they left behind. The shallow waters of lagoons, estuaries, rivers and lakes warm up rapidly. As water becomes warmer its oxygen content is reduced, making it difficult for fish to breathe. Many suffocate, but among the early colonizers of the shallow warm waters some fish developed the trick of gulping air from the surface and holding it in the gullet, where fine blood vessels absorbed the oxygen direct. With the passage of time and countless generations, this system of air-breathing evolved to find its most lasting expression in the lungfish, which could survive in oxygen-developed waters while gill-breathing fish died in their thousands. To stay alive the hungfish needed only to rise to the surface of the water for air. Later it developed nostrils, which further facilitated inhalation, and it even acquired the habit of retiring into the mud as watercourses dried up, waiting there in a state of aestivation (a hot weather version of hibernation) until the next rains washed them out again.

While the evolutionary process was leading the lungfish towards aestivation as a means of surviving drought, it was also refining an alternative answer to the same environmental problem in another group of fish. The alternative answer was mobility. Instead of sitting out the drought, the fish in question would retreat from it, across dry land if need be, and it was this strategic response to the problems of drought that finally endowed the vertebrates with the means of moving from the water to the land.

The strategy evolved in a group known as the lobe-finned fish. The basic equipment they employed was common to all the placoderms' descendants: two pairs of fins, one pair just behind the head and another near the tail. But the lobefins were powered with a greater mass of muscle than their cousins, and their fins were longer, stronger, and constructed in a way that was quite new. Where the fin joined the body a single bone formed a 'shoulder' joint with the main skeleton; in the next part of the fin there were two bones, and finally there was an irregular branching of small bones.

The limbs that have carried the vertebrates through the story of life evolved from the fins of the lobe-finned fishes. The basic structure of shoulder, elbow and wrist, and hip, knee and ankle, was already present in the fish that lived 350 million years ago, although its initial significance lay only in enabling the lobefins to support their own weight and to waddle from pond to pond down the riverbed and between lakes as the drought advanced. This also of course gave them access to the food that the terrestrial environment had on offer—a vast unexploited niche. The lobefins could already breathe air out of water; once they had evolved fins capable of lifting their bellies off the ground they were poised to lead the vertebrates' invasion of the land.

Chapter—4

Water

Water is the only common liquid on our planet. Next to air it is the substance with which we are most intimately acquainted. Because it is so familiar, we are apt to overlook the fact that water is an altogether peculiar substance. Its properties and behaviour are quite unlike those of any other liquid. To take just one example, water has the rare property of being denser as a liquid than as a solid, and it is probably the only substance that attains its greatest density at a few degrees above the freezing point (four degrees centigrade). The consequences of this behaviour are of great importance to life on our planet. When ice forms on a lake, for example, its lower density (only nine tenths that of liquid water) keeps it on top and it acts as an insulating blanket to retard cooling of the underlying water. As a result lakes in the temperate zones do not freeze solid to the bottom but leave a zone for the winter survival of aquatic life. On the other hand, the same peculiar property of water has fatal consequences for living cells. When water in the cells freezes and becomes less dense, its expansion damages or breaks up the cells.

Even the elements of which water is composed—oxygen and hydrogen—are chemically exceptional. Both are unusually reactive. Oxygen is our chief source of energy, being responsible for the respiration of living organisms and the combustion of fuels. Hydrogen unique in the fact that it has no enclosing shell but only a single electron, is able to attach itself to other atoms not only by means of its electron (a valence bond) but also by virtue of the attraction of its unoccupied, positively charged side for an electron in a second atom. This attachment is known as the *hydrogen bond.* In water the two hydrogen atoms attached to each oxygen atom can become linked to other atoms as well by means of these so-called hydrogen bonds. As a consequence the H_2O molecules are joined together, so that water should be considered not a collection of separate molecules but a united association. In effect the whole mass of water in a vessel is a single molecule.

The best method of detecting hydrogen bonds is to study water with the infrared spectrograph. We have found that the hydrogen bond absorbs radiation most strongly at a wavelength of about three microns, which is in the near infra-red region of heat radiation—i.e., close to the visible light spectrum. Liquid water absorbs this radiation so powerfully that if our eyes were sensitive to the infra-red, water would appear jet black. There is some absorption even in the visible spectrum at the red end. The fact that water absorbs red light accounts for its characteristic blue-green colour.

Water's 33 Components

One of the shocks to our familiar notions about water is that its formula is not simply H_2O. Nor is it a single substance. The beginning of this disillusionment came in 1934 when *Harold Urey discovered* "heavy water." Urey found that the purest water contained besides hydrogen and oxygen another substance like hydrogen but with an atomic weight of two, or twice that of hydrogen. This substance, which is now called deuterium, combines with oxygen to form the compound D_2O. By now, we know, of course, that there is a third isotope of hydrogen, called tritium, and three isotopes of oxygen: 0-16, 0-17 and 0-18. Thus the purest water that can be prepared in the laboratory is made up of six different isotopes, which may be combined in 18 different ways. If we add the various kinds of ions into which the addition or removal of an electron may transform water's atoms, we find that pure water contains no fewer than 33 substances.

Of course the amounts of the isotopes other than common hydrogen and common oxygen (0.16) are tiny. Tritium and oxygen 17 appear only in the minutest traces, and deuterium is present to the extent of about 200 parts per million and oxygen 18 about 1,000 parts per million. However, the properties of heavy water, particularly the D_2O variety, have attracted wide interest and have been extensively studied.

D_2O has a slightly higher boiling point than H_2O (101.4 degrees C.), freezes at a substantially higher temperature (3.8 degrees C.), and is somewhat more viscous than ordinary water. Its physiological properties are surprising. In animals and plants it

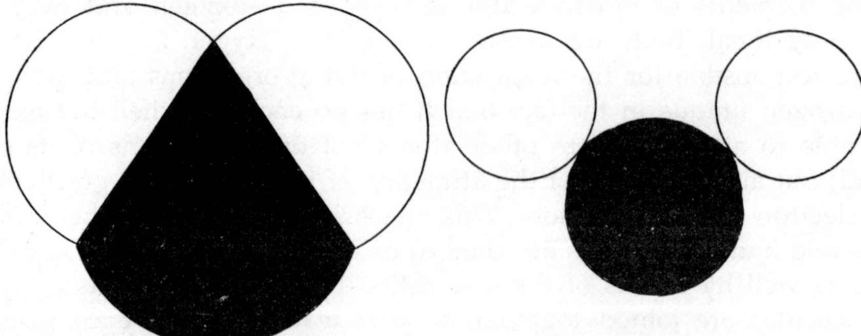

Fig. 4.1: Molecule of water consists of one oxygen atom (*black*) and two hydrogen atoms (*white*). The distance between the center of the oxygen atoms and the centre of each of the hydrogen atoms is .9 Angstrom unit (one Angstron unit: .00000001 centimetre). The angle formed by the two hydrogen atoms is 105 degrees. These dimensions are fitted together in the drawing at left. In the more schematic drawing at right the size of the atoms has been reduced. This representation of the molecule is used in the following drawings.

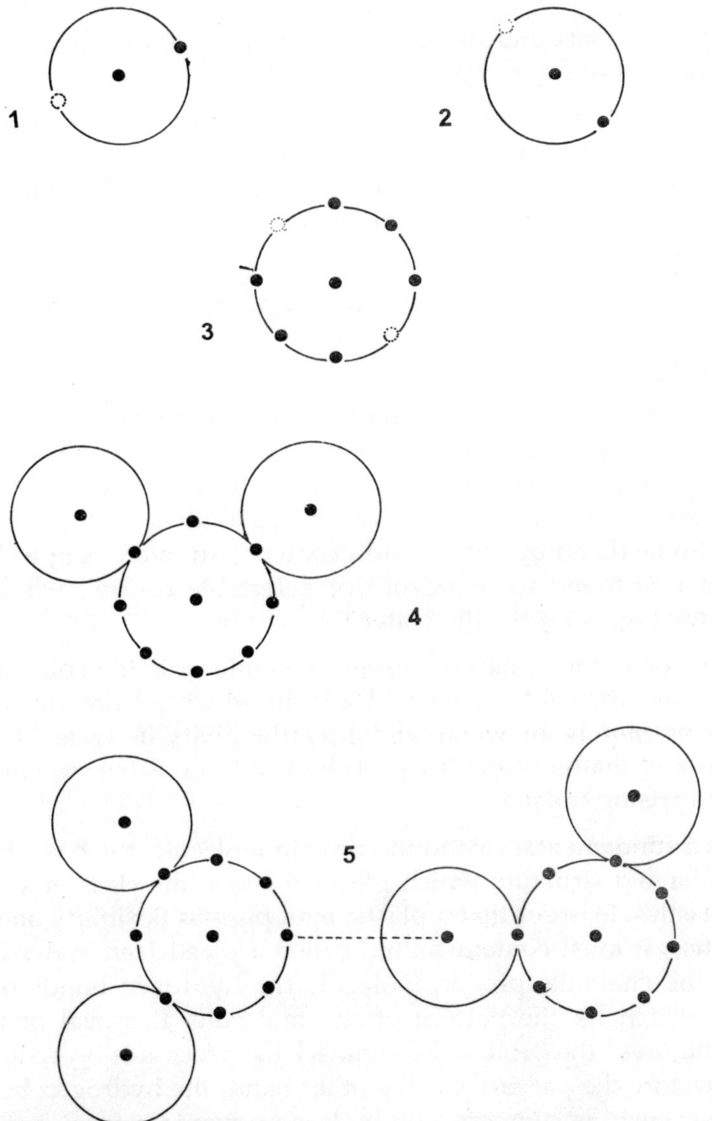

Fig. 4.2: Electrons and Protons of the water molecule account for most of its physical and chemical properties. The hydrogen atom (*1 and 2 in this highly schematic picture*) consists of a positively charged proton (*black dot*) and a negatively charged electron (*coloured dot*). The oxygen atom (3) has eight electrons, six of which are arranged in an outer shell. Because hydrogen shell has room for one more electron (*broken coloured circle*), and the outer shell of oxygen has room for two more electrons, the atoms have an affirmity for each other. In the water molecule (4) the electrons of the hydrogen atoms are shared by the oxygen atom. Because the positively charged proton of the hydrogen atom now sticks out from the water molecule, it has an attraction for the negatively charged electrons of a neighbouring water molecule (5). This relatively weak force (*broken line*) is called a hydrogen bond.

appears to be entirely inert and useless. Seeds will not sprout in D_2O, and rats given only D_2O to drink will die of thirst.

The largest use of heavy water is as a moderator in nuclear reactors, but it is also widely employed in theoretical research, especially in organic and biological chemistry. If compounds containing active hydrogen are treated with D_2O, deuterium will replace the hydrogen and the compound will show changes in chemical properties resulting from the lesser reactivity of deuterium.

It is interesting to find that the amount of D_2O in natural water appears to be the same whether the water comes from an alpine glacier or the bottom of the ocean, from willow wood or mahogany.

Tritium is more ephemeral and more variably distributed. It is formed in the highest layers of the atmosphere by the bombardment of cosmic rays, and falls in rain and snow. Since tritium is radioactive, with a half-life of 12.5 years, it disappears after a time from water which has been out of contact with the atmosphere. Wines, and water in wells, can be dated by their tritium content. An interesting well in the Urbana-Champaign area was found to be devoid of detectable tritium, which means that at least 50 years have elapsed since the water fell as rain.

The functions of water in nature are innumerable. It is the solvent par excellence. It is the medium in which life originated and in which all organisms still exist. The living cell consists largely of water and literally floats in water. Considering how predominantly living matter is made up of the fluid, the extent to which it takes on a solid shape is surprising indeed.

Water plays a fundamental role in the protein molecule, the basic material of living matter. Proteins have a structure which places them in the class of substances known to chemists as plastics. In order that a plastic man possess flexibility and other desirably physical properties, it must contain a fluid called a plasticizer. Water is a "plasticizer" for proteins. In the chainlike protein molecule the hydrogen bonds of water provide secondary links which fix the pattern of the molecule. Removal of water alters the pattern and "denatures" the protein. Fortunately the process is reversible, so that water in the cells can restore the pattern. On the other hand, the hydrogen bonds of hydrides other than water, such as ammonia or hydrogen cyanide, form a stable denatured configuration which freezes the protein in a dead pattern. This is why all the hydrides except water are extremely toxic. Their action is somewhat like that of a virus, which corrupts the true protein structure into a strange distorted pattern.

Water's Structure

To understand the behaviour of water we must understand its structure. It is far from simple. The best approach is a study of the structure of ice. The arrangement of the oxygen and hydrogen atoms in the ice crystal has been determined by X-rays and other means. The two hydrogens are bonded to the oxygen approximately at right angles to each other, more exactly, at an angle of 105 degrees. If the angle were 109

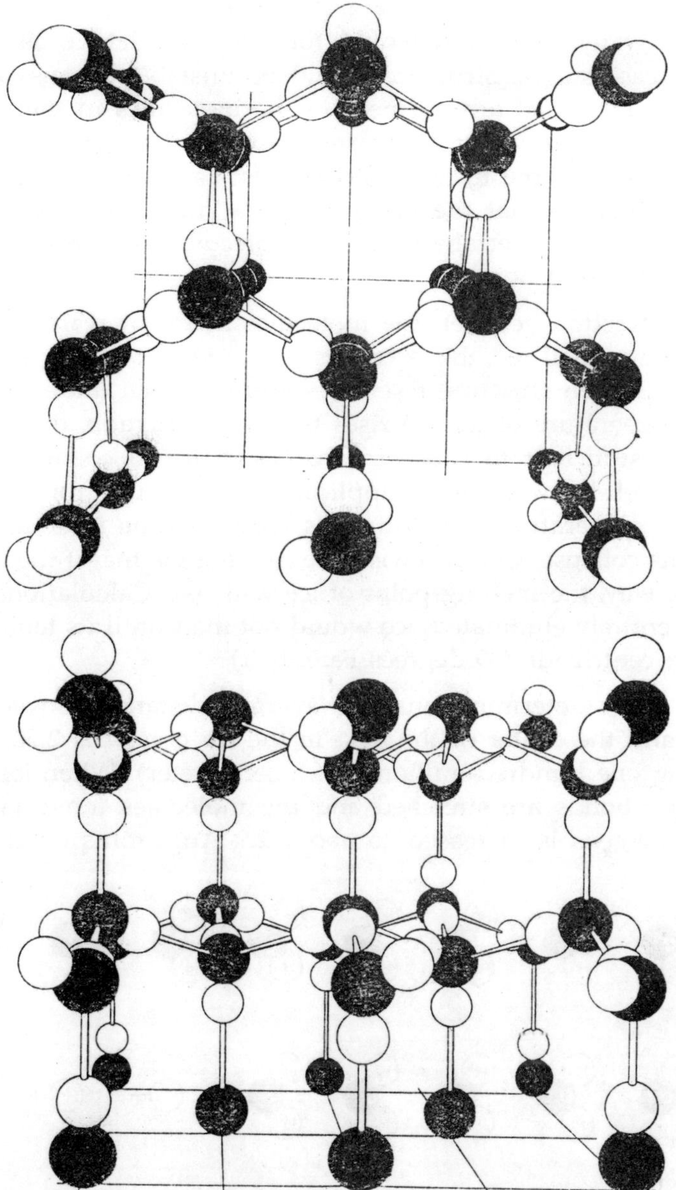

Fig. 4.3: ICE consists of water molecules in this arrangement. The top drawing shows a model of ice seen from one direction. The bottom drawing shows the same model seen as if the reader had turned the top drawing forward on a horizontal axis in the plane of the page. Some hydrogens have been omitted from the molecules which touch the grid. Each hydrogen in each molecule is joined to an oxygen in a neighbouring molecule by a hydrogen bond (*rods*). In actuality the molecules of ice are packed more closely together; here they have been pulled apart to show the structure. In a similar model of liquid water the molecules would be much more loosely organized, farther apart and joined by more hydrogen bonds.

degrees, the frozen water molecules would form a cubic lattice as in the diamond crystal. But in this case such a structure would be unstable because of the strain on the distorted bonds. The exact arrangement of the molecules in the ice crystal is not known with certainty; we know that they form a hexagonal structure, which is exhibited on the macroscopic scale in the form of snowflakes. Each molecule is surrounded by four nearest neighbours, so that the group has one molecule at the center and the other four at the corners of a tetrahedron. The molecules and group of molecules are joined together by hydrogen bonds.

The forces of attraction between the molecules in ice or water produce a strong inward pressure. As we shall see, this accounts for some of water's peculiar properties. In the form of ice, its open structure resembles a bridge arch under heavy downward stress. When the temperature of the ice rises to zero centigrade, the thermal agitation of the molecules is sufficient to cause the ice structure to collapse, and the water becomes fluid. It is well-known that the application of pressure from outside will make ice melt at a lower temperature; evidently this reinforces the internal pressure within the ice and assists its collapse. Contrarywise, we can assume that if the internal pressure is reduced in some way, the melting point of ice will rise. Calculations indicate that if this pressure were entirely eliminated, ice would not melt until its temperature reached 15 degrees or more centrigade (59 degrees Farenheit).

According to X-ray determinations, the average distance between the center of one oxygen atom and the center of the next in the ice crystal is 2.72 Angstrom units (an Angstrom being one hundred-millionth of a centimeter). When ice melts to liquid water, the hydrogen bonds are stretched and the molecules move farther apart: the distance between oxygens is increased to about 2.9 Angstroms on the average. This

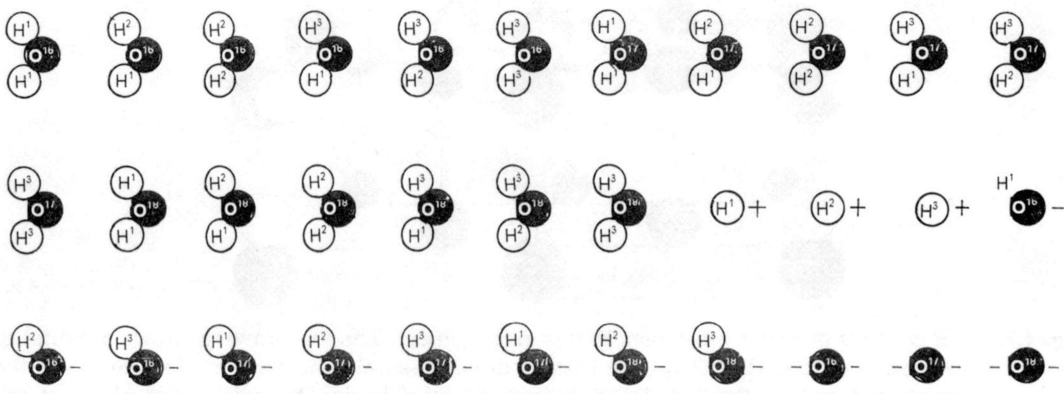

Fig. 4.4: Water is not H_2O but a mixture of 33 different substances. Eighteen of these are combinations of three isotopes of hydrogen and three of oxygen. The three hydrogen isotopes are ordinary hydrogen (H^1), deuterium (H^2) and tritium (H^3). the three oxygen isotopes are ordinary oxygen (O^{16}), oxygen 17 and oxygen 18. The remaining substances are various ions (*plus or minus signs*).

stretching would open the structure further and make water less dense were it not for the fact that in the fluid the molecules crowed together in more compact groups. Each molecule is now surrounded by five or more neighbours instead of only four.

The chaotic disorder in which water molecules exist in the liquid state is difficult to picture. Their arrangement shifts continually. The angle between the two hydrogen atoms in the water molecule no longer remains fixed near a right angle but becomes variable, so that the molecule is flexible. Each oxygen atom now attracts by electrical forces not two extra hydrogen atoms as in ice, but three or more. Thus we may find an oxygen atom surrounded by five or six hydrogens and a hydrogen atom surrounded by as many as three oxygens. In the closely knit, flexible structure the hydrogens constantly shift their positions and displace one another. Each such displacement is propagated in a chain or zipper fashion throughout the liquid. This has consequences which affect the viscosity, dielectric constant and electrical conductivity of water.

Water's Properties

In an ordinary unassociated liquid such as benzene the molecules flow by sliding around one another. In water the motion is rolling rather than sliding. Since the molecules are connected by hydrogen bonds, at least one bond must be broken before any flow can occur. From the fact that the molecules are bonded together, it might be expected theoretically that the viscosity of water should be comparatively high. However, each hydrogen bond in water is shared on the average between two other molecules, and one of these weakened bonds is easily broken. The greater viscosity of ice is due to the fact that each hydrogen is boded to only a single oxygen atom from another molecule, and this firmer bond must be broken before any movement can occur.

The dielectric constant of a liquid is a measure of its capacity to neutralize the attractions between electrical charges. For example, when sodium chloride dissolves in water, the positively charged sodium and negatively charged chlorine ions are separated. They are kept apart because water has a high dielectric constant—the highest of any common liquid. It reduces the force of attraction between the oppositely charged ions in solution to not much more than 1 per cent of the original value. The reason for water's strong neutralizing action lies in the arrangement of its molecules. In an aggregation of water molecules a hydrogen atom does not share its electron equally with the oxygen atom to which it is attached: the electron is closer to the oxygen atom than to the hydrogen. As a result the hydrogen atoms are positively charged and the oxygen negatively charged. Now when a substance is dissolved, separating into ions, the oxygen atoms are attracted to the positive ions and the hydrogens to the negative ones. Consequently water molecules surrounding a positive ion are oriented with their oxygens next to the ion, and molecules around a negative ion turn their hydrogens toward the ion. Thus the water molecules act as cages which separate and neutralize the ions. This explains why water is so effective a solvent for electrolytes (substances which dissociate into ions) such as sodium chloride.

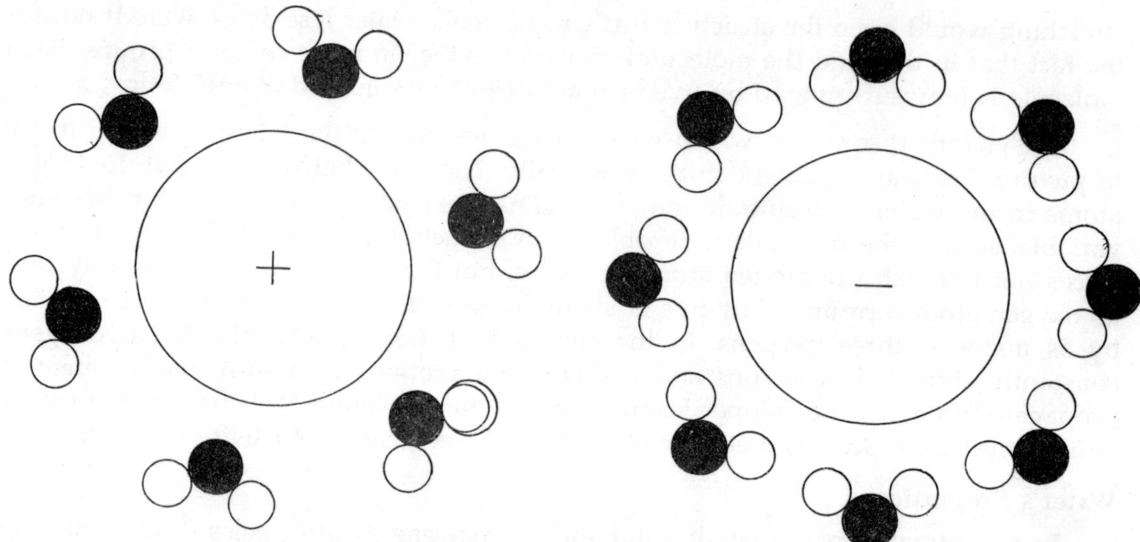

Fig. 4.5: Ions (*circles labelled with plus and minus signs*) in water are kept apart because they polarize the water molecules around them. Because the oxygen atom of the water molecule has more negative charge than the hydrogen atoms, it is attracted to a positive ion (*left*). Because the hydrogen atoms have more positive charge than the oxygen atom, they are attracted to a negative ion (*right*).

Water is generally supposed to be a good conductor of electricity. Every lineman knows the danger of handling high-voltage electrical lines when standing on a moist surface. Actually the conductivity is due to impurities dissolved in the water. Water is such a good solvent for electrolytes, including carbon dioxide from the atmosphere, that any moist surface may be assumed to be a good conductor. But pure water (which is difficult to keep pure—it must be kept out of contact with air and in a vessel of an inert material such as quartz) is a very good insulator indeed. The reason is that while the hydrogen and oxygen atoms in a water molecule are in a sense charged, or ionized, they cannot move about separately because they are attached to each other, and hence cannot carry an electric current.

One of the anomalous properties of liquid water is its high specific heat, or heat-holding capacity. The specific heat of a substance is the quantity of heat required to raise the temperature of one gram one degree centigrade. The specific heat of liquid water is more than twice as great as that of ice. The explanation is that the liquid's ionized oxygen and hydrogen atoms, though held together behave like free ions in their capacity to vibrate in response to heat. Thus they can absorb as much energy as if the ions were really free.

The strong bonding of water molecules accounts for the fact that water has unusually high melting and boiling points. It also explains why it is so difficult to vapourize ice. To do this we must break all the hydrogen bonds holding the molecules together. Calculations indicate that the total energy of the hydrogen bonds in one mole of water (18 grams) is equivalent to 6,000 calories.

Hydrates

For more than 60 years physical chemists have studied water largely in terms of solutions of electrolytes. This study has produced considerable information about electrolytes and ions, but not a great deal about the properties of water itself. Strangely enough, in recent years we have learned much more about water by examining its behaviour with substances which for all practical purposes are insoluble in water!

This behaviour was called to the attention of chemists in a dramatic fashion by certain surprising natural phenomena. One was the fact that corn sometimes showed frost effects when the temperature was 40 degree F., well above freezing. Another was the discovery that pipelines carrying natural gas often became clogged with a slushy "snow", containing water, at temperatures as high as 68 degree F. The plain indication was that these freeze-ups were due to the water. But this raised some startling and

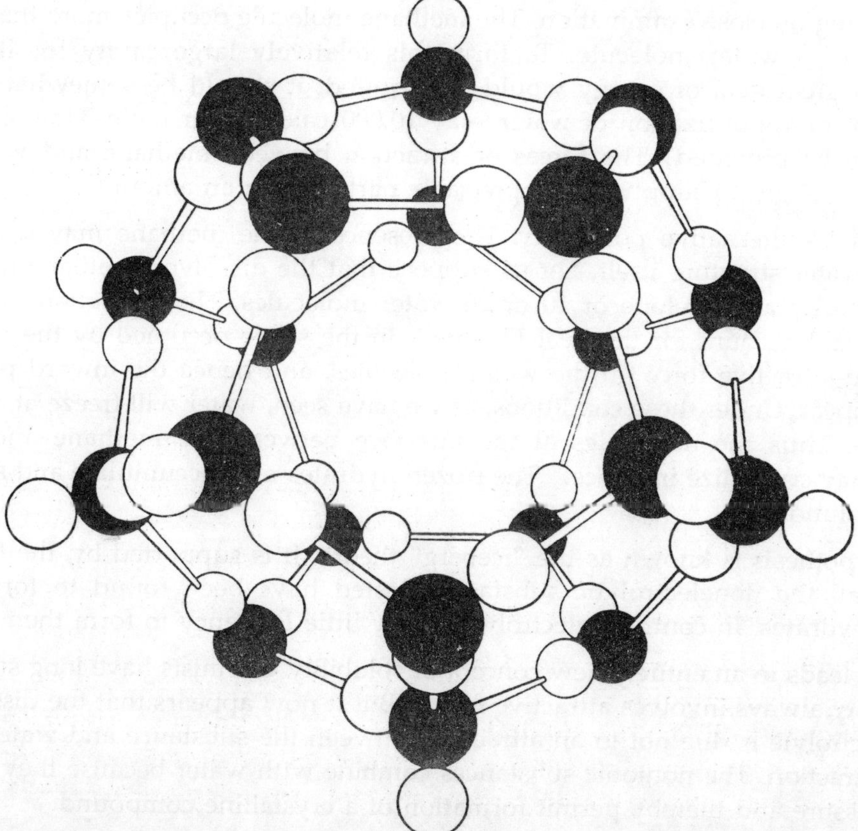

Fig. 4.6: Hydrate is formed when a foreign molecule in water is electrically neutral and just the right size for the water molecules to collect around it is crystalline cage. This cage can then grow to a much large crystal. It is part of a repeating unit of 136 molecules.

interesting questions. What made water freeze at these high temperatures? How could water combine, or become "bound," with substances which were all but insoluble in it? The mystery was not lessened when it was discovered that even the noble gases such as argon and krypton, which refuse all chemical reactions, could join with water to form a quasi compound.

Let us look at these questions in the light of what we have learned about water's structure and properties.

The methane molecule for example, does not form ions in water, nor does it accept the hydrogen bonds. There is very little attraction between it and the water molecule. It is, however, slightly soluble in water, and the dissolving methane molecules form compounds with water—"hydrates"—in which several water molecules are joined to one of methane.

The reaction liberates 10 times as much heat as when methane dissolves in hexane, although it is much more soluble in hexane than in water. This fact becomes even more surprising on close examination. The methane molecule occupies more than twice the volume of a water molecule. To form this relatively large cavity for itself on dissolving, a great deal of energy would be required: it should be somewhat greater than the heat of vapourization of water—say 10,000 calories per mole. How could so much energy be provided? The forces of attraction between methane and water are apparently too slight to supply any appreciable part of such an amount.

There is an alternative possibility. The presence of the methane may drastically change the water structure itself. Let us suppose that the dissolved methane molecule is surrounded by an envelope of 10 or 20 water molecules. The formation of such a structure would account for the heat liberated. In the space occupied by the methane molecule the attractive force on the water molecules, and hence the inward pressure, would disappear. Under these conditions, as we have seen, water will freeze at a higher temperature. Thus the molecules at the interface between the methane and water molecules may crystallize into "ice." The frozen hydrates may accumulate and separate out of the solution.

This hypothesis is known as the "iceberg" theory. It is supported by, the fact that practically all the nonelectrolytic substances tested have been found to form solid crystalline hydrates. In contrast, electrolytes show little tendency to form them.

All this leads to an entirely new concept of solubility. Chemists have long supposed that solubility always involves attractive forces. But it now appears that the dissolution of a nonelectrolyte is due not to an attraction between the substance and water but to a lack of attraction. The nonionic substances combine with water because they remove internal pressure and thereby permit formation of a crystalline compound.

In order to understand the formation of these hydrates, it is necessary to consider their molecular structure in detail. They tend to fall into groups according to the number of water molecules they contain.

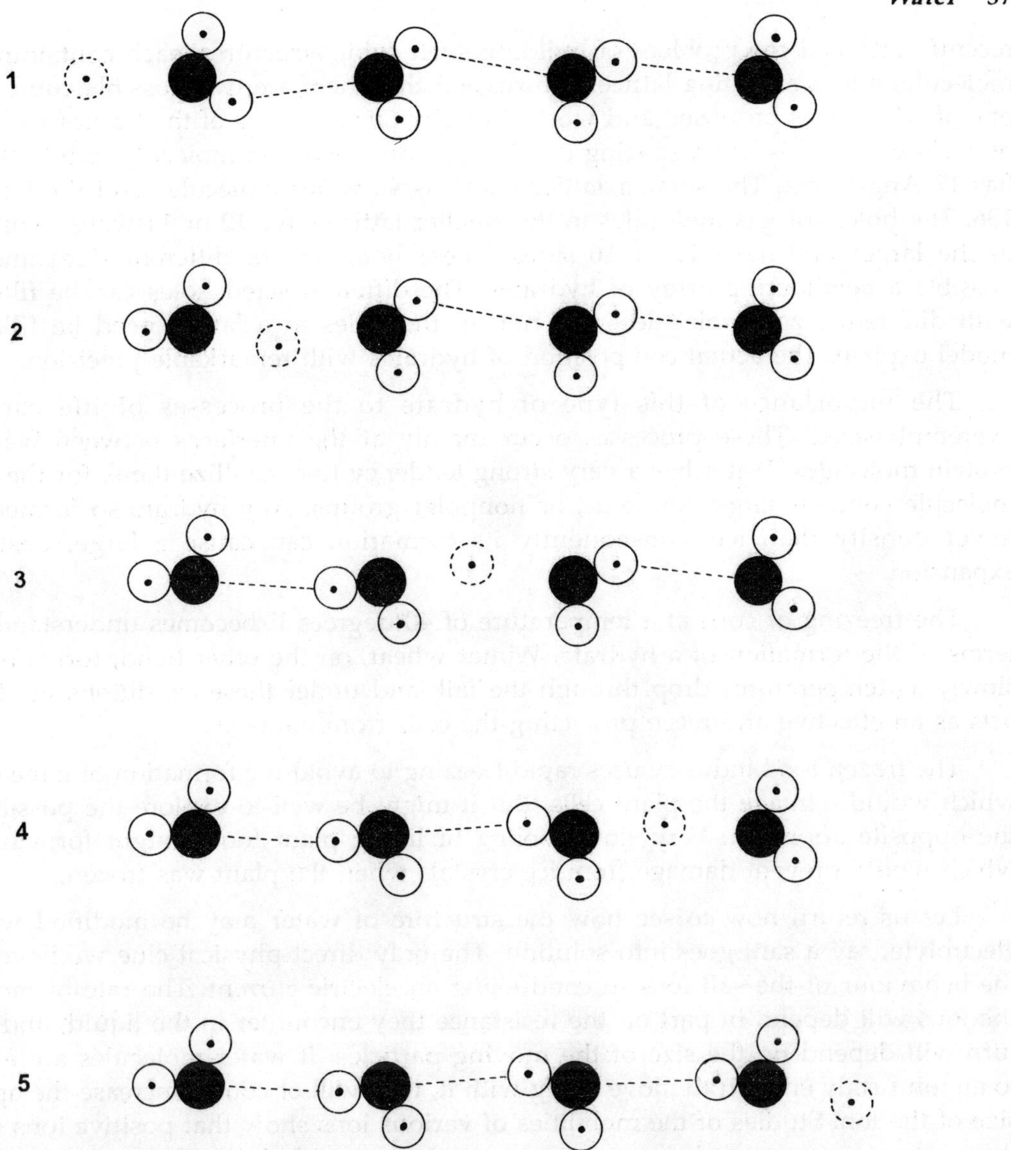

Fig. 4.7: Rapid Migration of Hydrogen Ions through water is explained by the assumption that the ions do not actually travel through the water but are passed from one molecule to the next by a process of exchange. Here hydrogen ion is represented by coloured dot surrounded by a broken circle. In the first horizontal row the hydrogen ion approaches a water molecule. In the second row the ion has taken the place of one hydrogen atom of the molecule, expelling the atom as a new ion. In the third row the new ion repeats this process.

The ground work for the study of the hydrates structure was laid by *M. von Stackelberg* in Germany 10 years ago. He showed by X-ray studies that this structure was cubic, in contrast to the hexagonal structure of ice. *W.F. Claussen* of our laboratory recently attacked the problem of building such cubic structures, each containing a gas molecule, into a repeating lattice. It turns out that there are two possible cubic lattices, one of which was proposed and worked out by *Linus Pauling* of the California Institute of Technology. This has a spacing of 12 Angstroms between molecules while the other has 17 Angstroms. The smaller lattice contains 46 water molecules and the larger one 136. The holes for gas molecules in the smaller lattice have 12 or 14 walls, while those in the larger one have 12 or 16 sides. These holes are of different sizes and make possible a bewildering array of hydrates. The different sized holes can be filled only with different sized molecules, and not all the holes in a lattice need be filled. The model explains the actual composition of hydrates with remarkable precision.

The importance of this type of hydrate to the processes of life cannot be overemphasized. These processes occur mainly at the interfaces between water and protein molecules. Water has a very strong tendency to crystallize there, for the protein molecule contains large non-ionic, or nonpolar groups. Any hydrate so formed has a lower density than ice; consequently its formation can cause a large, destructive expansion.

The freezing of corn at a temperature of 40 degrees F. becomes understandable in terms of the formation of a hydrate. Winter wheat, on the other hand, forms hydrates slowly as temperatures drop through the fall, and under these conditions the hydrate acts as an effective antifreeze protecting the cells from damage.

The frozen food industry uses rapid freezing to avoid the formation of large crystals which would damage the plant cells. But it might be well to explore the possibility of the opposite approach. Very slow cooling of living plant foods might form hydrates which would prevent damage from ice crystals when the plant was frozen.

Let us return now to see how the structure of water may be modified when an electrolyte, say a salt, goes into solution. The only direct physical clue we have lies in the behaviour of the salt ions in conducting an electric current. The rate of motion of the ions will depend in part on the resistance they encounter in the liquid, and this in turn will depend on the size of the moving particles. If water molecules are attached to an ion firmly enough to move along with it, they will of course increase the apparent size of the ion. Studies of the mobilities of various ions show that positive ions smaller than potassium carry such a cage of water molecules with them. The positively charged ion attracts the oxygens of two water molecules quite strongly, and if the volume of the ion plus its two water molecules is not greater than that of a methane molecule, a cage of hydrogen-bonded water molecules will form around this group as a nucleus. Positive ions larger than potassium fail to pick up such a cage. The same is true of most, but not all, negative ions.

The positively charged hydrogen ion and the negatively charged hydroxyl ion (OH) are surrounded by cages, and yet they show the highest mobility in carrying a current. We must conclude that they manage in some way to escape from the cage. Actually the mechanism is not hard to picture: they continually form new cages as they travel by a process called proton transfer. Under the influence of an electric field a hydrogen ion may jump from one water molecule to the next. When this has occurred, the hydrogen on the farther side of the water molecule takes up its part in the race like a relay runner and jumps to the next water molecule. Thus a succession of protons, each doing its bit, carries the current. The motion is rapid, because each proton moves only a short step. Transfer of the proton also explains the conduction of electricity by hydroxyl ions. When a proton jumps toward the right, say, and joins a hydroxyl ion, it leaves a hydroxyl ion on its left. The effect is the same as if a hydroxyl itself moved to the left.

Water, then, is not simple H_2O but a unique and complicated material with distinct and varied chemical properties. It has a definite though changing physical structure which depends on the orientation of its molecules with respect to one another, and to the molecules of dissolved substances. Since the behaviour of all living nature and much of the inanimate world is inseparably linked to the peculiar characteristics of this liquid, the study of water substance can tell us a great deal about fundamental aspects of the world in which we live.

Chapter—5

The Chemical Elements of Life

How many of the 90 naturally occurring elements are essential to life? After more than a century of increasingly refined investigation, the question still cannot be answered with certainty. Only a year or so ago the best answer would have been 20. Since then four more elements have been shown to be essential for the growth of young animals: fluorine, silicon, tin and vanadium. Nickel may soon be added to the list. In many cases the exact role played by these and other trace elements remains unknown or unclear. These gaps in knowledge could be critical during a period when the biosphere is being increasingly contaminated by synthetic chemicals and subjected to a potentially harmful redistribution of salts and metal ions. In addition, new and exotic chemical form of metals (such as methyl mercury) are being discovered, and a complex series of competitive and synergistic relations among mineral salts has been encountered. We are led to the realization that we are ignorant of many basic facts about how our chemical milieu affects our biological fate.

Biologists and chemists have long been fascinated by the way evolution has selected certain elements as the building blocks of living organisms and has ignored others. The composition of the earth and its atmosphere obviously sets a limit on what elements are available. The earth itself is hardly a chip off the universe. The solar system, like the universe, seems to be 99 per cent hydrogen and helium. In the earth's crust helium is essentially nonexistent (except in a few rare deposits) and hydrogen atoms constitute only about .22 per cent of the total. Eight elements provide more than 98 per cent of the atoms in the earth's crust: oxygen (47 per cent), silicon (28 per cent), aluminium (7.9 per cent), iron (4.5 per cent), calcium (3.5 per cent), sodium (2.5 per cent), potassium (2.5 per cent) and magnesium (2.2 per cent). Of these eight elements only five are among the 11 that account for more than 99.9 per cent of the atoms in the human body. Not surprisingly nine of the 11 are also the nine most abundant elements in sea-water.

Two elements, hydrogen and oxygen, account for 88.5 per cent of the atoms in the human body; hydrogen supplies 63 per cent of the total and oxygen 25.5 per cent. Carbon accounts for another 9.5 per cent and nitrogen 1.4 per cent. The remaining 20 elements now thought to be essential for mammalian life account for less than .7 per cent of the body's atoms.

The Background of Selection

Three characteristics of the biosphere or of the elements themselves appear to have played a major part in directing the chemistry of living forms. *First and foremost* there is the ubiquity of water, the solvent base of all life on the earth. Water is a unique compound; its stability and boiling point are both unusually high for a molecule of its simple composition. Many of the other compounds essential for life derive their usefulness from their response to water fullness from their response to water: whether they are soluble or insoluble, whether or not (if they are soluble) they carry an electric charge in solution and, not least, what effect they have on the viscosity of water.

The second directing force involves the chemical properties of carbon, which evolution selected over silicon as the central building block for constructing giant molecules. Silicon is 146 times more plentiful than carbon in the earth's crust and exhibits many of the same properties. Silicon is directly below carbon in the periodic table of the elements; like carbon, it has the capacity to gain four electrons and form four covalent bonds.

The crucial difference that led to the preference for carbon compounds over silicon compounds seems traceable to two chemical features: the unusual stability of carbon dioxide, which is readily soluble in water and always monometric (it remains a single molecule), and the almost unique ability of carbon to form long chains and stable rings with five or six members. This versatility of the carbon atom is responsible for the millions of organic compounds found on the earth.

Silicon in contrast, is insoluble in water and forms only relatively short chains with itself. It can enter into longer chains, however, by forming alternating bonds with oxygen, creating the compounds known as silicones (-Si-O-Si-O-Si-). Carbon-to-carbon bonds are more stable than silicon-to-silicon bonds, but not so stable as to be virtually immutable, as the silicon-oxygen polymers are. Nevertheless, silicon has recently been shown to be essential in a way as yet unknown for normal bone development and full growth in chicks.

The third force influencing the evolutionary selection of the elements essential for life is related to an atom's size and charge density. Obviously the heavy synthetic elements from neptunium (atomic number 93) to lawrencium (No. 103), along with two lighter synthetic elements, technetium (No. 43) and promethium (No. 61), were never available in nature. (The atomic number expresses the number of protons in the nucleus of an atom or the number of electrons around the nucleus). The eight heavy elements in another group (No. 84 and 85 and Nos. 87 through 92) are too radioactive to be useful in living structures. Six more elements are inert gases with virtually no useful chemical reactivities: helium, neon, argon, krypton, xenon and radon. On various plausible grounds one can exclude another 24 elements, or a total of 38 natural elements, as being clearly unsatisfactory for incorporation in living organisms because of their relative unavailability (particularly the elements in the lanthanide and actinide series) or their high toxicity (for example mercury and lead). This leaves 52 of the 90 natural elements as being potentially useful.

Only three of the 24 elements known to be essential for animal life have an atomic number above 34. All three are needed only in trace amounts: molybdenum (No. 42), in (No. 50) and iodine (No. 53). The four most abundant atoms in living organisms—hydrogen, carbon, oxygen and nitrogen—have atomic numbers of 1, 6, 7 and 8. Their preponderance seems attributable to their being the smallest and lightest elements that can achieve stable electronic configurations by adding one to four electrons. The ability to add electrons by sharing them with other atoms is the first step in forming chemical bonds leading to stable molecules. The seven next most abundant elements in living organisms all have atomic numbers below 21. In the order of their abundance in mammals they are calcium (No. 20), phosphorus (No. 15), potassium (No. 19), sulphur (No. 16), sodium (No. 11), magnesium (No. 12) and chlorine (No. 17). The remaining 10 elements known to be present in either plants or animals are needed only in traces. With the exception of fluorine (No. 9) and silicon (No. 14), the remaining eight occupy positions between No. 23 and No. 34 in the periodic table. It is interesting that this interval embraces three elements for which evolution has evidently found no role: gallium, germanium and arsenic. None of the metals with properties similar to those of gallium (such as aluminum and indium) has proved to be useful to living organisms. On the other hand, since silicon and tin, two elements with chemical activities similar to those of germanium, have just joined the list of essential elements, it seems possible that germanium too, in spite of its rarity, will turn out to have an essential role. Arsenic, of course, is a well-known poison.

Functions of Essential Elements

Some useful generalizations can be made about the role of the various elements. Six elements—carbon, nitrogen, hydrogen, oxygen, phosphorus and sulphur—make up the molecular building blocks of living matter: amino acids, sugars, fatty acids, purines, pyrimidines and nucleotides. These molecules not only have independent biochemical roles but also are the respective constituents of the following large molecules: proteins, glycogen, starch, lipids and nucleic acids. Several of the 20 amino acids contain sulphur in addition to carbon, hydrogen and oxygen. Phosphorous plays an important role in the nucleotides such as adenosine triphosphate (ATP), which is central to the energetics of the cell. ATP includes components that are also one of the four nucleotides needed to form the double helix of deoxyribonucleic acid (DNA), which incorporates the genetic blueprint of all plants and animals. Both sulphur and phosphorous are present in many of the small accessory molecules called coenzymes. In bony animals phosphorous and calcium help to create strong supporting structures.

The electrochemical properties of living matter depend critically on elements or combinations of elements that either gain or lose electrons when they are dissolved in water, thus forming ions. The principal cations (electron-deficient, or positively charged,

Composition of Universe		*Composition of Earth's Crust*		*Composition of Seawater*		*Composition of Human Body*	
Per cent of Total Number of Atoms							
[H]	91	[O]	47	[H]	66	[H]	63
He	9.1	Si	28	[O]	33	[O]	25.5
[O]	.057	Al	7.9	[Cl]	.33	[C]	9.5
[N]	.042	Fe	4.5	[Na]	.28	[N]	1.4
[C]	.021	[Ca]	3.5	[Mg]	.033	[Ca]	.31
Si	.003	[Na]	2.5	[S]	.017	P	.22
Ne	.003	[K]	2.5	[Ca]	.006	[Cl]	.03
[Mg]	.002	[Mg]	2.2	[k]	.006	[K]	.06
Fe	.002	Ti	.46	[C]	.0014	[S]	.05
[S]	.001	[H]	.22	Br	.0005	[Na]	.03
		[C]	.19			[Mg]	.01
All Others <.01		All Others < .1		All Others <.1		All Others < .01	

Al	Aluminum	C	Carbon	Fe	Iron	O	Oxygen	S	Sulphur
B	Boron	Cl	Chlorine	Mg	Magnesium	K	Potassium	Ti	Titanium
Br	Bromine	He	Helium	Ne	Neon	Si	Silicon		
Ca	Calcium	H	Hydrogen	N	Nitrogen	Na	Sodium		

Fig. 5.1: Chemical Selectivity of Evolution can be demonstrated by comparing the composition of the human body with the approximate composition of seawater, the earth's crust and the universe at large. The percentage are based on the total number of atoms in each case; because of rounding the totals do not exactly equal 100. Elements in the coloured boxes in the last column appear in one or more columns at the left. Thus one sees that phosphorus, the sixth most plentiful element in the body, is a rare element in inanimate nature. Carbon, the third most plentiful element, is also very scarce elsewhere.

ions are provided by four metals: sodium, potassium, calcium and magnesium. The principal anions (ions with a negative charge because they have surplus electrons) are provided by the chloride ion and by sulphur and phosphorous in the form of sulfate ions and phosphate ions. These seven ions maintain the electrical neutrality of body fluids and cells and also play a part in maintaining the proper liquid volume of the blood and other fluid systems. Whereas the cell membrane serves as a physical barrier to the exchange of large molecules, it allows small molecules to pass freely. The electrochemical functions of the anions and cations serve to maintain the appropriate relation of osmotic pressure and charge distribution on the two sides of the cell membrane.

One of the striking features of the ion distribution is the specificity of these different ions. Cells are rich in potassium and magnesium, and the surrounding plasma is rich in sodium and calcium. It seems likely that the distribution of ions in the plasma of higher animals reflects the oceanic origin of their evolutionary antecedents. One would like to know how primitive cells learned to exclude the sodium and calcium ions in which they were bathed and to develop an internal milieu enriched in potassium and magnesium.

The third and last group of essential elements consists of the trace elements. The fact that they are required in extremely minute quantities in no way diminishes their great importance. In this sense they are comparable to the vitamins. We now know that the great majority of the trace elements, represented by metallic ions, serve chiefly as key components of essential enzyme systems or of proteins with vital functions (such as hemoglobin and myoglobin, which respectively transports oxygen in the blood and stores oxygen in muscle). The heaviest essential element, iodine, is an essential constituent of the thyroid hormones thyroxine and triiodothyronine, although its precise role in hormonal activity is still not understood.

The Trace Elements

To demonstrate that a particular element is essential to life becomes increasingly difficult as one lowers the threshold of the amount of a substance recognizable as a "trace." It has been known for more than 100 years, for example, that iron and iodine are essential to man. In a rapidly developing period of biochemistry between, 1928 and 1935 four more elements, all metals, were shown to be essential: copper, manganese, zinc and cobalt. The demonstration can be credited chiefly to a group of investigators at the University of Wisconsin led by C.A. Elvehjem, E.B. Hart and W.R. Todd. At that time it seemed that these four metals might be the last of the essential trace elements. In the next 30 years, however, three more elements were shown to be essential chromium, selenium and molybdenum. Fluorine, silicon, tin and vanadium have been added since 1970.

The essentiality of five of these last seven elements was discovered through the careful, paintstaking efforts of *Klaus Schwarz* and his associates, initially located at the National Institute of Health and now based at the Veterans Administration Hospital in Long Beach, Calif. For the past 15 years Schwarz's group has made a systematic study of the trace-element requirements of rates and other small animals. The animals are maintained from birth in a completely isolated sterile environment.

The apparatus is constructed entirely of plastics to eliminate the stray contaminants contained in metal, glass and rubber. Although even plastics may contain some trace elements, they are so tightly bound in the structural lattice of the material that they cannot be leached out or be picked up by an animal even through contact. A typical isolator system houses 32 animals in individual acrylic cages. Highly efficient air filters remove all trace substances that might be present in the dust in the air. Thus the animals only access to essential nutrients is through their diet. They receive chemically

Element	*Symbol*	*Atomimc Number*	*Comments*
Hydrogen	H	1	Required for water and organic compounds.
Helium	He	2	Inert and unused.
Lithium	Li	3	Probably unsued.
Beryllium	Be	4	Probably unused; toxic.
[Boron]	B	5	Essential in some plants, function unknown.
Carbon	C	6	Required for organic compounds.
Nitrogen	N	7	Required for many organic compounds.
Oxygen	O	8	Required for water and organic compunds.
Fluorine	F	9	Growth factor in rats; possible constituent of teeth and bone.
Neon	Ne	10	Inert and unused.
Sodium	Na	11	Principal extracellular cation.
Magnesium	Mg	12	Required for activity of many enzymes in chlorophyll.
Aluminum	Al	13	Essentiality under study.
Silicon	Si	14	Possible structural unit of diatoms; recently shown to be essential in chicks.
Phosphorus	P	15	Essential for biochemical synthesis and energy transfer.
Sulphur	S	16	Required for proteins and other biological compounds.
Chlorine	Cl	17	Principal cellular and extracellular anion.
Argon	A	18	Inert and unused.
Potassium	K	19	Principal cellular cation.
Calcium	Ca	20	Major component of bone; required for some enzymes.
Scandium	Sc	21	Probably unused.
Titanium	Ti	22	Probably unused.
Vanadium	V	23	Essential in lower plants, certain marine animals and rat.
Chromium	Cr	24	Essential in higher animals; related to action of insulin.
Manganese	Mn	25	Required for activity of several enzymes.
Iron	Fe	26	Most important transition metal ion essential for hemoglobin and many enzymes.
Cobalt	Co	27	Required for activity of several enzymes in vitamin B_{12}.
Nickel	Ni	28	"Essentiality under study.
Copper	Cu	29	Essential in oxidative and other enzymes and hemocyanin.
Zinc	Zn	30	Required for activity of many enzymes.

Element	*Symbol*	*Atomimc Number*	*Comments*
Gallium	Ga	31	Probably unused.
Germanium	Ge	32	Probably unused.
Arsenic	As	33	Probably unused; toxic
Selenium	Se	34	Essential for liver function
Molybdenum	Mo	42	Required for activity of several enzymes.
Tin	Sn	50	Essential in rats; function unknown.
Iodine	I	53	Essential constituent of the thyroid hormones.

Fig. 5.2: Some Two-thirds of Lightest Elements, or 21 out of the first 34 elements in the periodic table, are now known to be essential for animal life. These 21 plus molybdenum (No. 42), tin (No. 50) and iodine (No. 53) constitute the total list of the 24 essential elements, which are here enclosed in coloured boxes. It is possible that still other light elements will turn out to be essential. The most likely candidates are aluminum, nickel and germanium. The element boron already appears to be essential for some plants.

pure amino acids instead of natural proteins, and all other dietary ingredients are screened for metal contaminants.

Since the standards of purity employed in these experiments far exceed those for reagents normally regarded as analytically pure, Schwarz and his co-workers have had to develop many new alalytical chemical methods. The most difficult problem turned out to be the purification of salt mixtures. Even the purest commercial reagents were contaminated with traces of metal ions. It was also found that trace elements could be passed from mothers to their offspring. To minimize this source of contamination animals are weaned as quickly as possible, usually from 18 to 20 days after birth.

With these precautions Schwarz and his colleagues have within the past several years been able to produce a new deficiency disease in rates. The animals grow poorly, lose hair and muscle tone, develop shaggy fur and exhibit other detrimental changes. When standard laboratory food is given these animals, they regain their normal appearance. At first it was thought that all the symptoms were caused by the lack of one particular trace element. Eventually four different elements had to be supplied to complete the highly purified diets the animals had been receiving. The four elements proved to be fluorine, silicon, tin and vanadium. A convenient source of these elements is yeast ash or liver preparations from a healthy animal. The animals on the deficiency diet grew less than half as fast as those on a normal or supplemented diet. Growth alone, however, may not tell the entire story. There is some evidence that even the addition of the four elements may not reverse the loss of hair and skin changes resulting from the deficiency diet.

Functions of Trace Elements

The addition of tin and vanadium to the list of essential trace metals brings to 10 the total number of trace metals needed by animals and plants. What role do these metals play? For six of the eight trade metals recognized from earlier studies (that is, for iron, zinc, copper, cobalt, manganese and molybdenum) we are reasonably sure of the answer. The six are constituents of a wide range of enzymes that participate in a variety of metabolic processes.

In addition to its role in hemoglobin and myoglobin, iron appears in succinate dehydrogenase, one of the enzymes needed for the utilization of energy from sugars and starches. Enzymes incorporating zinc help to control the formation of carbon dioxide and the digestion of proteins. Copper is present in more than a dozen enzymes, whose roles range from the utilization of iron to the pigmentation of the skin. Cobalt appears in enzymes involved in the synthesis of DNA and the metabolism of amino acids. Enzymes incorporating managnese are involved in the formation of urea and the metabolism of pyruvate. Enzymes incorporating molybdenum participate in purine metabolism and the utilization of nitrogen.

These six meals belong to a group known as transition elements. They owe their uniqueness to their ability to form strong complexes with ligands, or molecular groups, of the type present in the side chains of proteins. Enzymes in which transition metals are tightly incorporated are called metalloenzymes, since the metal is usually embedded deep inside the structure of the protein. If the metal atom is removed, the protein usually loses its capacity to function as an enzyme. There is also a group of enzymes in which the metal ion is more loosely associated with the protein but is nonetheless essential for the enzyme's activity. Enzymes in this group are known as metal-ion-activiated enzymes. In either group the role of the metal ion may be to maintain the proper conformation of the protein, to bind the substrate (the molecule acted on) to the protein or to donate or accept electrons in reactions where the substate is reduced or oxydized.

In 1968 the complete three dimensional structure of the first metalloenzyme, cytochrome *c*, was published. Cytochrome *c*, a red enzyme containing iron, is universally present in plants and animals. It is one of a series of enzymes, all called cytochromes, that extract energy from food molecules by the stepwise addition of oxygen.

The complete amino acid sequence of cytochrome *c* obtained from the human heart was determined some 10 years ago by a group led by *Emil L. Smith* of the University of Califormia at Los Angeles and by *Emanuel Margoliash* of North-western University. The iron atom is partially complexed with an intricate organic molecule, protoporphyrin, to form a heme group similar to that in hemoglobin. Of the iron atom's six coordination sites, four are attached to the heme group through nitrogen atoms. The other two sites form bonds with the protein chain; one bond is through a nitrogen atom in the side chain of a histidine unit at site No. 18 in the protein sequence and

Metal	*Enzyme*	*Biological Function*
Iron	Ferredoxin Succinate Dehydrogenase	Photosynthesis Aerobic oxidation of carbohydrates
Iron in Heme	Aldehyde Oxidase Cytochromes Catalase (Hemoglobin)	Aldehyde oxidation Electron transfer Protection against hydrogen peroxide Oxygen transport
Copper	Ceruloplasmin Cytochrome Oxidase Lysine Oxidase Tyrosinase Plastocyanin (Hemocyanin)	Iron utilization Principal terminal oxidase Elasticity of aortic walls Skin pigmentation Photosynthesis Oxygen transport in invertebrates
Zinc	Carbonic Anhydrase Carboxypeptidase Alcohol Dehydrogenase	CO_2 formation; regulation of acidity Protein digestion Alcohol metabolism
Manganese	Arginase Pyruvate Carboxylase	Urea formation Pyruvate metabolism
Cobalt	Ribonucleotide Reductase Glutamate Mutase	DNA biosynthesis Amino acid metabolism
Molybdenum	Xanthine Oxidase Nitrate Reductase	Purine metabolism Nitrate utilization
Calcium	Lipases	Lipid digestion
Magnesium	Hexokinase	Phosphate transfer

Fig. 5.3: Wide Variety of Metalloenzymes is required for the successful functioning of living organisms. Some of the most important are given in this list. The giant oxygen-transporting molecules hemoglobin and hemocyanin are included in the list (in brackets) even though they are not strictly enzymes, that is, they do not act as biological catalysts.

the other bond is through a sulphur atom in the side chain of a methionine unit at site No. 80.

Although the cytochrome *c* molecule is complicated, it is one of the simplest of the metalloenzymes. Cytochrome oxidase, probably the single most important enzyme in most cells, since it is responsible for transferring electrons to oxygen to form water, is far more complicated. Each molecule contains about 12 times as many atoms as cytochrome *c*, including two copper atoms and two heme groups, both of which participate in transferring the elections.

More complicated yet is cysteamine oxygenase, which catalyzes the addition of oxygen to a molecule of cysteamine; it contains one atom each of three different metals: iron, copper and zinc. There are many other combinations of metal ions and unique molecular assemblies. An extreme example is xanthine oxidase, which contains eight iron atoms, two molybdenum atoms and two molecules incorporating riboflavin (one of the B vitamins) in a giant molecule more than 25 times the size of cytochrome *c*.

The metal-containing proteins of another group, the metalloproteins, closely resemble the metalloenzymes except that they lack an obvious catalytic function. Hemoglobin itself is an example. Others are hemocyanin, the copper-containing blue protein that carries oxygen in many invertebrates, metallothionein, a protein involved in the absorption and storage of zinc, and transferrin, a protein that transports iron in the bloodstream. There may be many more such compounds still unrecognized because their function has escaped detection.

The Newest Essential Elements

Much remains to be learned about the specific biochemical role of the most recently discovered essential elements. In 1957 Schwarz and Calvin M. Foltz. working at the National Institutes of Health, showed that selenium helped to prevent several serious deficiency diseases in different animals, including liver necrosis and muscular dystrophy. Rats were protected against death from liver necrosis by a diet containing one-tenth of a part per million of selenium. Comparably low doses reversed the white muscle disease observed in cattle and sheep that happen to graze in areas where selenium is scarce.

In April a group at the University of Wisconsin under *J.T. Rotruck* reported a direct biochemical role for selenium. Oxidative damage to red blood cells was detected in rats kept on a selenium deficient diet. This damage was related to reduced activity of an enzyme, glutathione peroxidase, that helps to protect hemoglobin against the injurious oxidative effects of hydrogen peroxide. The enzyme uses hydrogen peroxide to catalyze the oxidation of glutathione, thus keeping hydrogen peroxide from oxidizing the reduced state of iron in hemoglobin. Oxidized glutathione can readily be converted to reduced glutathione by a variety of intracellular mechanisms. There is some reason to believe glutathione peroxidase may even contain some form of selenium acting as an integral part of the functional enzyme molecule.

The physiological importance of chromium was established in 1959 by *Schwarz* and *Walter Mertz*. They found that chromium deficiency is characterized by impaired growth and reduced life-span, corneal lesions and a defect in sugar metabolism. When the diet is deficient in chromium, glucose is removed from the bloodstream only half as fast as it is normally. In rats the deficiency is relieved by a single administration of 20 micrograms of certain trivalent chromic salts. It now appears that the chromium ion works in conjunction with insulin, and that in at least some cases diabetes may reflect faulty chromium metabolism.

After developing the all-plastic trace-element isolator described above, *Schwarz, David B. Milne* and *Elizabeth Vineyard* discovered that tin, not previously suspected as being essential, was necessary for normal growth. Without one or two parts per million of tin in their diet, rats grow at only about two-thirds the normal rate.

The next element shown to be essential in mammals by the Schwarz group was vanadium, an element that had been detected earlier in certain marine invertebrates but whose essentiality had not been demonstrated. On a diet in which vanadium is totally excluded rats suffer a retardation of about 30 per cent in growth rate. *Schwarz* and *Milne* found that normal growth is restored by adding one-tenth of a part per million of vanadium to the diet. At higher concentrations vanadium is known to have several biological effects, but its essential role in trace amounts remains to be established. A high dose of vanadium blocks the synthesis of cholesterol and reduces the amount of phospholipid and cholesterol in the blood. Vanadium also promotes the mineralization of teeth and is effective as a catalyst in the oxidation of many biological substances.

The third element most recently identified as being essential is fluorine. Even with tin and vanadium added to highly purified diets containing all other elements known to be essential, the animals in Schwarz's plastic cages still failed to grow at a normal rate. When up to half a part per million of potassium fluoride was added to the diet, the animals showed a 20 to 30 per cent weight gain in four weeks. Although it had appeared that a trace amount of fluorine was essential for building sound teeth, Schwarz's study showed that fluorine's biochemical role was more fundamental than that. In any case fluoridated water provides more than enough fluorine to maintain a normal growth rate.

Although there were earlier clues that silicon might be an essential life element, firm proof of its essentiality, at least in young chicks, was reported only three months ago. *Edith M. Carlisle* of the School of Public Health at the University of California at Los Angeles finds that chicks kept on a silicon-free diet for only one or two weeks exhibit poor development of feathers and skeleton, including markedly thin leg bones. The addition of 30 parts per million of silicon to the diet increases the chicks' growth more than 35 per cent and makes possible normal feathering and skeletal development. Considering that silicon is not only the second most abundant element in the earth's crust but is also similar to carbon in many of its chemical properties, it is hard to see how evolution could have totally excluded it from an essential biochemical role.

Nickel, nearly always associated with iron in natural substances, is another element receiving close attention. Also a transition element, it is particularly difficult to remove from the food used in special diets. Nickel seems to influence the growth of wing and tail feathers in chicks but more consistent data are needed to establish its essentiality. One incidental result of Schwarz's work has been the discovery of a previously unrecognized organic compound, which will undoubtedly prove to be a new vitamin.

Synergism and Antagonism

The interaction of the various essential metals can be extremely complicated. The absence of one metal in the diet can profoundly influence, either positively or negatively, the utilization of another metal that may be present. For example, it has been known for nearly 50 years that copper is essential for the proper metabolism of iron. An animal deprived of copper but not iron develops anaemia because the biosynthetic machinery fails to incorporate iron in hemoglobin molecules. It has only recently been found in our laboratories at Florida State University that ceruloplasmin, the copper-containing protein of the blood, is a direct molecular link between the two metals. Ceruloplasmin promotes the release of iron from animal liver so that the iron-binding protein of the serum, transferring, can complex with iron and transfer it to the developing red blood cells for direct utilization in the biosynthesis of hemoglobin. This represents a synergistic relation between copper and iron.

As an example of antagonism between elements one can cite the instance of copper and zinc. The ability of sheep or cattle to absorb copper is greatly reduced if too much zinc or molybdenum is present in their diet. Evidently either of the two metals can displace copper in an absorption process that probably involves competition for sites on a metal-binding protein in the intestines and liver.

The recent discoveries present many fresh challenges to biochemists. One can expect the discovery of previously unsuspected metalloenzymes containing vanadium, tin, chromium and selenium. New compounds or enzyme systems requiring fluorine and silicon may also be uncovered. The multiple and complex interdependencies of the elements suggest many hitherto unrecognized and important facts about the role and inter-relations of metal ions in nutrition and in health and disease.

Chapter—6
Oxygen

For convenience of description and reference science has divided its account of the Earth's history into four broad sections called eras, and named each according to the status of life at the time—as revealed by the fossil record. Going back from the present the eras are: the Cenozoic, meaning recent life; the Mesozoic, middle life; the Paleozoic, early life, and the Precambrian, which was known as the Azoic—without trace of life—until it was discovered that life had existed even then.

By far the greatest part of the information available on the Earth and life comes from the Cenozoic era; less comes from the Mesozoic, even less from the Paleozoic, and least of all from the Precambrian. And yet the Precambrian encompasses nearly 90 per cent of the Earth's history, stretching from 4.5 billion years ago, when the Earth condensed to become a solid planet, to 570 million years ago, when the fossil evidence first reveals living cells combined in the form of multicellular organisms.

The *Precambrian era*—so vast, so distant, so veiled by the knowledge of more recent times--is easiest to imagine in terms of what it lacked, like a distant empty desert lying beyond the consciousness of a dynamic city.

But deserts have their dynamics too. For the first thousand million years or so of the Precambrian era these dynamics were simply the physical forces of the planet itself, but once some enigmatic molecules had acquired the knack of replicating themselves around 3.5 billion years ago, the process of life became the dynamic force most responsible for the way the *Earth* looks today. Without the advent of life the Earth would have taken on an entirely different aspect—most probably resembling that of the *Moon* or of *Mars*.

The effects of life on the Earth have been accumulative, but slow moving. For the best part of its first 2.5 billion years life would have been microscopic in scale. A host of organisms, each barely one thousandth of a millimetre in diameter, teeming through the oceans and all the 'warm little ponds' of the world.

The fossil record provides firm evidence that single-cell organism have existed on Earth for a very long time, and gives an idea of their increasing size, shape and functional complexity through time, but the story of their evolutionary development from single cell to more complex organism comes largely from the evidence of modern living organisms.

All living things share four fundamental and interrelated characteristics: they are all cellular in structure; the protein in all cells is made up from basically the same 20 amino acid units; all cells use basically the same nucleotides in their genes; and all cells use ATP (adenosine triphosphate) as the molecule that energizes their life systems. These four characteristics are in effect a definition of life. Each is a very complex phenomenon and that they should be common to every living thing from microbe to man cannot be an accident. Such universality can only result from a common origin. It must mean that all living things have evolved from a single ancestral form. And that ancestral form was probably very similar to the smallest and simplest organisms alive on Earth today.

The smallest and simplest living things are single-cell organisms with a simple DNA molecule (their genetic material) floating free within the cell. Since these cells do not keep their DNA in a separate enclosed nucleus, they are called prokaryotes, from the Greek *pro,* meaning 'before', and *karyon,* meaning 'kernel' or 'nucleus'. Prokaryotes are the most widely dispersed living things, inhabiting environments of all sorts—from the depths of the oceans to the vents of volcanoes; from the Polar ice-caps to the near boiling water of natural hot springs. Some can remain frozen, or desiccated, or otherwise dormant for years and return to life afterwards as though nothing had happened. Others can survive several hours of boiling in water.

Among the prokaryotes, the very simplest are a group called the methanogens, bacteria which are found in marshes, lake beds and the digestive tracts of animals. The methanogens have been known for a long time but they were not studied extensively until the 1970s—mainly because oxygen is poisonous to them, which made it impossible to rear them under ordinary laboratory conditions. Studied were eventually instigated, however, and it became known that the methanogens are members of a small and exclusive group of organisms that are able to live off inorganic chemicals without the aid of any other source of energy. In the process they give off methane, marsh gas. The life process employed by the methanogens is called fermentation, well-known as the means of leavening bread and making alcohol—prokaryotic bacteria are the active fermenting agents in both cases.

Two other kinds of bacteria that have the capacity to live off inorganic chemicals under extreme conditions and the halobacteria that live in very salty environments, and the thermoacidophiles, which thrive in hot acidic environments. These three kinds of bacteria would have been well equipped for life in the harsh environments that prevailed on the early Earth, when there was no oxygen in the atmosphere but plenty of heat and inorganic chemicals about. They could be survivors of a lineage that arose then. Certainly the methanogens and their kind can be considered good analogues for the forms of life that existed on Earth at the earliest stages of evolution.

The inorganic chemical-consuming bacteria had the early Earth all to themselves for some considerable time—hundreds of millions of years. Eventually the materials they took from the environment to build and energize themselves would have been finished, and at that point life on Earth might have come to a lingering end. But the crisis was avoided, in fact it was never even a threat, because from the beginning life has been ruled by its inherent capacity to adapt and change as the environment dictates. The cell form, the amino acids, the ATP and the DNA remain the same, but their arrangements are adapted to take advantage of new circumstances. There is nothing conscious or directed, or even survival-oriented about the procedure. It is simply a matter of the chance mutations between generations producing individual organisms that are better suited than their antecedents to the changing environment. The mutants and their offspring thrive. The old guard either becomes extinct or retreats to some place where the environment has remained unchanged—this process is called natural selection. Thus the dinosaurs disappeared off the face of the Earth following the climatic changes that occurred 65 million years ago, and the methanogens and their kind are found living today only in very special circumstances.

Adaptive change is the basis of evolutionary theory. Evolution is not a fight for survival as is often supposed, with success meaning the direct elimination of competing organisms. It is a process by which new organisms and systems constantly arise to exploit new resources and opportunities.

Mutation is the key factor, Experiments have shown that by the time a culture of the modern prokaryote *Escherichia coli* (an inhabitant of the human gut and similar environments) has divided 30 times, no less than 1.5 per cent of its number are mutants. And the generation times can be very short; given ideal conditions *E. coli* can double in number every 20 minutes.

So, with the densities of their living populations probably exceeding millions of cells per cubic centimetre, and with the passage of millions of years, the methanogens and their kind must have come up with countless mutations during the first few hundred million years of life on Earth. Quite early on these would have produced organisms variously adapted to thrive in every environment that the Earth had to offer. But the process did not stop there. Eventually, somewhere among the teeming microscopic hordes, mutation produced an organism that was able to harness the radiant energy of the Sun to its life processes. This was an innovation of immense significance that has marked the course of life ever since.

The principle was that instead of using chemicals absorbed from its surroundings both as the materials needed for growth *and* as the fuel to process those materials, as every prokaryote had done until then, the new mutant used the energy of sunlight to power its life processes. Energy coming in from outside left more available inside, with the consequence that the organism was more energy-efficient than its antecedents. Life now had the means to exploit a widened realm of resources. There has always been plenty of sunlight—harnessing it to life is called *photosynthesis*.

The essence of photosynthesis is the ability to convert the energy of light to another form, and this ability is related to the phenomenon of colour. Black objects warm up very quickly when exposed to sunlight because they convert all the light striking them into heat; white objects remain cool because they reflect all the light that strikes them; red objects warm up just to the extent that they absorb yellow and blue light, blue objects to the extent that they absorb yellow and red light, and so on, through all the colours of the spectrum and the intensities of light, energy that they represent. Put at its simplest, colour is a visual representation of the energy left over when an object has absorbed all that it is tuned to take from the visible spectrum.

Studies of light and colour and of how photosynthesis works in organisms living today—from the simplest to the most complex—have provided the evidence for a hypothesis explaining how photosynthesis may have evolved on the early Earth.

In the course of those counltess mutations, some prokaryotes evolved with the capacity to produce and use porphyrins in their life processes. *Porphyrins* are light-absorbing compounds; their advent and incorporation in living matter was probably fortuitous, so too perhaps was the fact that they were coloured. Until then life had been colourless, but now among the millions of organisms there were, for the first time, macroscopic grains of colour that today are called chlorophyll—the wonder ingredient that takes light from the Sun and converts it into a tiny bank of energy in the core of the cell. The first chlorophylls were probably purplish—some living forms among the simplest bacteria still are—but they have subsequently evolved into a variety of other hues as well, including the numerous shades of green that colour the Earth today.

Photosynthesis requires a source of hydrogen. The first photosynthesizers would have taken this very basic ingredient from the compound molecules they absorbed from the water around them, just as their antecedents had done. But time and evolution brought further refinement. Eventually, about three billion years ago, an organism arose that was able to absorb hydrogen direct by splitting the water molecule into its component parts; two parts hydrogen and one part oxygen. The hydrogen was used in photosynthesis; the oxygen was released as waste.

It is thought likely that these first photosynthesizers lived mostly in dense, mat-like communities on the floor of shallow seas. Diversifying, limited only by the balance between water shallow enough to allow them sufficient light but deep enough to protect them from the destructive effects of the Sun's ultraviolet radiation, the photosynthesizing prokaryotes spread around the Earth. By around 2.2 billion years ago they were the dominant life-form, and the oxygen they released as waste began to have telling effects.

There was *no oxygen* in the atmosphere surrounding the early Earth, indeed the gas would have been lethal to all its first inhabitants. Experiments have shown that even the chemical processes that formed the first amino acids and other organic compounds in the 'warm little pond' could not have occurred if there had been even very small concentrations of molecular oxygen around. It seems certain that life as we know it could not have arisen in the presence of oxygen; and yet today life could not exist without it. How has this paradox come about?

In the early days the oxygen given off by the photosynthesizers would have been only a very small drop in the ocean—literally. The small amounts produced would have been dispersed and chemically neutralized before they could have much effect on the surroundings. But the trend towards an oxygenerated environment had been set. Millions of years passed. The photosynthesizers proliferated slowly, but eventually the amounts of oxygen they were giving off approached critical levels. Until that moment the evolution of the life had been subject to only the physical characteristics of the Earth—life had adapted to the environment—but now life itself because a formative factor. With the advent of oxygen life began to change the face of the Earth.

From among the diverse variety of single-cell organisms then existing, the methanogens and their kind retreated to the oxygen-free muds and other places similar to the habitats in which their descendants still flourish; any organisms that found no safe haven became extinct. Among the photosynthesizers natural selection favoured forms that could tolerate increasing concentrations of oxygen.

Meanwhile, the increasing amount of oxygen released by the photosynthesizers was also bringing about considerable changes to the environment. In the first instance it turned the oceans rusty.

Iron in a major component of the Earth. In the absence of oxygen it can exist in a soluble state, and the early oceans would have contained huge quantities of dissolved iron. But once the concentration of oxygen in the oceans reached a critical level, it began to react with the dissolved iron, forming iron oxides, which are insoluble. The iron oxides were precipitated out of solution and sank to the ocean floor.

In the relatively short space of 200 million years or so, billions of tonnes of iron oxides were laid down on ocean floors around the world. The deposits are remarkably regular in form, but they are not solid masses, they are made up of bands of iron oxide lying between bands of a red, silica-rich rock. The iron bands vary in thickness from a few millimetres to several centimetres. In section the bands of iron-grey and red give the rock a most distinctive appearance and the deposits have been named the Banded Ironstone Formations.

The banded nature of the deposits must surely be related to a pattern of increase and decrease of numbers among the photosynthesizers themselves. During warmer periods there were more photosynthesizers at work, producing more oxygen, precipitating more iron. In cooler periods their numbers would have declined, and for a time silica and other materials (including the remains of dead photosynthesizers) were the dominant components of the accumulating deposits. And so it went on for millions of years.

Eventually, the oxygen produced by the microscopic *photosynthesizers* swept the oceans clean of iron. Now for the first time free oxygen escaped from the oceans and began to accumulate in the atmosphere. This development brought further irreversible change to the environment of life on earth.

Oxygen was becoming a fact of life. Many more oxygen-tolerant forms had arisen, and some had even gone so far as to add a few molecules of oxygen to the fermentation process that alone had powered life until then. The result was very beneficial to the organisms concerned, releasing 18 times more energy than fermentation alone could release from the same amount of material. In effect the innovators held on to the waste product of fermentation and combined it with oxygen for another round of energy-producing reactions. This is called cellular respiration; basically it is why we breathe.

The respirators used the available resources more efficiently than any of their predecessors. They proliferated. At the same time there came the single-cell *cyanobacteria* (also known as the blue-green algae), still very small but more inclined to a communal life than anything that had come before. The cyanobacteria joined themselves together end-to-end to form strands of green living matter.

Photosynthesizers, respirators and cyanobacteria were now the dominant life-forms, churning out more and more oxygen, but still confined to those bands of water shallow enought to transmit sufficient sunlight for their life processes, and deep enough to transmit sufficient sunlight for their life processes, and deep enough to filter out the Sun's lethal ultraviolet radiation. For some time the progression of life on Earth seems to have halted at this critical point of balance between the good and the bad effects of the Sun.

Some sort of evolutionary innovation might have been the device most expected to break the impasse, but in the event it was not needed. Life was finally released from the tyranny of the ultraviolet rays by the indirect effects of life itself. Without doubt this was one of the most significant interactions between life and the global environment that has ever occurred. Once again the active agent was oxygen.

Like so much else, oxygen gas is susceptible to the destructive effects of sunlight. When subjected to intense solar radiation the gaseous oxygen molecule O_2 is split into two single oxygen atoms, some of which get together in groups of three to form the O_3 molecule, which is another gas called *ozone*, and *ozone* has the capacity of absorbing ultraviolet radiation.

The process was continuous, *ozone* accumulating above the oxygen as it escaped from the oceans. And as the *ozone* layer thickened, life could move closer and closer to the surface of the water, it could even, perhaps, survive for a time where warm little ponds had dried out completely. The surfaces of water and land presented new habitats, new opportunities for the evolution of life and, as is abundantly clear from the state of the Earth today, they were not neglected. Furthermore, evolutionary development took a great leap forward at this stage: after the prokaryotes had been in existence for possibly two billion years, a major transition took place, hardly of less significance than the origin of life itself—a new type of cell appeared.

The fossil record shows that until about 1.45 billion years ago prokaryotic cells were the only living forms. The prokaryotes had carried life from its very beginnings to a state of really quite advanced diversity. They laid the ground-plan of life and set the environmental scene for all that it had produced—and then their development stopped. The prokaryotes living today are still the smallest and the simplest organisms that life has ever known. Every other living thing has come into being through the medium of another kind of cell, the eukaryote (*eu* meaning 'true').

The four fundamentals of life—*cellular structure, amino acids* for *proteins,* DNA for genes, and ATP for energy transmission—link the eukaryotes and the prokaryotes as surely as a flower to its stem. The evidence of the fossil record leaves no doubt that the prokaryotes came first, and this conclusion is further supported by the evidence from living organisms, amongst whom it is found that although the prokaryotes are variably tolerant of oxygen—all the way from being unable to live *with* it to being unable to live *without* it—virtually all the eukaryotes have an absolute need of oxygen, and even the exceptions seem to be descendants of oxygen-dependent forms. All this leads to the conclusion that the eukaryotes must have arisen after the oceans and the atmosphere had been oxygenated by the prokaryotes.

However, although the evidence strongly suggests that the eukaryotes must have evolved from the prokaryotes, the differences between the two groups are of a scale that seems to call for more than just time and mutation to have brought forth the really rather complex eukaryotes from the so much simpler prokaryotes.

The eukaryote cell is generally larger, with several distinct parts, called organelles, within the cell, and their genetic material is contained within a nucleus, where it is associated with protein structures known as chromosomes. A typical eukaryote has about 1,000 times more DNA than any prokaryote and this has an important bearing on its mode of reproduction and its evolutionary potential.

Prokaryote reproduction is conducted by the relatively simple procedure of binary fission. First the DNA is replicated, then the cell elongates, pinches in the middle and divides, leaving two cells, each containing a copy of the original DNA.

In eukaryotes reproduction is essentially a process of division too, but it is rendered much more complex by the number of different elements and larger quantities involved. Each organelle has to be duplicated and apportioned, and as the nucleus divides each part must receive exactly the right number of chromosomes. Sometimes the component parts of the genetic material may be slightly reshuffled, so that duplication is not necessarily exact. The complexity of their reproductive process broadens the potential effects of mutation in eukaryote cells.

So how did the eukaryotes come into being? The most plausible hypothesis yet put forward to explain their origin suggests that the evolutionary development involved may have been more behavioural than physical. The idea is that it was not so much alterations to the prokaryote cell as such that created the eukaryotes, rather that some among them began to behave differently, living together in a mutually beneficial relationship that is common enough in the living world today, called symbiosis.

The hypothesis suggests that the eukaryote organelles were once independent cells that evolved the behavioural capacity to enter and live inside another prokaryote, and subsequently became part of a give-and-take relationship that bestowed benefits on both parties. This explanation was first mooted in the nineteenth century. It surfaced again in the 1920s, in a book by an American physician, J.E. Wallin, but achieved more notoriety than support when it was found that Wallin had fudged his experiments and distorted the data. In the 1970s the role of symbiosis in the evolution of the eukaryote cell was examined again by *Lynn Margulis* of Boston University. As a result of her work the suggestion that organelles were once independent organisms has gained in credibility.

The hypothesis is strongly supported by evidence from the living world. Lichen, for instance, is actually algae and fungus living together in a symbiotic relationship. The photosynthesizing algae fuel the partnership while the fungus provides a home base and security. Together the partners can inhabit a far wider range of environments and live much longer than either could alone. Scientists have demonstrated the symbiotic nature of the lichens by separating their algal and fungal components, and rearing them independently, but the partners have been together for so long that their symbiosis is now hereditary. The algae and fungi do not have to find each other with each generation as their earliest antecedents must have done; when they reproduce, the symbiotic lichen is produced—ready-made.

Hereditary symbiosis such as the lichens have developed occurs in other organisms too. There is a green plant, *Psychotria bacteriophila,* for instance, whose seeds contain not only the genetic material and other essential material usually passed on from one generation to the next, but also a batch of the symbiotic bacteria the plant must have at its roots to convert atmospheric nitrogen to a form the plant can utilize.

Yet more complex is the case of *Myxotricha paradoxa,* a single-cell organism that has found a home in the gut of certain Australian termites, where it feeds itself and at the same time performs the essential task of releasing the nutrients that nourish its host. The small termites eat dead wood, but they could not survive without the microscopic *Myxotricha* in their gut to help them digest it.

Myxotricha acquired the name *paradoxa* because it swims in an usual and puzzling way. A joint study by *A.V. Grimstone* of Cambridge University and *L.R. Cleveland* of the University of Georgia has solved the puzzle of how it swims and shows that *Myxotricha* may itself be the key to the solution of yet other puzzles.

The hair-like appendages called *flagella* and *clia* that propel *Myxotricha* are in fact independent bacteria of an elongated form, living symbiotically on the surface of their host. Somehow, in fulfilling their life processes, they drive *Myxotricha* where it needs to go. Lynn Margulis believes that something like the symbiotic relationship that *Myxotricha* has developed was the origin of the eukaryote cell.

She suggests that as oxygen began to accumulate around the Earth a fermenting bacterium would have found considerable advantage in taking on some smaller respirators as symbiotic partners, and natural selection would have favoured the trend. With time the symbiosis would have become hereditary and a new kind of organism evolved. Meanwhile smaller, more mobile single cells discovered the advantage of living in close association with larger organisms—much as pilot fish align themselves with sharks. Eventually that association too became symbiotic. The smaller organisms attached themselves to their larger partners and became their means of propulsion—the flagella and cilia. Again, the association brought considerable advantage to both parties and the symbiosis eventually became hereditary. So through time, with mutation and evolution, some simple single-cell prokaryotes became first the symbionts of other larger prokaryotes and then the organelles of the ancestral eukaryote. Margulis believes it may have taken one billion years to perfect the eukaryote cell. And the perfect eukaryote cell is the foundation upon which the living world is built.

Standing tall, sharing the Earth with trees, elephants, whales and so many other living things, large and small, it may seem perverse to insist that the perfection of life is represented by the microscopic eukaryote cell that arose in the primordial seas some 1.45 billion years ago, but it is a fact. In a sense nothing has changed since then; the Earth is still populated by prokaryote and eukaryote cells. The only difference is that while the prokaryotes have remained largely invisible, the eukaryotes have assembled themselves into some impressively visible structures. How has this come about?

The fossil evidence of the eukaryotes' early development gives some clues to its timing and progress. Measurements of some 8,000 fossil cells from 18 Precambrian locations around the world have shown that all cells from rocks older than 1.45 billion years are within the size range of modern prokaryotes. Then there is a distinct jump; cells larger than modern prokaryotes become increasingly abundant, and there is evidence of increasingly complex structure and internal organisation among these larger microfossils.

Filaments of large cells have come from rocks in California that are more than 1.2 billion years old; spiny cells of unquestionable eukaryote affinity have been found in Siberian shales 950 million years old, and cells containing small dense bodies thought to be preserved organelles have come from formations in central Australia that are about 850 million years old.

It is thought likely that the eukaryotes began to reproduce sexually around one billion years ago. This was an especially significant innovation for, by sharing the genetic material of two parents among the offspring, both the chance of mutuation and the potential degree of mutation between generations was increased. In effect, sexual reproduction enhanced the mechanism by which evolution operates and thereby accelerated the process. Increased diversity and proliferation of life-forms was inevitable.

Meanwhile, lifestyles were changing too. For a long time all the single-cell inhabitants of the Earth's seas fed off the same, basically chemical, resources of the planet. Eventually, however, some began to eat others and predation began. Here was a particularly dynamic force that could well have accelerated the evolutionary process still further. Predation brought advantage to the predators and applied strong selection pressure on their prey. Some prey populations probably became extinct; others evolved avoidance equipment or strategies. Among other things, the advent of predation in the story of life may have encouraged some cells to find refuge inside larger cells, where they eventually became the symbionts of the eukaryote cell.

At the same time the concentration of oxygen in the atmosphere was slowly but steadily increasing. The scene was ripe with potential: after some 2.5 billion years life was at last poised to move beyond the microscopic stage of its existence. The key to this step was the advent of the multicellular organism, when some cells developed the characteristic of living together as a colony of cells that itself was a distinct organism, bestowing advantages of nutrition, security and mobility on each of its component parts.

The oldest known multicellular fossil comes from deposits that are about 750 million years old, but their evolution probably began before then. With time and mutation and the pressure of natural selection, some component cells evolved special functions or features that were specific to themselves but beneficial to the whole. New organisms arose, creating and exploiting new opportunities in environments that were becoming increasingly interactive with life itself.

Once the trend was set its progress was inexorable, perhaps even inevitable, but the cellular diversity required to produce the modern living world is not as great as might be supposed. The human body, for instance, is composed of millions upon millions of individual cells, but among them there are hardly more than 100 different kinds of cell. And every single one is a variation on that microscopic, self-sustaining unit—the *eukaryote cell.*

Chapter—7
The Origin of Life

About a century ago the question, How did life begin? Which has interested men throughout their history, reached an impasse. Up to that time two answers had been offered: one that life had been created supernaturally, the other that it arises continually from the nonliving. The first explanation lay outside science; the second was now shown to be untenable. For a time scientists felt some discomfort in having no answer at all. Then they stopped asking the question.

Recently ways have been found again to consider the origin of life as a scientific problem—as an event within the order of nature. In part this is the result of new information. But a theory never rises of itself, however rich and secure the facts. It is an act of creation. Our present ideas in this realm were first brought together in a clear and defensible argument by the Russian biochemist A.I. Oparin in a book called *The Origin of Life,* published in 1936. Much can be added now to Oparin's discussion, yet it provides the foundation upon which all of us who are interested in this subject have built.

The attempt to understand how life originated raises a wide variety of scientific questions, which lead in many and diverse directions and should end by casting light into many obscure corners. At the center of the enterprise lies the hope not only of explaining a great past event—important as that should be—but of showing that the explanation is workable. If we can indeed come to understand how a living organism arises from the nonliving, we should be able to construct one—only of the simplest description, to be sure, but still recognizably alive. This is so remote a possibility now that one scarcely dares to acknowledge it; but it is there nevertheless.

One answer to the problem of how life originated is that it was created. This is an understandable confusion of nature with technology. Men are used to making things; it is a ready thought that those things not made by men were made by a superhuman being. Most of the cultures we know contain mythical accounts of a supernatural creation of life. Our own tradition provides such an account in the opening chapters of *Genesis.* There we are told that beginning on the third day of the Creation, God brought forth living creatures—first plants, then fishes and birds, then land animals and finally man.

Theories of Origin. Four theories have been advanced to account for the existence of the varied kinds of animals and plants on earth to-day—theories in some respects diametrically opposed to one another, in other respects somewhat in accord. They are:

1. Eternity of Present Conditions.
2. Special Creation or Creationism.
3. Catastrophism with—
   (a) Repopulation by immigration.
   (b) Repopulation by successive creations.
4. Organic Evolution

Theory of Eternity of Present Conditions

The first theory argues for the unchangeableness of the universe, holding not only that organisms have been unalterable throughout their existence, but that they have always existed and will continue to exist in the same unchanging state throughout eternity. This was apparently the belief of very few authorities, for one finds almost no allusion to it in the literature of science, although Hutton wrote: "The result of this physical enquiry is that we find no vestige of a beginning—no prospect of an end." Whether this should be interpreted as a statement that the world has neither beginning nor end is, however, open to question.

Theory of Special Creation

The second theory, that of Special Creation, or Creationism, is the literal interpretation of the Mosaic account of creation set forth in the first chapter of Genesis—a simple story, beautifully told, derived from the Hebrew tradition and well suited to the state of knowledge of the times and of the people for whom it was written. This account, strictly interpreted, has been the teaching, not alone of the Hebrew, but of the Christian church authorities for many centuries, although the increase of zoological knowledge made it harder and harder to reconcile with observed facts, until it was replaced by the doctrine of Evolution.

Suarez—One of the greatest advocates of the Special Creation doctrine during Christian times was Father Suarez (1548-1617), a Spanish Jesuit priest, who taught emphatically that "the world was made in six natural days. On the first of these days the *materia prima* was made out of nothing, to receive afterwards those 'substantial forms' which moulded it into the universe of things; on the third day, the ancestors of all living plants suddenly came into being, full-grown, perfect, and possessed of all the properties which now distinguish them; while, on the fifth and sixth days, the ancestors of all existing animals were similarly caused to exist in their complete and perfect state, by the infusion of their appropriate material substantial forms into the matter which had already been created. Finally, on the sixth day, the *anima rationalis*—that rational and immortal substantial form which is peculiar to man—was created out of nothing, and 'breathed into' a mass of matter which, till then, was mere dust of the earth, and so man arose. But the species man was presented by a solitary made individual, until the Creator took out one of his ribs and fashioned it into a female" (Huxley).

So profound was Suarez' influence upon European Catholic thought that his teaching continued to be the only orthodox belief in Europe until the middle of the nineteenth century. In a similar manner John Milton (1608-1674) influenced Protestant thought in England by the wondrously written story of the creation in *Paradise Lost.*

Some advocates of the theory claimed that none of the forms had changed in the several thousand years which had elapsed since the beginning; but that the latter-day descendants were in every way precisely similar to the original pair when they issued from the hands of their Creator. Other keen observers, like Linnaeus, thought that all the species of one genus constituted at the creation but one form, *ab initio unam constituerint speciem;* their number being subsequently increased through intercrossing with other species, and the hybrids thus produced forming additional species to those originally created. Linnaeus also held that certain forms had lost their pristine character through degeneracy—the result of climate and environment.

Theory of spontaneous Generation or abiogenesis

According to this theory, life has originated from non-living organic matter *abiogenetically (Gr. A, not; bios, life; genesis,* origin) from time to time. Greek philosophers of prechristian era, like *Thales, Anaximander, Anaximenes, Xenophanes, Empedocles, Plato, Aristotle* etc. are the followers of this theory.

According to *Epicurus* (342-271 B.C.) worms and numerous other animals were generated from the soil or manure by the action of moist warmth of sun and air. *Anaxagoras* (510-428 B.C.) thought that life come in tiny seeds (spermeia) with the rain water to fructifly the earth. According to *Aristotle* (384-322 B.C.) living creatures are born from like species no doubt, but they also arise spontaneously.

Thus, it was regarded a normal thing for worms, larvae of bees, wasps, mites, glow-worms and other insects to be formed from due, rotting dung and mud, from dry wood, from sweat from meat Flies, moths, butterflies, midges, dungbeetles, fleas, bugs and lice arise from field humus, from moulds and dung, from decaying wood and fruits, (impurities) of vineger and from old wool. *Aristotle* was of the opinion that not only insects and worms but also other highly organized creatures could be spontaneously formed, for example, crabs and various molluscs from foul earth and mud; eels and many other fish from the mud of lakes, from sand and from decaying water, plants; even frogs and salamanders could arise from moist earth.

Belief in abiogenesis was so strong that *Van Helmont* (1652) claimed the formation of mice in 21 days from dirty, sweat-soaked shirt put in wheat barn in the dark.

Till the middle of the 17th century, the theory of spontaneous generation was widely accepted. But later on several scholars viz., *Francesco Redi* in 17th century and *Abbe Spallanzani* in 18th century disproved the idea of spontaneous generation.

Redi (1621-1691) described a series of experiments in which he placed meat or fish (eel) under clean muslin coverings and demonstrated that while flies laid eggs on muslin, maggots or larvae appeared only when those eggs were transferred to the meat and allowed to hatch. He concluded that maggots come only from pre-existing flies and were not spontaneously generated by any other form of material. The discovery of microorganisms by *Leeuwenhock* (1632-1723) brought the question into the fore-front of biological though once agains.

Spallanzani (1729-1799) in his experiment sealed the necks of the flasks containing infusion so that even air could not enter inside after boiling the infusion for four hours. Some of the flasks kept open or loosely-corked. Micro-organisms appeared in those flasks which were open or loosely-corked while the sealed flasks remained sterile. According to him air carried microorganisms which germinate when get moisture and food. They were not formed abiogenetically.

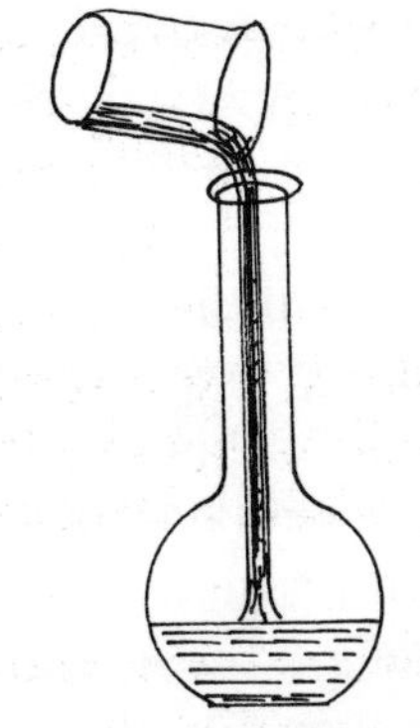

1. hay solution added to flask

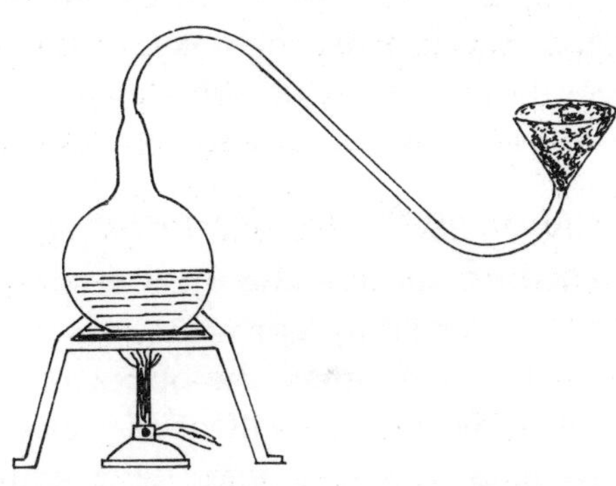

3. solution boiled vigorously for several minutes

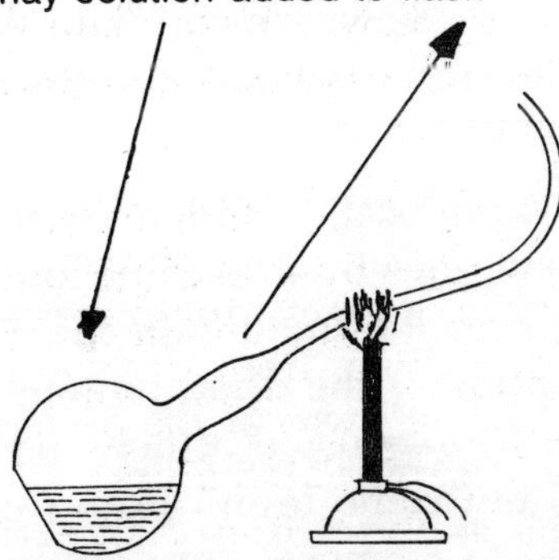

2. neck of flask bent into s shaped curve using heat.

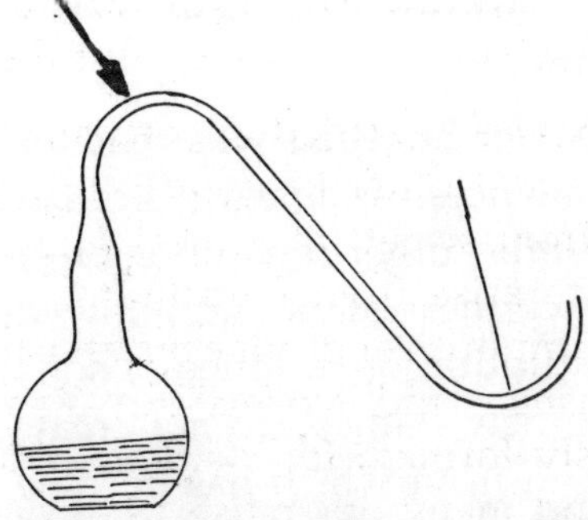

4. solution is colled slowly and remains sterile for many months.

Fig. 7.1: Pasteur's experiment with band-necked flask to disproove Theory of Spontaneous Generation.

In the beginning of the 17th century, spontaneous generation of more highly developed animals such as reptiles, insects, worms was proved to be impossible. *Needham* in 1745 described experiments in which sealed vessels having meat extracts were swarmed with microorganisms. *Spallanzani* (1765) challenged the views and results of *Needham.*

Louis Pasteur (1862) disputed the theory of spontaneous generation. He boiled *bouillon* (strong broth) in a special flask with a sealed neck, in which the nutritive medium was sterile for a long time. No bacterial growth appeared. When the flask was opened the micro-organisms entered into the bouillon where they started reproducing. Thus, it was demonstrated that self-generation of organisms is impossible.

Experiments performed by *Spallanzani* and *Pasteur* proved beyond doubt the fallacy of spontaneous generations.

Theory of Catastrophism

Cuvier—A new complication arose through the discovery of older faunas, the remains of which were preserved in the form of fossils and which seemed to represent creatures whose existence antedated that of the living types. Cuvier (1769-1832), one of the founders of the science of Paleontology, became interested in the bones which lay buried in the gypsum quarries in the hill of Montmartre within the present limits of the city of Paris. His studies of these forms, and especially his reconstructions of their skeletons, showed the great anatomist that he was dealing with extinct animals which had no existing representatives. Cuvier also had, because of his official position in the Jardin des Plantes, the opportunity to study hosts of specimens from all parts of the earth, and as a result of his research, gave to the world a new theory, that of Catastrophism or Catalclysm, to account for the extinction of these forms. He is generally accredited with the belief that the cataclysms were world-wide and that the slaughter of the older fauna necessitated the creation of a new one to take its place. That belief, however, was held by later scholars of the same school, but apparently not by Cuvier.

What Cuvier believed was that the catastrophes were local, "sudden revolutions, such as subsidences of the earth's crust, followed by invasions by the sea of continents once dry;" while "other revolutions resulting in the upheaval of mountain chains have again cast back the waters and allowed, on the foundation of the dried bottom of the sea, the constitution of continental soils favourable to the expansion of new terrestrial faunas; these new faunas are not created on the spot, but come from distant regions, their migration from which has become possible owing to temporary bridges between continents" (Depéret). Cuvier's belief has a great deal of truth in it, except that the "revolutions," with resulting climatic change and consequent extinctions and immigrations, have been rapid only in proportion to the length of geologic time, but very, very slow as morals note the flight of years.

D'Orbigny—Further knowledge of historical geology led to an expansion of the catastrophic belief far beyond the teaching of Cuvier, and postulated a re-creation following each cataclysm and corresponding to the principal geologic periods. Alcide d'Orbigny (1802-1857), writing in the year 1848, expounded this theory as follows:

> "The first creation shows itself in the Silurian stage. After its annihilation through some geological cause or other, a second creation took place a considerable time after in the Devonian stage, and, twenty-seven times in succession, *distinct creations* have come to re-people the whole earth with its plants and animals after each of the geological disturbances which destroyed everything in living nature. Such is the fact, certain but incomprehensible, which we confine ourselves to stating, without endeavoring to solve the superhuman mystery which envelops it" (Depéret).

Cosmozoic theory or Theory of Panspermia

According to *Richter*, the protoplasm reached the earth in the form of sports or germ or other simple particle with cosmic duct from other planet.

Arrehenius proposed *cosmic panspermia theory,* according to him the life existed throughout the Universe and their spores etc. could freely travel through the space from one planet to other.

According to *Helmholtz,* the life was brough on the earth with falling meteorites.

However, this theory has been rejected due to two reasons. First, due to intense cold, extreme dryness and intense radiation of interstellar space. Even the most resistant living spores cannot withstand exposure to interstellarspace. Moreover, this theory does not answer the basic question of 'origin of life' upon earth.

Modern Concepts on the origin of Life

Scientists are now of the opinion that spontaneous generation is not possible at present on Earth and that specific conditions are necessary for the appearance of life.

T.H. Huxley (1869) and *John Tyndall* (1874) asserted that life could be generated from inorganic chemicals. But their ideas were vague as the knowledge of biochemistry was not available at that time.

In 1920s, a new interest in the origin of life arose from independent speculation of two biologists, *A.I. Oparin* and *J.B.S. Haldane.* Both these scientists got the ideas from the new biochemistry that was founded by *Sir F.G. Hopkins.* They were of the opinion that the evolution is purely chemical—the gradual transmutation of inorganic compounds into organic ones.

After 1920s, the subject of the origin of life attracted ever increasing interest. *Pirie* coined the term *Biopoiesis* for the study of the origin of life.

Oparin first of all gave the scientific account of the origin of life in his book entitled. *"The Origin of Life"*, which was published in 1936. His main postulates are as follows:

(a) ***Origin of earth.*** The earth is presumed to have originated about 5-6 billion years ago. There are two hypothesis for its origin (i) *Planetesimal hypothesis* believes that earth is originated as a part broken off from the modern mass of sub (ii) *Nebular hypothesis.* According to this hypothesis the earth is originated by gradual condensation of interstellar dust from which entire solar system is supposed to be formed. At that time temperature was extremely high which served as basis of primary origin of organic compounds on the earth.

(b) ***Presence of elements on earth.*** In the begining, the earth was fiery spinning ball of hot gases and vapours of elements. Gradually, through hundreds of millions of years, the gases condensed into a molten core and different element got stratified according to their density. The original temperature of the earth was very high, about 5000-6000°C. Due to this high temperature carbon elements existed in three major forms, dicarbon (C_2), cynogen (CN), methane (CH_4), metal carbides, carbon dioxide etc. Oxygen was not existing in free state but oxides of aluminium, boron, hydrogen etc. were present. Nitrogen existed in combination with metals to form nitrides.

The earth cooled gradually and some of the gases liquefied and some solidified. Steam condensed into water and resulted in rain. The water again returned to atmosphere because of superheated earth. This cycle continued for millions of years and resulted in the cooling of earth. Due to cooling down of earth first ocean came in existence. The oceanic water contained methane, ammonia of atmosphere in dissolved condition.

Theory of Organic Evolution

Evolution is the gradual development from the simple unorganized condition of primal matter to the complex structure of the physical universe; and in like manner, from the beginning of organic life on the habitable planet, a gradual unfolding and branching out into all the varied forms of beings which constitute the animal and plant kingdoms. The first is called Inorganic, the last Organic Evolution, or descent with modification.

Early Greek Theories—Organic Evolution is often imagined to be a nineteenth century contribution to biologic science, whereas the idea is itself the product of an evolution of thought and is the fruition of no fewer than twenty-four centuries of speculation and research. The germ of the evolutionary idea had its inception with the Greeks, whose wonderful fertility of mind has so enriched the world, the first writer to deal with the problem, Anaximander, living five and a half centuries before the Christian era. Empedocles (495-435 B.C.) may be called the father of Evolution, though the Evolution that he taught implied no succession of related animals, gradually improving in successive generations, but a series of attempts on the part of nature to produce more perfect forms, the unfit being eliminated. He is the first to show the possibility of the origin in the fittest forms through chance rather than through design.

Another Greek, Democritus (460-?357 B.C.), went further than Empedocles in that he taught the adaptations of single structures and organs, whereas the latter applied the idea to entire organisms. But by far the most notable figure in Greek philosophy was Aristotle (384-322 B.C.), whose versatility as a writer upon all aspects of human knowledge was remarkable. In view of the limited opportunities for observation possible in those days when the teeming host of microscopic forms as well as the extinct creatures were utterly unknown, the deductions of Aristotle, even where he appears to retrogress from the truth, are highly logical. He did not believe in Special Creation, nevertheless he postulates an intelligent design as the primary cause of the changes which has been wrought in nature, and the central thought in his evolutionary theory, if such it was, is an *internal perfecting tendency* impelling organisms to greater and greater perfection. As a result of this, he saw a complete gradation in nature from the mineral to the plant, the plant-like animal, the animal with senses and hence locomotor powers, and finally man.

Aristotle considered life a function of the organism, not a separate principle, and had an understanding of adaptations and of heredity, even of the atavistic heredity wherein an ancestral trait reappears in a later descendant after having lain dormant for several generations. Osborn says of him:

Aristotle's argument for "operation of natural law, rather than of chance, in the lifeless and in the living world, is a perfectly logical one, and his consequent rejection of the hypothesis of the Survival of the Fittest, a sound induction from his own limited knowledge of Nature... If he had accepted Empedocles' hypothesis [of the origin of the fittest through chance rather than through design] he would have been the literal prophet of Darwinism."

To summarize, then, the Greeks offered as causes of evolutionary change three explanations:

1. Intelligent design,
2. The operation of natural laws implanted by intelligent design.
3. The operation of natural causes due to laws of change—no evidence of design, even in origin.

Middle Ages—And now, for hundreds of years, owing largely to the repressive measures of the church authorities, though some, like Saint Augustine, would have taught otherwise, the progress of the evolutionary idea virtually ceased until the coming of the philosophers Bacon, Descartes, Leibnitz, and Kant, and the naturalists Linnaeus, Buffon, Erasmus and Charles Darwin, E. Geofisms which lead, during their larval state, a planktonic existence. These forms are as a rule extremely small and have feeble powers of locomotion, generally by means of cilia. Nevertheless they are so numerous that the upper strata of the oceans are literally crowded with them and they form a great source of food supply for the more aggressive forms of life. They pass their short existence floating about in the sea, in swarms. Sooner or later, however, they sink to the bottom, and if they fall upon the proper sort of substratum, they develop into the benthonic adult; but if they fall upon an unfavourable bottom, or if food supply is scarce, they perish.

The necessity of some means of dispersal or for the repopulation of an area wherein accident has destroyed the original inhabitants is imperative, and in sedentary adults can only be attained by this means. Mero-planktonic larvae are found in every group of aquatic sedentary benthonic animals. The young of the sedentary scale insects, on the other hand, which are active for a brief time, are vagrant benthos. Germs, spores, and many seeds like those of the maple and dandelion constitute about all that can possibly be included in aerial mero-plankton.

Pseudo-plankton (Gr. ψεῦδος, false) is a term proposed for organisms such as the sargassum or gulf sea-weed which is normally or in early life an attached benthonic oganism but which becomes planktonic. The meaning of the term has been extended to include plants or animals living as sedentary or vagrant benthos upon floating objects, such as the algae, hydroids, or bryozoans, which may be attached to the floating sargassum, and the crustaceans, molluscs, or other animals which dwell among them. In many instances the pseudo-planktonic existence of these forms is due to accident, but on the other hand it seems to be habitual with certain forms, which, like the goose-barnacle, *Lepas,* rarely occur except attached to floating objects such as timber or the bottoms of ships, especially when the latter are derelict. Many of the animals found on floating sargassum seem to be characteristic of it in this condition, as they do not occur when it is attached.

The Task

To make an organism demands the right substances in the right proportions and in the right arrangement. We do not think that anything more is needed—but that is problem enough.

The substances are water, certain salts—as it happens, those found in the ocean—and carbon compounds. The latter are called *organic* compounds because they scarcely occur except as products of living organisms.

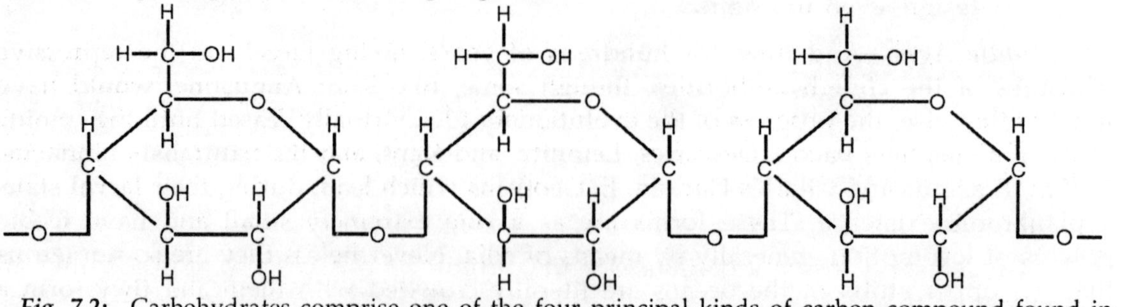

Fig. 7.2: Carbohydrates comprise one of the four principal kinds of carbon compound found in living matter. This structural formula represents parts of a characteristic carbohydrate. It is a polysaccharide consisting of six-carbon sugar units, three of which are shown.

Organic compounds consist for the most part of four types of atoms: carbon, oxygen, nitrogen and hydrogen. These four atoms together constitute about 99 per cent of living material, for hydrogen and oxygen also form water. The organic compounds found in organisms fall mainly into four great classes: carbohydrates, fats, proteins and nucleic acids. The illustrations on this and the next three pages give some notion of their composition and degrees of complexity. The fats are simplest, each consisting of three fatty acids joined to glycerol. The starches and glycogens are made of sugar units strung together to form long straight and branched chains. In general only one type of sugar appears in a single starch or glycogen; these molecules are large, but still relatively simple. The principal function of carbohydrates and fats in the organism is to serve as fuel—as a source of energy.

The *nucleic acids* introduce a further level of complexity. They are very large structures, composed of aggregates of at least four types of unit—the nucleotides—brought together in a great variety of proportions and sequences. An almost endless variety of different nucleic acids is possible, and specific differences among them are believed to be of the highest importance. Indeed, these structures are thought by many to be the main constituents of the genes, the bearers of hereditary constitution.

Variety and specificity, however, are most characteristic of the proteins, which include the largest and most complex molecules known. The units of which their structure is built are about 25 different amino acids. These are strung together in chains hundreds to thousands of units long, in different proportions, in all types of sequence, and with the greatest variety of branching and folding. A virtually infinite number of different proteins is possible. Organisms seem to exploit this potentiality, for no two species of living organism, animal or plant, possess the same proteins.

Organic molecules therefore form a large and formidable array, endless in variety and of the most bewildering complexity. One cannot think of having organisms without them. This is precisely the trouble, for to understand how organisms originated we must first of all explain how such complicated molecules could come into being. And that is only the beginning. To make an organism requires not only a tremendous variety of these substances, in adequate amounts and proper proportions, but also just the right arrangement of them. Structure here is as important as composition—and what a complication of structure! The most complex machine man has devised—say an electronic brain-is child's play compared with the simplest of living organisms. The especially trying thing is that complexity here involves such small dimensions. It is on the molecular level; it consists of a detailed fitting of molecule to molecule such as no chemist can attempt.

The Possible and Impossible

One has only to contemplate the magnitude of this task to concede that the spontaneous generation of a living organism is impossible. Yet here we are—as a result of spontaneous generation. It will help to digress for a moment to ask what one means by "impossible."

With every event one can associate a probability—the chance that it will occur. This is always a fraction, the proportion of times the event occurs in a large number of trials. Sometimes the probability is apparent even without trial. A coin has two faces; the probability of tossing a head is therefore 1/2. A die has six faces; the probability of throwing a deuce is 1/6. When one has no means of estimating the probability beforehand, it must be determined by counting the fraction of successes in a large number of trials.

Our everyday concept of what is impossible, possible or certain derives from our experience: the number of trials that may be encompassed within the space of a human lifetime, or at most within recorded human history.

But even within the bounds of our own time there is a serious flaw in our judgement of what is possible. It sounds impressive to say that an event has never been observed in the whole of human history. We should tend to regard such an event as at least "practically" impossible, whatever probability is assigned to it on abstract grounds. When we look a little further into such a statement, however, it proves to be almost meaningless. For men are apt to reject reports over very improbable occurrences. Persons of good judgement think it safer to destrust the alleged observer of such an event than to believe him. The result is that events which are merely very extraordinary acquire the reputation of never having occurred at all. Thus the highly improbable is made to appear impossible.

To give an example: Every physicist knows that there is a very small probability, which is easily computed. The event requires no more than that the molecules of which the table is composed, ordinarily in random motion in all directions, should happen by chance to move in the same direction. Every physicist concedes this possibility; but try telling one that you have seen it happen.

A final aspect of the problem is very important. When we consider the spontaneous origin of a living organism, this is not an event that need happen again and again. It is perhaps enough for it to happen once. The probability with which we are concerned is of a special kind; it is the probability that an event occur *at least once*. To this type of probability a fundamentally important thing happens as one increases the number of trials. However improbable the event in a single trial, it becomes increasingly probable as the trials are multiplied. Eventually the event becomes virtually inevitable. For instance, the chance that a coin will not fall head up in a single toss is ½. The chance that no head will appear in a series of tosses is ½ x ½ x ½ . . . . as many times over as the number of tosses. In 10 tosses the chance that no head will appear is therefore ½ multiplied by itself 10 times, or 1/1,000. Consequently the chance that a head will appear at least once in 10 tosses is 999/1,000. Ten trials have converted what started as a modest probability to a near certainty.

The same effect can be achieved with any probability, however small, by multiplying sufficiently the number of trials. Consider a reasonably improbable event, the chance of which is 1/1,000. The chance that this will not occur in one trial is 999/

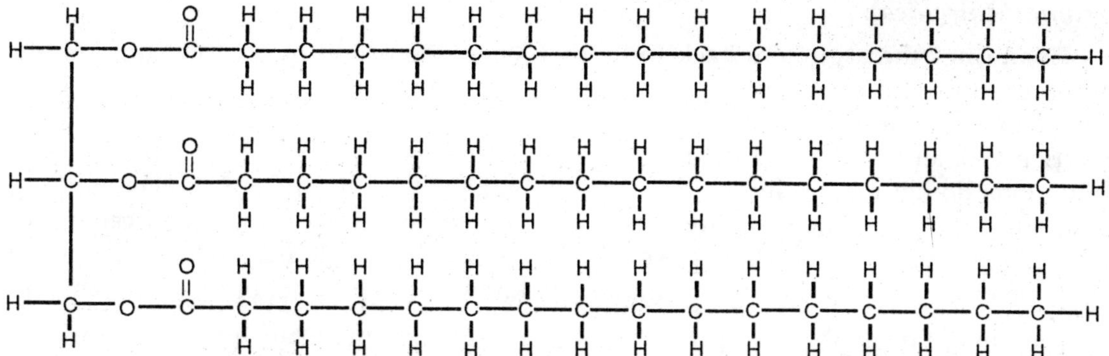

Fig. 7.3: FATS are a second kind of carbon compound found in living matter. This formula represents the whole molecule of palmitin, one of the commonest fats. The molecule consists of glycerol (*11 atoms at the far left*) and fatty acids (*hydrocarbon chains at the right)*

1,000. The chance that it won't occur in 1,000 trials is 999/1,000 multiplied together 1,000 times. This fraction comes out to be 37/100. The chance that it will happen at least once in 1,000 trials is therefore one minus this number—63/100—a little better than three chances out of five. One thousand trials have transformed this from a highly improbable to a highly probable event. In 10,000 trials the chance that this event will occur at least once comes out to be 19,999/20,000. It is now almost inevitable.

It makes no important change in the argument if we assess the probability that an event occur at least two, three, four or some other small number of times rather than at least once. It simply means that more trials are needed to achieve any degree of certainty. Otherwise everything is the same.

In such a problem as the spontaneous origin of life we have no way of assessing probabilities beforehand, or even of deciding what we mean by a trial. The origin of a living organism is undoubtedly a stepwise phenomenon, each step with its own probability and its own conditions of trial. Of one thing we can be sure, however: whatever constitutes a trial, more such trials occur the longer the interval of time.

The important point is that since the origin of life belongs in the category of at least once phenomena, time is on is side. However improbable we regard this event, or any of the steps which it involves, given enough time it will almost certainly happen at least once. And for life as we know it, with its capacity for growth and reproduction, once may be enough.

Time is in fact the hero of the plot. The time with which we have to deal is of the order of two billion years. What we regard as impossible on the basis of human experience is meaningless here. Given so much time, the "impossible" becomes possible, the possible probable, and the probable virtually certain. One has only to wait: time itself performs the miracles.

Organic Molecules

This brings the argument back to its first stage: the origin of organic compounds. Until a century and a quarter ago the only known source of these substances was the stuff of living organisms. Students of chemistry are usually told that when, in 1828, Friedrich Wöhler synthesized the first organic compound, urea, he proved that organic compounds do not require living organisms to make them. Of course it showed nothing of the kind. Organic chemists are alive; Wöhler merely showed that they can make organic compounds externally as well as internally. It is still true that with almost negligible exceptions all the organic matter we know is the product of living organisms.

The almost negligible exceptions, however, are very important for our argument. It is now recognized that a constant, slow production of organic molecules occurs without the agency of living things. Certain geological phenomena yield simple organic compounds. So, for example, volcanic eruptions bring metal carbides to the surface of the earth, where they react with water vapour to yield simple compounds of carbon and hydrogen. The familiar type of such a reaction is the process used in old-style bicycle lamps in which acetylene is made by mixing iron carbide with water.

Recently harold Urey, Nobel laureate in chemistry, has become interested in the degree to which electrical discharges in the upper atmosphere may promote the formation of organic compounds. One of his students, S.L. Miller, performed the simple experiment of circulating a mixture of water vapour, methane (CH_4), ammonia (NH_3) and hydrogen—all gases believed to have been present in the early atmosphere of the earth—continuously for a week over an electric spark. The circulation was maintained by boiling the water in one limb of the apparatus and condensing it in the other. At the end of the week the water was analyzed by the delicate method of paper chromatography. It was found to have acquired a mixture of amino acids! Glycine and alanine, the simplest amino acids and the most prevalent in proteins, were definitely identified in the solution, and there were indications it contained aspartic acid and two others. The yield was surprisingly high. This amazing result changes at a stroke our ideas of the probability of the spontaneous formation of amino acids.

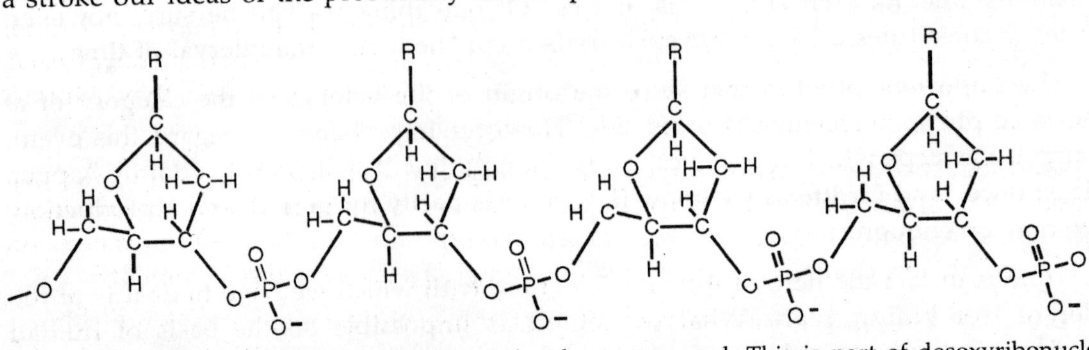

Fig. 7.4: Nucleic Acids are a third kinds of carbon compound. This is part of desoxyribonucleic acid, the backbone of which is five-carbon sugars alternating with phosphoric acid. The letter R is any one of four nitrogenous bases, two purines and two pyrimidines.

A final consideration, however, seems to me more important than all the special processes to which one might appeal for organic synthesis in inanimate nature.

It has already been said that to have organic molecules one ordinarily needs organisms. The synthesis of organic substances, like almost everything else that happens in organisms, is governed by the special class of proteins called enzymes—the organic catalysts which greatly accelerate chemical reactions in the body. Since an enzyme is not used up but is returned at the end of the process, a small amount of enzyme can promote an enormous transformation of material.

Enzymes play such a dominant role in the chemistry of life that it is exceedingly difficult to imagine the synthesis of living material without their help. This poses a dilemma, for enzymes themselves are proteins, and hence among the most complex organic components of the cell. One is asking, in effect, for an apparatus which is the unique property of cells in order to form the first cell.

This is not, however, an insuperable difficulty. An enzyme, after all, is only a catalyst; it can do not more than change the *rate* of a chemical reaction. It cannot make anything happen that would not have happened, though more slowly, in its absence. Every process that is catalyzed by an enzyme, and every product of such a process, would occur without the enzyme. The only difference is one of rate.

Once again the essence of the argument is time. What takes only a few moments in the presence of an enzyme or other catalyst may take days, months or years in its absence; but given time, the end result is the same.

Indeed, this great difficulty in conceiving of the spontaneous generation of organic compounds has its positive side. In a sense, organisms demonstrate to us what organic reactions and products are *possible*. We can be certain that, given time, all these things must occur. Every substance that has ever been found in an organism displays thereby the finite probability of its occurrences. Hence, given time, it should arise spontaneously. One has only to wait.

It will be objected at once that this is just what one cannot do. Everyone knows that these substances are highly perishable. Granted that, within long spaces of time, now a sugar molecule, now a fat, now even a protein might form spontaneously, each of these molecules should have only a transitory existence. How are they ever to accumulate; and, unless they do so, how form an organism?

We must turn the question around. What, in our experience, is known to destroy organic compounds? Primarily two agencies: decay and the attack of oxygen. But decay is the work of living organisms, and we are talking of a time before life existed. As for oxygen, this introduces a further and fundamental section of our argument.

It is generally conceded at present that the early atmosphere of our planet contained virtually no free oxygen. Almost all the earth's oxygen was bound in the form of water and metal oxides. If this were not so, it would be very difficult to imagine how organic matter could accumulate over the long stretches of time that

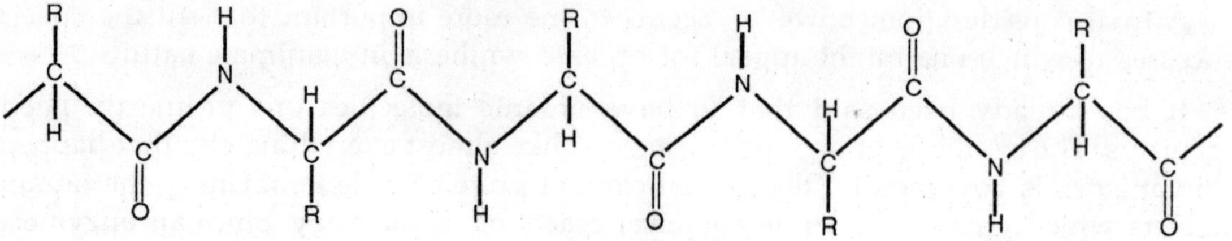

Fig. 7.5: Protlins are a fourth kind of carbon compound found in living matter. This formula represent part of a polypeptide chain, the backbone of a protein molecule. The chain is made up of amino acids. Here the letter R represents the side chains of these acids.

alone might make possible the spontaneous origin of life. This is a crucial point, therefore, and the statement that the early atmosphere of the planet was virtually oxygen-free comes forward so opportunely as to raise a suspicion of special pleading.

Apparently something similar was true also for another common component of our atmosphere—carbon dioxide. It is believed that most of the carbon on the earth during its early geological history existed as the element or in metal carbides and hydrocarbons; very little was combined with oxygen.

This situation is not without it irony. We tend usually to think that the environment plays the tune to which the organisms must dance. The environment is given; the organism's problem is to adapt to it or die. It has become apparent lately, however, that some of the most important features of the physical environment are themselves the work of living organisms. Two such features have just been named. The atmosphere of our planet seems to have contained no oxygen until organisms placed it thereby the process of plant photosynthesis. It is estimated that at present all the oxygen of our atmosphere is renewed by photosynthesis once in every 2,000 years, and that all the carbon dioxide passes through the process of photosynthesis once in every 300 years. In the scale of geological time, these intervals are very small indeed. We are left with the realization that all the oxygen and carbon dioxide of our planet are the products of living organisms, and have passed through living organisms over and over again.

Forces of Dissolution

In the early history of our planet, when there were no organisms or any free oxygen, organic compounds should have been stable over very long periods. This is the crucial difference between the period before life existed and our own. If one were to specify a single reason why the spontaneous generation of living organisms was possible once and is so no longer, this is the reason.

We must still reckon, however, with another destructive force which is disposed of less easily. This can be called spontaneous dissolution—the counter-part of spontaneous generation. We have noted that any process catalyzed by an enzyme can occur in time without the enzyme. The trouble is that the processes which synthesize an organic substance are reversible: any chemical reaction which an enzyme may catalyze will go backward as well as forward. We have spoken as though one has only to wait to achieve synthesis of all kinds; it is true to say that what one achieves by waiting is *equilibria* of all kinds—equilibria in which the synthesis and dissolution of substances come into balance.

In the vast majority of the processes in which we are interested the point of equilibrium lies far over toward the side of dissolution. That is to say, spontaneous dissolution is much more probable, and hence proceeds much more rapidly, than spontaneous synthesis. For example, the spontaneous union, step by step, of amino acid units to form a protein has a certain small probability, and hence might occur over a long stretch of time. But the dissolution of the protein or of an intermediate product into its component amino acids is much more probable, and hence will go ever so much more rapidly. The situation we must face is that of patient Penelope waiting for Odysseus yet much worse: each night she undid the weaving of the preceding day, but here a night could readily undo the work of a year or a century.

How do present-day organisms manage to synthesize organic compounds against the forces of dissolution? They do so by a continuous expenditure of energy. Indeed, living organisms commonly do better than oppose the forces of dissolution; they grow in spite of them. They do so, however, only at enormous expense to their surroundings. They need a constant supply of material and energy merely to maintain themselves, and much more of both to grow and reproduce. A living organism is an intricate machine for performing exactly this function. When, for want of fuel or through some internal failure in its mechanism, an organism stops actively synthesizing itself in opposition to the processes which continuously decompose it, it dies and rapidly disintegrates.

Forces of Integration

At present we can make only a beginning with this problem. We know that it is possible on occasion to protect molecules from dissolution by precipitation or by attachment to other molecules. A wide variety of such precipitation and "trapping" reactions is used in modern chemistry and biochemistry to promote synthesis. Some molecules appear to acquire a degree of resistance to disintegration simply through their size. So, for example, the larger molecules composed of amino acids—polypeptides and proteins—seem to display much less tendency to disintegrate into their units than do smaller compounds of two or three amino acids.

Again, many organic molecules display still another type of integrating force—a spontaneous impulse toward structure formation. Certain types of fatty molecules—lecithins and cephalins—spin themselves out in water to form highly oriented and well-shaped structures—the so-called myelin figures. Proteins sometimes orient even in solution, and also may aggregate in the solid state in highly organized formations. Such spontaneous architectonic tendencies are still largely unexplored, particularly and they may occur in complex mixtures of substances, and they involve forces the strength of which has not yet been estimated.

What we are saying is that possibilities exist for opposing *intra*molecular dissolution by *inter*molecular aggregations of various kinds. The equilibrium between union and disunion of the amino acids that make up a protein is all to the advantage of disunion, but the aggregation of the protein with itself or other molecules might swing the equilibrium in the opposite direction: perhaps by removing the protein from access to the water which would be requried to disintegrate it or by providing some particularly stable type of molecular association.

In such a scheme the protein appears only as a transient intermediate, an unstable way-station, which can either fall back to a mixture of its constituent amino acids or enter into the formation of a complex structural aggregate: amino acids $\leftrightharpoons$ protein $\rightarrow$ aggregate.

Such molecular aggregates, of various degrees of material and architectural complexity, are indispensable intermediates between molecules and organisms. We have no need to try to imagine the spontaneous formation of an organism by one grand collision of its component molecules. The whole process must be gradual. The molecules form aggregates, small and large. The aggregates add further molecules, thus growing in size and complexity. Aggregates of various kinds interact with one another to form still larger and more complex structures. In this way we imagine the ascent, not by jumps or master strokes, but gradually, piecemeal, to the first living organisms.

First Organisms

Where may this have happened? It is easiest to suppose that life first arose in the sea. Here were the necessary salts and the water. The latter is not only the principal component of organisms, but prior to their formation provided a medium which could dissolve molecules of the widest variety and ceaselessly mix and circulate them. It is this constant mixture and collision of organic molecules of every sort that constituted in large part the "trials" of our earlier discussion of probabilities.

The sea in fact gradually turned into a dilute broth, sterile and oxygen-free. In this broth molecules came together in increasing number and variety, sometimes merely to collide and separate, sometimes to react with one another to produce new combinations, sometimes to aggregate into multimolecular formations of increasing size and complexity.

What brought order into such complexes? For order is as essential here as composition. To form an organism, molecules must enter into intricate designs and connections; they must eventually form a self-repairing, self-constructing dynamic machine. For a time this problem of molecular arrangement seemed to present an almost insuperable obstacle in the way of imagining a spontaneous origin of life, or indeed the laboratory synthesis of a living organism. It is still a large and mysterious problem, but it no longer seems insuperable. The change in view has come about because we now realize that it is not altogether necessary to *bring* order into this situation; a great deal of order is implicit in the molecules themselves.

The epitome of molecular order is a crystal. In a perfect crystal the molecules display complete regularity of position and orientation in all planes of space. At the other extreme are fluids—liquids or gases—in which the molecules are in ceaseless motion and in wholly random orientations and positions.

Lately it has become clear that very little of a living cell is truly fluid. Most of it consists of molecules which have taken up various degrees of orientation with regard to one another. That is, most of the cell represents various degrees of approach to crystallinity—often, however, with very important differences from the crystals most familiar to us. Much of the cell's crystallinity involves molecules which are still in solution—so-called liquid crystals—and much of the dynamic, plastic quality of cellular structure, the capacity for constant change of shape and interchange of material, derives from this condition. Our familiar crystals, furthermore, involves only one or a very few types of molecule, while in the cell a great variety of different molecules come together in some degree of regular spacing and orientation—*i.e.*, some degree of crystallinity. We are dealing in the cell with highly mixed crystals and near-crystals, solid and liquid. The laboratory study of this type of formation has scarcely begun. Its further exploration is of the highest importance for our problem.

In a fluid such as water the molecules are in very rapid motion. Any molecules dissolved in such a medium are under a constant barrage of collisions with water molecules. This keeps small and moderately sized molecules in a constant turmoil; they are knocked about at random, colliding again and again, never holding any position or orientation for more than an instant. The larger a molecule is relative to water, the less it is disturbed by such collisions. Many protein and nucleic acid molecules are so large that even in solution their motions are very sluggish, and since they carry large numbers of electric charges distributed about their surfaces, they tend even in solution to align with respect to one another. It is so that they tend to form liquid crystals.

We have spoken above of architectonic tendencies even among some of the relatively small molecules; the lecithins and cephalins. Such molecules are insoluble in water yet possess special groups which have a high affinity for water. As a result they tend to form surface layers, in which their water-seeking groups project into the water phase, while their water-repelling portions project into the air, or into an oil phase, or unite to form an oil phase. The result is that quite spontaneously such molecules, when exposed to water, take up highly oriented positions to form surface membranes, myelin figures and other quasicrystalline structures.

Recently several particularly striking examples have been reported of the spontaneous production of familiar types of biological structure by protein molecules. Cartilage and muscle offer some of the most intricate and regular patterns of structure to be found in organisms. A fibre from either type of tissue presents under the electron microscope a beautiful pattern of cross striations of various widths and densities, very regularly spaced. The proteins that form these structures can be coaxed into free solution and stirred into completely random orientation. Yet on precipitating, under proper conditions, the molecules realign with regard to one another to regenerate with extraordinary fidelity the original patterns of the tissues.

We have therefore a genuine basis for the view that the molecules of our oceanic broth will not only come together spontaneously to form aggregates but in doing so will spontaneously achieve various types and degrees of order. This greatly simplifies our problem. What it means is that, given the right molecules, one does not have to do everything for them; they do a great deal for themselves.

Oparin has made the ingenious suggestion that natural selection, which *Darwin* proposed to be the driving force of organic evolution, begins to operate at this level. He suggests that as the molecules come together to form colloidal aggregates, the latter begin to compete with one another for material. Some aggregates, by virtue of especially favourable composition or internal arrangement, acquire new molecules more rapidly than others. They eventually emerge as the dominant types. Oparin suggests further than considerations of optimal size enter at this level. A growing colloidal particle may reach a point at which it becomes unstable and breaks down into smaller particles, each of which grows and redivides. All these phenomena lie within the bonds of known processes in nonliving systems.

The Sources of Energy

We suppose that all these forces and factors, and others perhaps yet to be revealed, together give us eventually the first living organism. That achieved, how does the organism continue to live?

It has already noted that a living organism is a dynamic structure. It is the site of a continuous influx and outflow of matter and energy. This is the very sign of life, its cessation the best evidence of death. What is the primal organism to use as food, and how derive the energy it needs to maintain itself and grow?

For the primal organism, generated under the conditions we have described, only one answer is possible. Having arisen in an oceanic broth of organic molecules, its only recourse is to live upon them. There is only one way of doing that in the absence of oxygen. It is called fermentation: the process by which organisms derive energy by breaking organic molecules and rearranging their parts. The most familiar example of such a process is the fermentation of sugar by yeast, which yields alcohol as one of the products. Animal cells also ferment sugar, not to alcohol but to lactic acid. These are two examples from a host of known fermentations.

The yeast fermentation has the following over-all equation: $C_6H_{12}O_6 \rightarrow 2CO_2 + 2C_2H_5OH$ + energy. The result of fragmenting 180 grams of sugar into 88 grams of carbon dioxide and 92 grams of alcohol is to make available about 20,000 calories of energy for the use of the cell. The energy is all that the cell derives by this transaction;

the carbon dioxide and alcohol are waste products which must be got rid of somehow if the cell is to survive.

The cell, having arisen in a broth of organic compounds accumualted over the ages, must consume these molecules by fermentation in order to acquire the energy it needs to live, grow and reproduce. In doing so, it and its descendants are living on

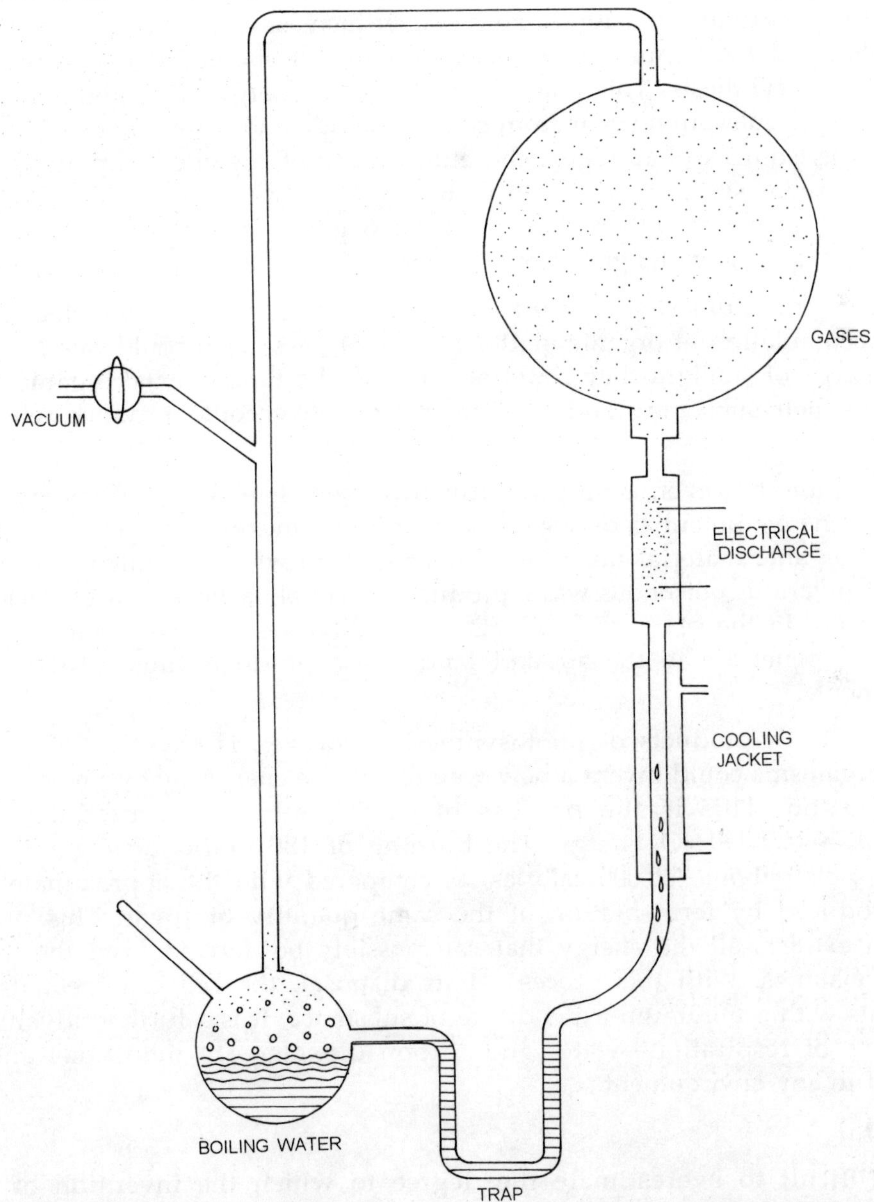

Fig. 7.6: Experiment of S.L. Miller made amino acids by circulating methane (CH_4), ammonia (NH_3), water vapour (H_2O) and hydrogen (H_2) past an electrical discharge. The amino acids collected at the bottom of apparatus and were detected by paper chromatography.

borrowed time. They are consuming their heritage, just as we in our time have nearly consumed our heritage of coal and oil. Eventually such a process must come to an end, and with that life also should have ended. It would have been necessary to start the entire development again.

Fortunately, however, the waste product carbon dioxide saved this situation. This gas entered the ocean and the atmosphere in ever-increasing quantity. Some time before the cell exhausted the supply of organic molecules, it succeeded in interventing the process of photosynthesis. This enabled it, with the energy of sunlight, to make its own organic molecules; first sugar from carbon dioxide and water, then, with ammonia and nitrates as sources of nitrogen, the entire array of organic compounds which it requires. The sugar synthesis equation is: $6\ CO_2 + 6\ H_2O + \text{sunlight} \rightarrow C_6H_{12}O_6 + 6\ O_2$. Hence 264 grams of carbon dioxide *plus* 108 grams of water *plus* about 700,000 calories of sunlight yield 180 grams of sugar and 192 grams of oxygen.

This is an enormous step forward. Living organisms no longer needed to depend upon the accumulation of organic matter from past ages; they could make their own. With the energy of sunlight they could accomplish the fundamental organic synthesis that provide their substance, and by fermentation they could produce what energy they needed.

Fermentation, however, is an extraordinarily inefficient source of energy. It leaves most of the energy potential of organic compounds unexploited; consequently huge amounts of organic material must be fermented to provide a modicum of energy. It produces also various poisonous waste products—alcohol, lactic acid, acetic acid, formic acid and so on. In the sea such products are readily washed away, but if organisms were ever to penetrate to the air and land, these products must prove a serious embarrassment.

One of the by-products of photosynthesis, however, is oxygen. Once this was available, organisms could invent a new way to acquire energy, many times as efficient as fermentation. This is the process of cold combustion called respiration: $C_6H_{12}O+6O_2 \rightarrow 6CO_2+6H_2O+\text{energy}$. The burning of 180 grams of sugar in cellular respiration yields about 700,000 calories, as compared with the approximately 20,000 calories produced by fermentation of the same quantity of sugar. This process of combustion extracts all the energy that can possibly be derived from the molecules which it consumes. With this process at its disposal, the cell can meet its energy requirements with a minimum expenditure of substance. It is a further advantage that the products of respiration—water and carbon dioxide—are innocuous and easily disposed of in any environment.

Life's Capital

It is difficult to overestimate the degree to which the invention of cellular respiration released the forces of living organisms. No organism that relies wholly upon fermentation has ever amounted to much. Even after the advent of photosynthesis, organisms could have led only a marginal existence. They could indeed produce their own organic materials, but only in quantities sufficient to survive. Fermentation is so profligate a way of life that photosynthesis could do little more than keep up with it. Respiration used the material of organisms with such enormously greater efficiency as for the first time to leave something over. Coupled with fermentation, photosynthesis made organisms self-sustaining; coupled with respiration, it provided a surplus. To use an economic analogy, photosynthesis brought organisms to the subsistence level; respiration provided them with capital. It is mainly the capital that they invested in the great enterprise of organic evolution.

The entry of oxygen into the atmosphere also liberated organisms in another sense. The sun's radiation contains ultraviolet components which no living cell can tolerate. We are sometimes told that if this radiation were to reach the earth's surface, life must cease. That is not quite true. Water absorbs ultraviolet radiation very effectively, and one must conclude that as long as these rays penetrated in quantity to the surface of the earth, life had to remain under water. With the appearance of oxygen, however, a layer of ozone formed high in the atmosphere and absorbed this radiation. Now organisms could for the first time emerge from the water and begin to populate the earth and air. Oxygen provided not only the means of obtaining adequate energy for evolution but the protective blanket of ozone which alone made possible terrestrial life.

This is really the end of our story. Yet not quite the end. Our entire concern in this argument has been to bring the origin of life within the compass of natural phenomena. It is of the essence of such phenomena to be repetitive, and hence, given time, to be inevitable.

This is by far our most significant conclusion—that life, as an orderly natural event on such a planet as ours, was inevitable. The same can be said of the whole of organic evolution. All of it lies within the order of nature, and apart from details all of it was inevitable.

Astronomers have reason to believe that a planet such as ours—of about the earth's size and temperature, and about as well-lighted—is a rare event in the universe. Indeed, filled as our story is with improbable phenomena, one of the least probable is to have had such a body as the earth to begin with. Yet though this probability is small, the universe is so large that it is conservatively estimated at least 100,000 planets like the earth exist in our galaxy alone. Some 100 million galaxies lie within the range of our most powerful telescopes, so that throughout observable space we can count apparently on the existence of at least 10 million planets like our own.

What it means to bring the origin of life within the realm of natural phenomena is to imply that in all these places life probably exists—life as we know it. Indeed, there can be no way of composing and constructing living organisms which is fundamentally different from the one we know—though this is another argument, and must await another occasion. Wherever life is possible, given time, it should rise. It should then ramify into a wide array of forms, yet including many which should look familiar to us—perhaps even men.

We are not alone in the universe, and do not bear alone the whole burden of life and what comes of it. Life is a cosmic event—so far as we know the most complex state of organisation that matter has achieved in our cosmos. It has come many times, in many places—places closed off from us by impenetrable distances, probably never to be crossed even with a signal. As men we can attempt to understand it, and even somewhat to control and guide its local manifestations. On this planet that is our home, we have every reason to wish it well. Yet should we fail, all is not lost. Our kind will try again elsewhere.

Chapter—8

The Evolution of the Earliest Cells

When *On the Origin of Species* appeared in 1859, the history of life could be traced back to the beginning of the Cambrian period of geologic time, to the earliest recognized fossils, forms that are now known to have lived more than 500 million years ago. A far longer prehistory of life has since been discovered: it extends back through geologic time almost three billion years more. During most of that long Precambrian interval the only inhabitants of the earth were simple microscopic organisms, many of them comparable in size and complexity to modern bacteria. The conditions under which these organisms lived differed greatly from those prevailing today, but the mechanisms of evolution were the same. Genetic variations made some individuals better fitted than others to survive and to reproduce in a given environment, and so the heritable traits of the better-adapted organisms were more often represented in succeeding generations. The emergence of new forms of life through this principle of natural selection worked great changes in turn on the physical environment, thereby altering the conditions of evolution.

One momentous event in *Precambrian* evolution was the development of the biochemical apparatus of oxygen-generating photosynthesis. Oxygen released as a by-product of photosynthesis accumulate in the atmosphere and effected a new cycle of biological adaptation. The first organisms to evolve in response to this environmental change could merely tolerate oxygen; later cells could actively employ oxygen in metabolism and were thereby enabled to extract more energy from foodstuff.

A second important episode in Precambrian history led to the emergence of a new kind of cell, in which the genetic material is aggregated in a distinct nucleus and is bounded by a membrane. Such nucleated cells are more highly organized than those without nuclei. What is most important, only nucleated cells are capable of advanced sexual reproduction, the process whereby the genetic variations of the parents can be passed on to the off spring in new combinations. Because sexual reproduction allows novel adaptations to spread quickly through a populationists development accelerated the pace of evolutionary change. The large, complex, multicellular forms of life that have appeared and quickly diversified since the beginning of the *Cambrian period* are without exception made up of nucleated cells.

The history of life in its later phases, since the start of the *Cambrian period,* has been reconstructed mainly from the study of fossils preserved in sedimentary rocks. In the 18th and 19th centuries it gradually became apparent that the fossil record has appreciable chronological and geographical continuity. The fossil deposits form recognizable layers which can be identified in widely separated geological formations. Boundaries between such layers, where one characteristic suite of fossils gives way to another, provide the basis for dividing geologic time into eras, periods and epochs.

One of the most dramatic boundaries in the rock record is the one that separates the Cambrian period from all that came before. The 11 periods of geologic time since the start of the Cambrian are referred to collectively as the *Phanerozoic era,* which might be translated from the Greek as the era of manifest life. The preceding era is called simply the Precambrian.

By itself the geologic time scale cannot provide dates for fossil deposits; it only lists their sequence. Ages can be calculated, however, from the constant rate of decay of radioactive isotopes in the earth's crust. By determining how much of an isotope has decayed since the minerals in a rock unit crystallized, a date can be assigned to that unit and to nearby strata containing fossils. Radioactive-isotope studies of this kind, carried out on rocks from many parts of the world, have established a rather well-defined date for the start of the Phanerozoic era: it began about 570 million years ago. The same method indicates that the earth itself and the rest of the solar system are 4.6 billion years old. Thus the Precambrian era encompasses some seven-eighths of the earth's entire history.

The boundary between the Precambrian era and the Cambrian period has traditionally been viewed as a sharp discontinuity. In Cambrian strata there are abundant fossils of marine plants and animals: seaweeds, worms, sponges, mollusks, lampshells and, what are perhaps most characteristic of the period, the early arthropods called trilobites. It was thought from many years that fossils were entirely absent in the underlying Precambrian strata. The Cambrian fauna seemed to come into existence abruptly and without known predecessors.

Life could not have begun with organisms as complex as *trilobites.* In *on the Origin of Species* Darwin wrote: "To the question why we do not find rich fossiliferous deposits belonging to....periods prior to the Cambrian system. I can give no satisfactory answer... The case at present must remain inexplicable; and may be truly urged as a valid argument against the views here entertained." The argument is no longer valid, but it is only in the past 20 years or so that a definitive answer to it has been found.

One part of the answer lies in the discovery of primitive fossil animals in rocks below the earliest Cambrian strata. The fossils include the remains of jellyfishes, various kinds of worms and possibly sponges, and they make up a fauna quite distinct from that of the predominantly shelled animals of the Cambrian period. These discoveries, however, extend the fossil record by only about 100 million years, less than four per cent of the Precambrian era. It can still be asked: What came before?

Since the 1950's a far-reaching explanation has emerged. It has come to be recognized that not only are many Precambrian rocks fossil-bearing but also Precambrian fossils can be found even in some of the most ancient sedimentary deposits known. These fossils had escaped notice earlier largely because they are the remains only of microscopic forms of life.

An important clue in the search for Precambrian life was discovered in the early years of the 20th century, but its significance was not fully appreciated until much later. The clue came in the form of masses of thinly layered limestone rock discovered by *Charles Doolittle Walcott* in Precambrian strata from western North America. Walcott found numerous moundlike or pillar like structures made up of many draped horizontal layers, like, tall stacks of pancakes. These structures are now called stromatolites, from the Greek *stroma* meaning bed or coverlet, and *lithos,* meaning stone.

Walcott interpreted the stromatolites as being fossilized reefs that had probably been formed by various types of algae. Other workers were skeptical, and for many years the stromatolites were widely attributed to some non biological origin. The first convincing evidence substantiating Walcott's hypothesis came in 1954, when *Stanley A. Tyler* of the University of Wisconsin and *Elso S. Barghoorn* of Harvard University reported the discovery of fossil microscopic plants in an outcropping of Precambrian rocks called the Gunflint Iron formation near Lake Superior in Ontario. Most of the Gunflint fossils, which form the layers of dome-shaped and pillar-like stromatolites, resemble modern blue-green algae and bacteria. More recently, living stromatolites have been identified in several coastal habitats, most notably in a lagoon at Shark Bay on the western coast of Australia. They are indeed built up by communities of blue-green algae and bacteria, and they are strikingly similar in form to the fossilized Precambrian structures

Today microfossils have been identified in some 45 stromatolitic deposits (All but three of these fossilized communities have been found in the past 10 years). The fossils are often well preserved, the cell walls being petrified in three-dimensional form, and they have become a prime source of documentation for the early history of life. In recent years the search for Precambrian microfossils in other kinds of sediments, such as shales deposited in offshore environments, has also been rewarded. These fossils are generally not as well preserved as the ones in stromatolites, most of them having been flattened by pressure; on the other hand, they supply information about Precambrian life in a habitat quite different from that of the shallow-water stromatolites.

A surprising amount of information can be derived from the fossil remains of a microorganism. Size, shape and degree of morphological complexity are among the most easily recognized features, but under favourable circumstances even details of the internal structure of cells can be discerned. In retracing the course of Precambrian evolution, however, there is no need to rely exclusively on the fossil record. An entirely independent archive has been preserved in the metabolism and the biochemical pathways of modern, living cells. No living organism is biochemically identical with its Precambrian antecedents, but vestiges of earlier biochemistries have been retained. By studying their distribution in modern forms of life it is sometimes possible to deduce when certain biochemical capabilities first appeared in the evolutionary sequence.

Still another independent source of information about the early evolutionary progression is based neither on living nor on fossil organisms but on the inorganic geological record. The nature of the minerals found there reflects physical conditions at the time the minerals were deposited, conditions that may have been influenced by biological innovations. In order to understand the introduction of oxygen into the early atmosphere, for example, all three fields of study must be called on to testify, the mineral record tells when the change took place the fossil record reveals the organisms responsible and the distribution of biochemical capabilities among modern organisms puts the development in its proper evolutionary context.

Since the 1960's it has become apparent that the greatest division among living organisms is not between plants and animals but between organisms whose cells have nuclei and those that lack nuclei. In terms of biochemistry, metabolism, genetics and intracellular organization, plants and animals are very similar; all such higher organisms, however, are quite different in these features from bacteria and blue-green algae, the principal types of non-nucleated life. Recognition of this discontinuity has been important for understanding the early stages of biological history.

Organisms whose cells have nuclei are called eukaryotes, from the Greek roots *eu-*, meaning well or true, and *karyon,* meaning kernel or nut. Cells without nuclei are prokaryotes, the prefix promeaning before. All green plants and all animals are eukaryotes. So are the fungi, including the molds and the yeasts, and protists such as *Paramecium* and *Euglena.* The prokaryotes include only two groups of organisms, the bacteria and the blue-green algae. The latter produce oxygen through photosynthesis like other algae and higher plants, but they have much stronger affinities with the bacteria than they do with eukaryotic forms of life. Later on several workers refer to blue-green algae by an alternative and more descriptive name, the *cyanobacteria.*

Several important traits distinguish eukaryotes from prokaryotes. In the nucleus of a eukaryotic cell the DNA is organized in chromosomes and is enclosed by an intracellular membrane, many prokaryotes have only a single loop of DNA which is loose in the cytoplasm of the cell. Prokaryotes reproduce asexually by the comparatively simple process of binary fission. In contrast, asexual reproduction in eukaryotic cells takes place through the complicated process of mitosis, and most eukaryotes can also reproduce sexually through meiosis and the subsequent fusion of sex cells. (The "parasexual" reproduction of some prokaryotes differs markedly from advanced eukaryotic sexuality). Eukaryotic cells are generally larger than prokaryotic ones, although the range of sizes overlaps, and almost all prokaryotes are unicellular

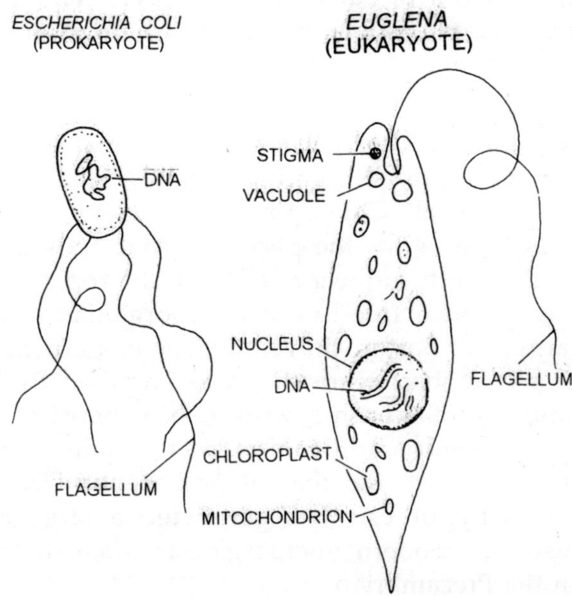

	Prokaryotes	*Eukaryotes*
Organisms Represented	Bacteria and Cyanobacteria	Protists, Fungi, Plants and Animals
Cell Size	Small, Generally 1 to 10 Micrometers	Large, Generally 10 to 100 Micrometers
Metabolism and Photosynthesis	Anaerobic or Aerobic	Aerobic
Motility	Nonmotile or with Flagella made of the Protein Flagellin	Usually Motile, Cilia or Flagella Constructed of Microtubues
Cell Walls	Of Characteristic Sugars and Peptides	Of cellulose or chitin but Lacking in Animals
Organelles	No Membrane-Bounded Organelles	Mitochondria and Chloroplasts
Genetic Organization	Loop of DNA in Cytoplasm	DNA organized in Chromosomes and bounded by Nuclear Membrane
Reproduction	By Binary Fission	By Mitosis or Meiosis
Cellular Organization	Mainly Unicellular	Mainly Multicellular with differenttiation of Cells

Fig. 8.1: Greatest Division among organisms is the one separating cells with nuclei (eukaryotes) from those without nuclei (prokaryotes). The only prokaryotes are bacteria and cyanobacteria, and here they are represented by the bacterium *Escherichia coli (top left).* all other organisms are eukaryotes, including higher plants and animals, fungi and protists such as *Euglena (top right.)* Eukaryotic cells are by far the more complex ones, and some of the organelles they contain, such as mitochondria and chloroplasts, may be derived from prokaryotes that established a symbiotic relationship with the host cell. Prokaryotes vary widely in their tolerance of or requirement for free oxygen, and they are thought to have evolved during a period of fluctuating oxygen. All eukaryotes require oxygen for metabolism and for the synthesis of various substances, and they must have emerged after an atmosphere rich in oxygen became established.

organisms whereas the majority of eukaryotes are large, complex and many-celled. A mammalian animal, for example, can be made up of billions of cells, which are highly differentiated in both structure and function.

An intriguing feature of eukaryotic cells is that they have within them smaller membrane-bounded subunits, or organelles, the most notable being mitochondria and chloroplasts. Mitochondra are present in all eukaryotes, where they play a central role in the energy economy of the cell. Chloroplasts are present in some protists and in all green plants and are responsible for the photosynthetic activities of those organisms. It has been suggested that both mitochondria and chloroplasts may be evolutionary derivatives of what were once free-living microorganisms, an idea discussed in particular by Lynn Margulis of Boston University. The modern chloroplast, for example, may be derived from a cyanobacterium that was engulfed by another cell and that later established a symbiotic relationship with it. In support of this hypothesis it has been noted that both mitochondria and chloroplasts contain a small fragment of DNA whose organization is somewhat like that of prokaryotic DNA. In the past several years the testing of this hypothesis has generated a large body of data on the comparative biochemistry of modern microorganism, data that also provide clues to the evolution of life in the Precambrian.

One further difference between prokaryotes and eukaryotes is of particular importance in the study of their evolution: the extent to which the two types of organisms tolerate oxygen. Among the prokaryotes oxygen requirements are quite variable. Some bacteria cannot grow or reproduce in the presence of oxygen: they are classified as obligate anaerobes. Others can tolerate oxygen but can also survive in its absence; they are facultative anaerobes. There are also prokaryotes that grow best in the presence of oxygen but only at low concentrations, far below that of the present atmosphere. Finally, there are fully aerobic prokaryotes forms that cannot survive without oxygen.

In contrast to this variety of adaptations the eukaryotes present a pattern of great consistency: with very few exceptions they have an absolute requirement for oxygen, and even the exceptions seem to be evolutionary derivatives of oxygen-dependent organisms. This observation leads to a simple hypothesis: the prokaryotes evolved during a period when environmental oxygen concentrations were changing, but by the time the eukaryotes arose the oxygen content was stable and relatively high.

One indication that eukaryotic cells have always been aerobic is provided by mitotic cell division, a process that can be considered a definitive characteristic of the group. Many eukaryotic cells can survive temporary deprivation of oxygen and can even carry on some metabolic functions; it appears that no cell, however, can undergo mitosis unless oxygen is available at least in low concentration .

The pathways of metabolism itself—the biochemical mechanisms by which an organism extracts energy from foodstuff—provide more detailed evidence. In eukaryotes the central metabolic process is respiration, which in overall terms can be

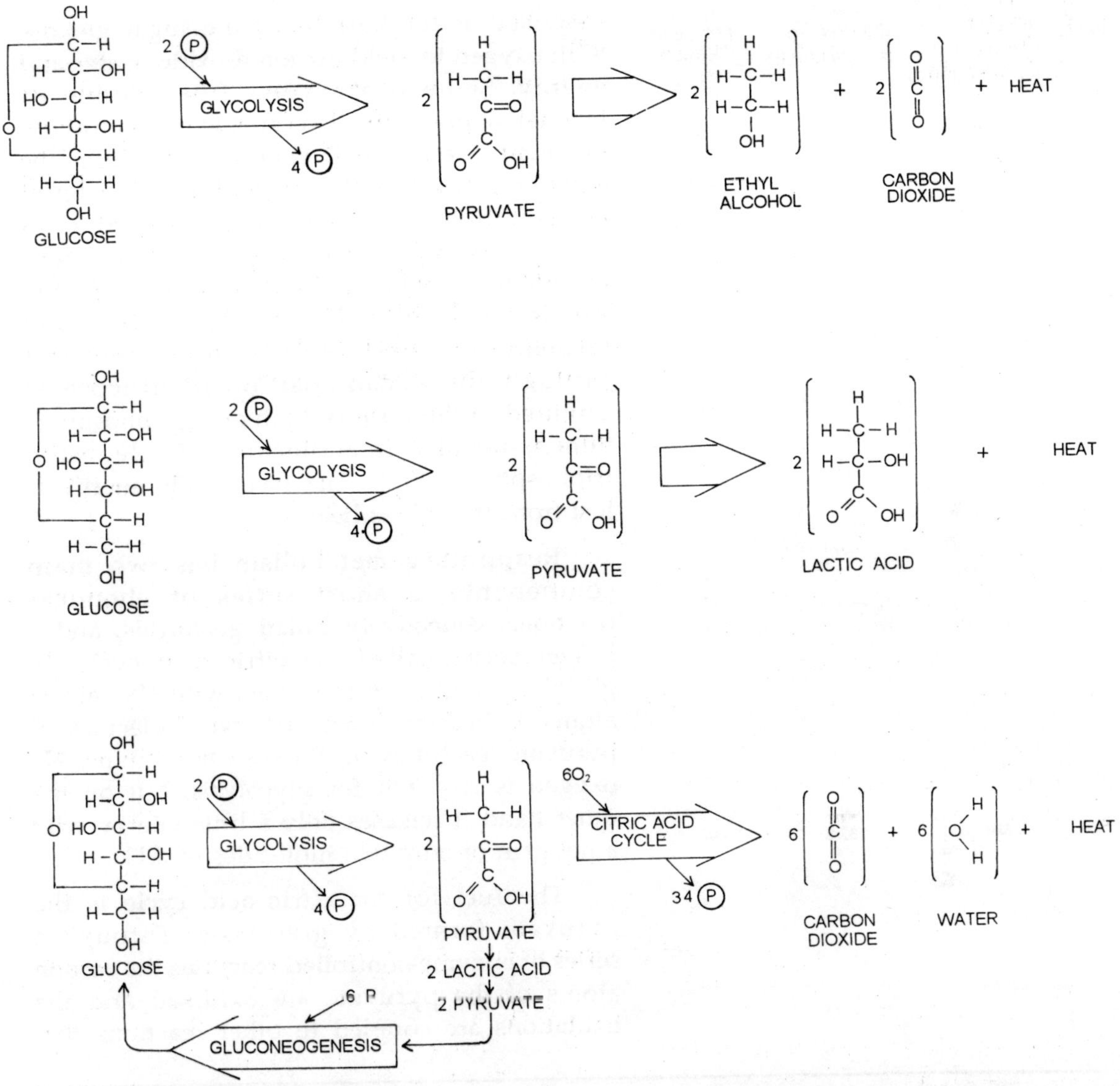

Fig. 8.2: Metabolic Pathways by which cells extract energy from foodstuff apparently evolved in response to an increase in free oxygen. In all organisms the only usable energy derived from the breakdown of carbohydrates such as glucose is the fraction stored high-energy phosphate bonds, denoted~P; the rest is lost as heat. In anaerobic organisms (those that live without oxygen) glucose is broken down through fermentation: each molecule of glucose is split into two molecules of pyruvate, the process called glycolysis, with a net gain of two phosphate bonds. In bacterial fermentation the pyruvate is converted, in a step that provides no usable energy, into products such as lactic acid or ethyl alcohol and carbon dioxide, which are excreted as wastes. The metabolic system of aerobic organisms (those that require oxygen) is respiration. It begins with glycolysis, but the

GAIN IN Ⓟ	TOTAL ENERGY RELEASED Ⓟ PLUS HEAT (KILOCALORIES)	AVAILABLE ENERGY Ⓟ (KILOCALORIES)	CALCULATED EFFICIENCY (PERCENT)
2	57	14.6	26
2	47	14.6	31
36	686	262.8	38

described as the burning of the sugar glucose with oxygen to yield carbon dioxide, water and energy. Some prokaryotes (the aerobic or facultative ones) are also capable of respiration, but many derive their energy solely from the simpler process of fermentation. In bacterial fermentation glucose is not combined with oxygen (or with any othersubstance from outside the cell) but is simply broken down into smaller molecules. In both respiration and fermentation part of the energy released through the decomposition of glucose is captured in the form of high-energy phosphate bonds, usually in molecules of adenosine triphosphate (ATP). The rest of the energy is lost from the cell as heat.

Respiratory metabolism has two main components: a short series of chemical reactions, collectively called glycolysis, and a longer series called the citric acid cycle. In glycolysis a glucose molecule, with six carbon atoms, is broken down into two molecules of pyruvate, each having three carbon atoms. No oxygen is required for glycolysis, but on the other hand it releases only a little energy with a net gain of only two molecules of ATP.

The fuel for the citric acid cycle is the pyruvate formed by glycolysis. Through a series of enzyme-controlled reactions the carbon atoms of the pyruvate are oxidized and the oxidations are coupled to other reactions that

---

pyruvate is treated not as a waste but as a substrate for a further series of reactions that make up the citric acid cycle. In these reactions pyruvate is decomposed one carbon atom at a time and combined with oxygen, the ultimate products being carbon dioxide and water. Respiration releases far more energy than fermentation, and the proportion of the energy recovered in useful form is also greater; as a result 36 phosphate bonds are formed instead of two. Respiratory metabolism could have evolved, however, only when free oxygen became readily available; it appears to have developed simply by appending the citric acid cycle to the glycolytic pathway. When aerobic cells are deprived of oxygen, many revert to fermentative metabolism, converting pyruvate into lactic acid. In vertebrates the lactic acid from muscle cells is exported to the liver, where it is retured to the form of pyruvate and converted into glucose.

result in the synthesis of ATP. For each two molecules of pyruvate (and hence for each molecule of glucose entering the sequence) 34 additional molecules of ATP are formed. The complete respiratory pathway is thus far more effective than glycolysis alone. In respiration the proportion of energy released that can be recovered in useful form (as ATP) is higher than it is in fermentation, about 38 per cent instead of only some 30 per cent, and in respiration the net energy yield to the cell is some 18 times greater. By breaking down the glucose to simple inorganic molecules (carbon dioxide and water) respiration liberates virtually all the biologically usable energy stored in the chemical bonds of the sugar.

The metabolism of the prokaryotes immediately suggests an evolutionary relationship between them and the eukaryotes; up to a point fermentation is indistinguishable from glycolysis. In bacterial fermentation a molecule of glucose is split into two molecules of pyruvate, with a net yield of two molecules of ATP. As in glycolysis, no oxygen is required for the process. In anaerobic prokaryotes, however, the metabolic pathway essentially ends at pyruvate. The only further reactions transform the pyruvate into such compounds as lactic acid, ethyl alcohol or carbon dioxide, which are excreted by the cell as wastes.

The similarity of fermentation in prokaryotes to glycolysis in eukaryotes seems too close to be a coincidence, and the assumption of an evolutionary relationship between the two groups provides a ready explanation. It seems likely that anaerobic fermentation became established as an energy-yielding process early in the history of life. When atmospheric oxygen became available for metabolism, it offered the potential for extracting 18 times as much useful energy from carbohydrate: a net yield of 36 molecules of ATP instead of only two molecules. The oxygen-dependent reactions did not however, simply replace the anaerobic ones: they were appended to the existing anaerobic pathway.

Further evidence for this proposed evolutionary sequence can be found in the behaviour of some eukaryotic cells under conditions of oxygen deprivation. In mammalian muscle cells, for example, prolonged exertion can demand more oxygen than the lungs and the blood can supply. The citric acid cycle is then disabled, but the cells continue to function, albeit at reduced efficiency, through glycolysis alone. Under such conditions of oxygen debt pyruvate is not consumed in the cell, but in the liver it can be converted back into glucose (at a cost in energy of six ATP molecules). Significantly the pyruvate itself is not transported to the liver but instead is converted into lactic acid, which in the liver must then be returned to the form of pyruvate. This use of lactic acid may represent a vestige of an earlier, bacterial pathway that under aerobic conditions has been suppressed. Indeed, the oxygen-starved muscle cell seems to revert to a more primitive, entirely anaerobic form of metabolism.

The development of an oxygen-dependent biochemistry can also be traced through a consideration of reaction sequences in the synthesis of various biological molecules. Once again stages in the synthetic pathway that emerged early in the Precambrian

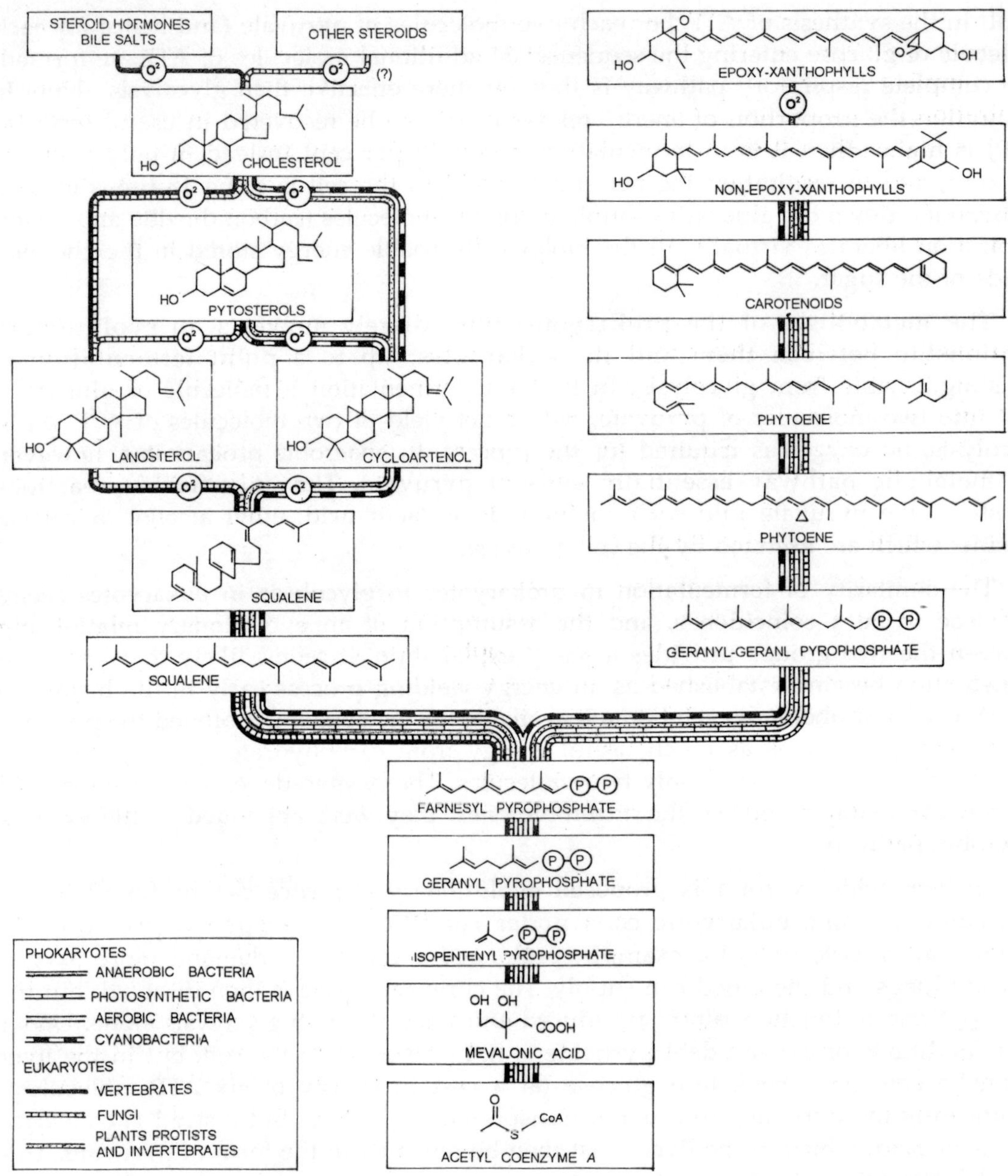

Fig. 8.3: Synthesis of Sterols and of related compounds such as the carotenoid pigments of plants requires molecular oxygen (O_2) only for steps near the end of the reaction sequence. Oxygen-dependent steps can be carried out only by aerobic organisms that evolved comparatively late in history of Precambrian life, Organic molecules are in schematic form with most carbon and hydrogen atoms omitted.

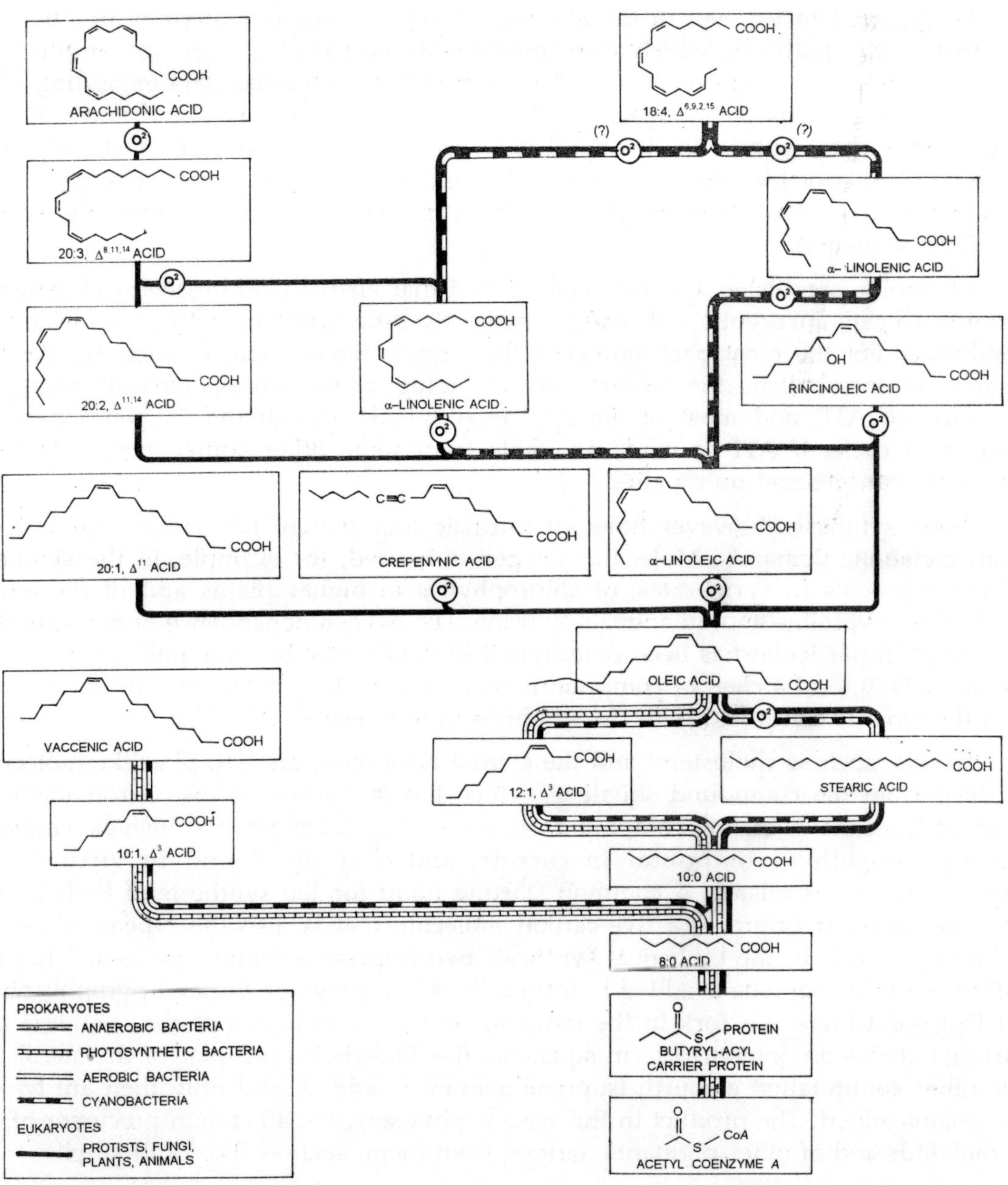

Fig. 8.4: Fatty-acid Synthesis also follows a pattern suggesting the late addition of oxygen-dependent steps. Most prokaryotes can make mono-unsaturated fatty acids (those having one double bond) by inserting double bond during elongation of molecule. Eukaryotes and some prokaryotes first make a fully saturated molecule, stearic acid, then introduce double bonds by the process of oxidative desaturation.

can be expected to proceed in the absence of oxygen. Reaction steps nearer the end product of the pathway, which were presumably added at a later age, might with increasing frequency require oxygen. The distribution of the oxygen-demanding steps among various kinds of organisms could also have evolutionary significance. If only one pathway has evolved for the synthesis of a class of biochemical substances, then primitive forms of life might be expected to exhibit only the initial, anaerobic steps. Organisms that arose later might exhibit progressively longer, oxygen-dependent synthetic sequences.

In aerobic organisms it might seem at first that virtually all biochemical synthesis require oxygen; eukaryotic cells exhibit relatively little synthetic activity under anoxic conditions. For the most part, however, the oxygen requirement of such synthesis is simply for metabolism: the construction of biological molecules demands energy in the form of ATP, and most of the ATP is supplied through the oxygen-dependent citric acid cycle. If ATP is made available from some other source many synthetic pathways can proceed unimpaired.

Some synthesis, however have an intrinsic requirement for oxygen, quite apart from metabolic demands. Molecular oxygen is needed, for example, in the synthesis of bile pigments in vertebrates, of chlorophyll *a* in higher plants and of the amino acids hydroxyproline and, in animals tyrosine. The oxygen dependence of two synthetic pathways in particular has been determined in detail. One of these pathways controls the manufacture of a class of compounds that includes the sterols and the carotenoids and the other is concerned with the synthesis of fatty acids.

Sterols, such as cholesterol and the steroid hormones, are flat, platelike molecules derived from the compound squalene, which has 30 carbon atoms. Carotenoids are derived from the 40-carbon compound phytoene: they are pigments, such as carotene, the orange-yellow compound in carrots, and they are found in virtually all photosynthetic organisms. A common starting point for the synthesis of both groups of compounds is isoprene, a five-carbon molecule that is also the repeating unit in synthetic rubber. In the biological synthesis two isoprene subunits are joined head to tail then a third isoprene is added to form a 15-carbon polymer, farnesyl pyrophosphate. At this point there is a fork in the pathway. In one continuaion of the synthesis two farnesyl chains are joined to form squalene, the 30-carbon precursor of the sterols. In the other continuation a fourth isoprene subunit is added, and only then are two of the chains joined. The product in this case is phytoene, the 40-carbon precursor of the carotenoids and of other pigments derived from them, such as the xanthophylls.

Up to this step in the synthetic pathway none of the reactions requires the participation of molecular oxygen. The next step in the synthesis of sterols, however, is the conversion of the linear squalene molecule to a 30-carbon ring, and this transformation does require oxygen; so do most of the subsequent steps in sterol synthesis. On the other branch of the pathway there are a few more anaerobic reactions, and indeed carotenoids can be made from phytoene without oxygen. Several further modifications of the carotenoids, however, such as the production of the pigments called epoxy-xanthophylls, are oxygen dependent.

Two observations about the evolution of these biosynthetic pathways are appropriate. Even in groups of organisms that have long been aerobic the first steps in the synthesis are independent of the oxygen supply; molecular oxygen enters the reaction sequence only at later stages. In a similar way the most primitive living organisms, the anaerobic bacteria, are capable only of the first segments of the pathway, the anaerobic segments. The more complex aerobic bacteria and the photosynthetic cyanobacteria have longer synthetic pathways, including some steps that require oxygen. Advanced eukaryotes, such as vertebrate animals and higher plants, have long, branched synthetic pathways, with many steps in which molecular oxygen is required.

A similar pattern can be discerned in the synthesis of fatty acids and their derivatives. The fatty acids are straight carbon-chain compounds that have a carboxyl group (COOH) at one end. A fatty acid is said to be saturated if there are no double bonds between carbon atoms in the chain: it is saturated with hydrogen, which fills all the available bonding positions. An unsaturated fatty acid has a double bond between two carbon atoms or it may have several such double bonds: for each double bond two hydrogen atoms must be removed from the molecule.

In the synthesis of fatty acids the molecule grows by the repeated addition of units two carbon atoms long. The first few steps in the synthesis are identical in all organisms, and they yield fully saturated fatty acids. The first branch in the pathway comes when the developing chain is eight carbons long. At that point many prokaryotes can introduce a double bond, which eukaryotes cannot. There is a second branch at the next step when the saturated chain is 10 carbons long: a double bond can similarly be introduced at that point by many prokaryotes but not by eukaryotes. No matter which branch is followed, elongation of the chain ends at 18 carbons. At that point the fatty acids produced by many prokaryotes contain a double bond, but in eukaryotes the product is always the fully saturated molecule, stearic acid. None of the steps in this sequence, whether in prokaryotes or eukaryotes, requires molecular oxygen.

If no subsequent transformations of fatty acids were possible, eukaryotic cells would be incapable of synthesizing any but the fully saturated forms. Actually extensive modifications can be accomplished through the process of oxidative desaturation, in which double bonds are formed by removing two hydrogen atoms and combining them with oxygen to form water, Oxidative desaturation can take place only in the presence of molecular oxygen (O_2). Through this mechanism cyanobacteria make unsaturated fatty acids with two, three and four double bonds, and eukaryotes form polyunsaturated fatty acids (with multiple double bonds).

As in the sterol-carotenoid synthesis, an analysis of the fatty-acid pathway argues for a pattern of biochemical evolution in which the increasing availability of atmospheric oxygen played a central role. The first steps in the synthetic sequence are common to all organisms capable of making fatty acids, and in the most primitive organisms those are the only steps. Hence the reactions that come first in the biochemical sequence apparently also developed early in the history of life; these first steps are all anaerobic. Organisms that presumably emerged somewhat later (such as aerobic bacteria and cyanobacteria) have longer pathways, including a few steps of oxidative desaturation. In advanced eukaryotes a substantial proportion of the steps are oxygen dependent.

Comparisons of the metabolism and biochemistry of prokaryotes and eukaryotes thus provide strong evidence that the latter group arose only after a substantial quantity of oxygen had accumulated in the atmosphere. Hence it is of interest to ask when eukaryotic cells first appeared. It seems apparent that an oxygen-rich atmosphere cannot have developed later than this signal evolutionary event.

The primary means of assigning a date to the origin of the eukaryotes is through the fossil record. Because this field of study is so new, however, the available information is scanty and often difficult to interpret. It is rarely a straightforward task

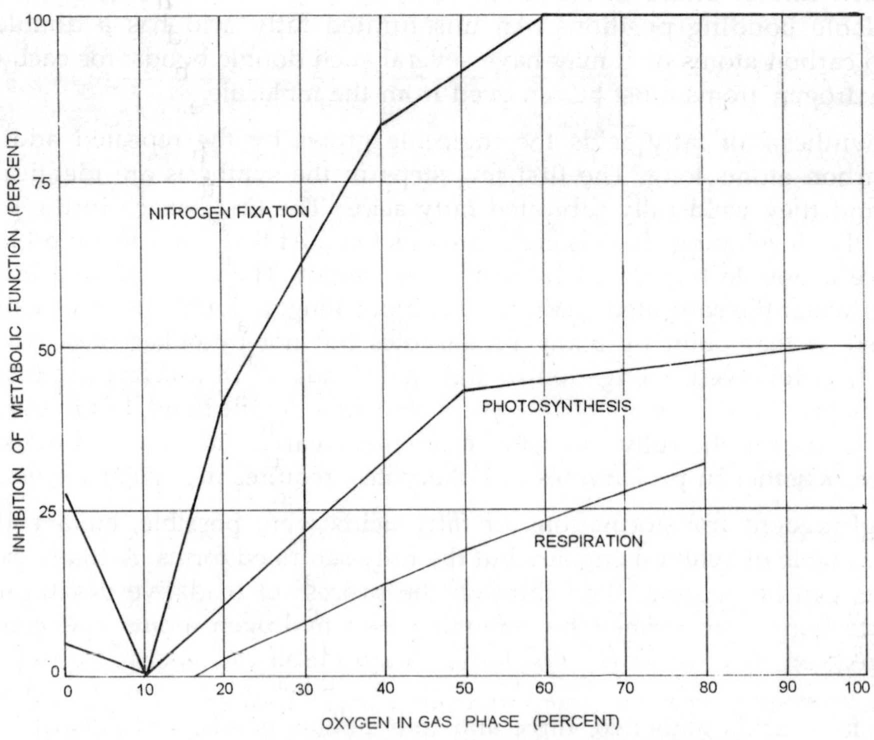

Fig. 8.5: Oxygen Inhibition of metabolic functions in cyanobacteria suggests that these aerobic prokaryotes are adapted to an optimum oxygen concentration of about 10 per cent, or roughly half the oxygen concentration of the earth's present atmosphere. Nitrogen fixation is completely halted by high oxygen levels, but even respiration, which requires oxygen, can be partially inhibited. Data are for the heterocyst-forming cyanobacterium *Anabaena flos-aquae.*

to identify a microscopic, single-cell organism as being eukaryotic merely from an examination of its fossilized remains. And even when a fossil has been identified as unequivocally eukaryotic the available radioactive-isotope methods of dating can rarely assign it a precise age. At best such methods have an accuracy of only about *plus* or minus 5 per cent. What is more, the age determinations are generally carried out on rocks that were once molten, such as volcanic lavas, whereas the fossils are found in sedimentary deposits. Consequently the stratum of the fossil itself usually cannot be dated: it is merely assigned an age somewhere between the ages of the nearest underlying and overlying datable rock units.

In spite of these difficulties there is now substantial evidence for the existence of eukaryotic fossils in rocks hundreds of millions of years older than the earliest Phanerozoic strata. The evidence is of two kinds: microfossils that display a morphological or organizational complexity judged to be of eukaryotic character and the presence of fossil cells whose size is typical only of eukaryotes.

The evidence from relatively complex microscopic fossils includes the following: (1) branched filaments, made up of the cells with distinct cross walls and resembling modern fungi or green algae, from the Olkhin formation of Siberia, a deposit thought to be about 725 million years old (but with a known age of between 680 and 800 million years); (2) complex, flask-shaped microfossils from the Kwagunt formation in the eastern Grand Canyon, thought to be about 800 (or 650 to 1,150) million years old; (3) fossils of unicellular algae containing intracellular membranes and small dense bodies that may represent preserved organelles, from the Bitter Springs formation of central Australia, dated at approximately 850 (or 740 to 950) million years; (4) a group of four sporelike cells in a tetrahedral configuration that may have been produced by mitosis or possibly meiosis, also from Bitter Springs rocks; (5) spiny cells or algal cysts several hundred micrometers in diameter and with unquestionable affinities to eukaryotic organisms, from Siberian shales that are reportedly 950 (or 750 to 1,050) million years old; (6) highly branched filaments of large diameter and with rare cross walls, similar in some respects to certain green or golden-green eukaryotic algae, from the Beck Spring dolomite of southeastern California (1,300, or 1,200 to 1,400 million years old) and from the Skillogalee dolomite of South Australia (850, or 740 to 867 million years old); (7) spheroidal microfossils described as exhibiting two-layered walls and having "medial splits" on their surface and which may represent an encystment stage of a eukaryotic alga from shales 1,400 (or 1,280 to 1,450) million years old in the McMinn formation of northern Australia; (8) a tetrachedral group of four small cells, resembling spores produced by mitotic cell division of some green algae, from the Amelia dolomite of northern Australia, approaching 1,500 (or 1,390 to 1,575) million years in age; (9) unicellular fossils that appear to be exceptionally well preserved and that are reported to contain small membrane-bounded structures that could be remnants of organelles, from the Bungle-Bungle dolomite in the same region as the Amelia dolomite and of approximately the same age.

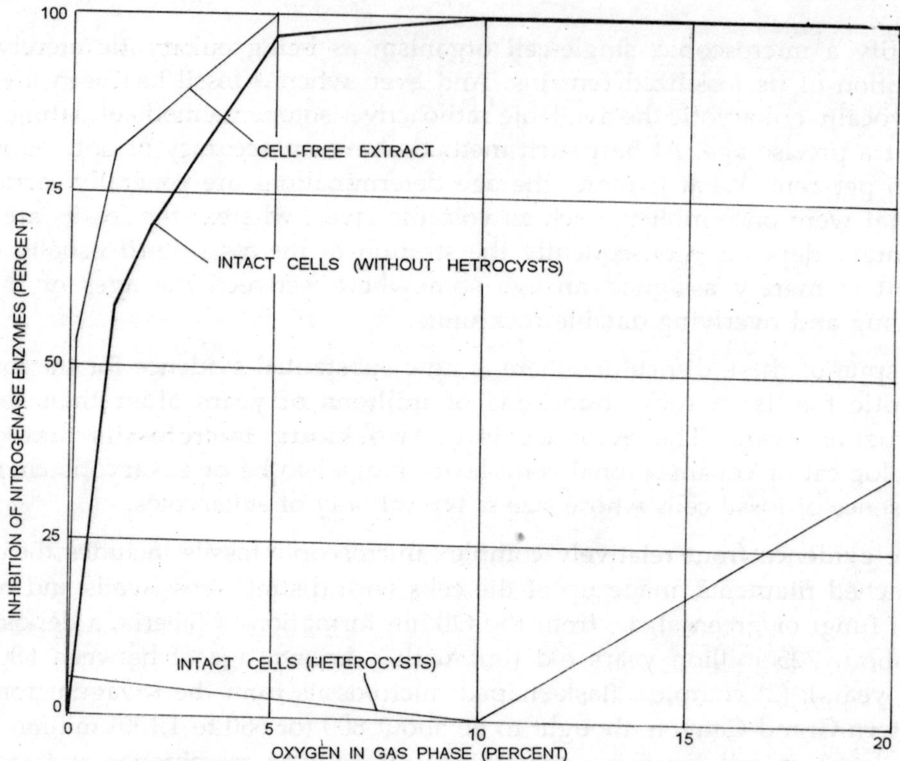

Fig. 8.6: Inhibition of Nitrogen Fixation in the presence of oxygen is caused by the deactivation of the nitrogenase enzymes. In cell-free extracts of the cyanobacterium *Plectonema boryanum* the nitrogenase enzymes are inhibited by even minute quantities of oxygen, and the intact cells of this species, which does not form heterocysts, offer little protection against the inhibition; such organisms can fix nitrogen only in an anoxic habitat. The thick cell walls and other special features of heterocyst cells, such as those formed by *Nostoc muscorum,* allow fixation to continue in a fully aerobic environment. Data suggest that the capability for nitrogen fixation evolved before significant quantities of oxygen had accumulated in the atmosphere.

Thus the earliest of these eukaryote-like fossils are probably somewhat less than 1,500 million years old. Numerous types of microfossils have been discovered in older sediments, but none of them seems to be a strong candidate for that in overall effect (although not in mechanism) is the reverse of respiration. The energy of sunlight is employed to make carbohydrates from water and carbon dioxide, and molecular oxygen is released as a by-product. The cyanobacteria can tolerate the oxygen they produce and can make use of it both metabolically (in aerobic respiration) and in synthetic pathways that seem to be oxygen dependent (as in the synthesis of chlorophyll *a*). Nevertheless, the biochemistry of the cyanobacteria differs from that of green eukaryotic plants and suggests that the group originated during a time of fluctuating oxygen concentration. For example, although many cyanobacteria can make unsaturated fatty acids by oxidative desaturation some of them can also employ the anaerobic mechanism of adding a double bond during the elongation of the chain. In a similar manner oxygen-dependent synthesis of certain sterols can be carried out be some cyanobacteria, but the amounts of the sterols made in this way are minuscule compared with the amounts typical of eukaryotes. In other cyanobacteria those sterols are not found at all, the biosynthetic pathway being terminated after the last anaerobic step: the formation of squalene. Hence in their biochemistry the cyanobacteria seem to occupy a middle ground between the anaerobes and the eukaryotes.

In metabolism too the cyanobacteria occupy an intermediate position. They flourish today in fully oxygenated environments, but physiological experiments indicate that for many species optimum growth is obtained at an oxygen concentration of about 10 per cent. Which is only half that of the present atmosphere. Both photosynthesis and respiration are increasingly inhibited when the oxygen concentration exceeds that optimum level. It has recently been discovered that some cyanobacteria can switch the cellular machinery of aerobic metabolism on and off according to the availability of oxygen. Under anoxic conditions these species not only halt respiration but also adopt an anaerobic mode of photosynthesis, employing hydrogen sulfide (H_2S) instead of water and releasing sulphur instead of oxygen. This capability for anaerobic metabolism is probably a relic of an earlier stage in the evolutionary development of the group.

Another activity of some cyanobacteria that seems to reflect an earlier adaptation to anoxic conditions is nitrogen fixation. Nitrogen is an essential element of life, but it is biologically useful only in 'fixed" form, for example combined with hydrogen in ammonia (NH_3). Only prokaryotes are capable of fixing nitrogen (although they often do so in symbiotic relationships with higher plants). The crucial complex of enzymes for fixation, the nitrogenases, is highly sensitive to oxygen. In cell-free extracts nitrogenases are partially inhibited by as little as ·1 per cent of free oxygen, and they are irreversibly inactivated in minutes by exposure to oxygen concentrations of only about 5 per cent.

Such a complex of enzymes could have originated only under anoxic conditions, and it can operate today only if it is protected from exposure to the atmosphere. Many nitrogen-fixing bacteria provide at protection simply by adopting an anaerobic habitat, but among the cyanobacteria a different strategy has developed; the nitrogenase enzymes are protected in specialized cells, called heterocysts, whose internal milieu is anoxic. The heterocysts lack certain pigments essential for photosynthesis, and so they generate no oxygen of their own. They have thick cell walls and are surrounded by a mucilaginous envelope that retards the diffusion of oxygen into the cell. Finally, they are equipped with respiratory enzymes that quickly consume any uncombined oxygen that may leak in.

Because of the thick cell walls heterocysts should be comparatively easy to recognize in fossil material. Indeed, possible heterocysts have been reported from several Precambrian rock units, the oldest being about 2.2 billion years in age. If these cells are indeed heterocysts, they may be taken as a sign that free oxygen was present by then, at least in small concentrations.

Nitrogen fixation has a high cost in energy, and the capability for it would therefore seem to confer a selective advantage only when fixed nitrogen is a scarce resource. Today the main sources of fixed nitrogen are biological and industrial, but biologically usable nitrate (NO_3-) is formed by the reaction of atmospheric nitrogen and oxygen. In the anoxic atmosphere of the early Pre-identification as eukaryotic for example, the well-studied Canadian fossils of the Gunflint and Belcher Island iron formations, which are about two billion years old, have been interpreted as exclusively prokaryotic.

The testimony of these as yet rare and unusual specimens can be checked through statistical studies of the sized of known Precambrian microfossils. The size ranges of prokaryotes and eukaryotes overlap, so that a particular fossil cannot always be classified unambiguously on the basis of size alone; by cataloguing the measured sized in a large sample of fossils, however, it may be possible to determine whether or not eukaryotic cells are present. Among modern species of spheroidal cyanobacteria about 60 per cent are very small, less than five micrometers in diameter: of the remaining species only a few are larger than 20 micrometers and none is larger than 60 micrometers. Unicellular eukaryotes, such as green or red algae, can be much larger. Typically they fall in the range between 5 and 60 micrometers, but several per cent of living species are larger than 60 micrometers and a few are larger than 1,000 micrometers (one millimeter).

Systematic size measurements have been made on some 8,000 fossil cells from 18 widely dispersed Precambrian deposits. On the basis of those data certain tentative conclusions can be drawn. Cells larger than 100 micrometers, and hence of distinctly eukaryotic dimensions, are unknown in rocks older than about 1,450 million years. Virtually all the unicellular fossils from rocks of that age, whether they grew in shallow-water stromatolites or were deposited in offshore shales, are of prokaryotic size.

Cells larger than modern prokaryotes (greater than 60 micrometers in diameter) first become abundant in rocks about 1,400 million years old. Algae of this type were apparently free-floating rather than mat-forming species, and they are therefore particularly common in shales, sediments deposited in deeper water. Such eukaryote-size fossils have been known for several years from shales of this age in China and in the U.S.S.R. Recently cells more than 100 micrometers in diameter have also been discovered in the Newland limestone of Montana, and cells more than 600 micrometers in size (10 times the size of the largest spheroidal prokaryote) have been found in the McMinn formation of Australia; the age of both of these fossil-bearing deposits is about 1,400 million years.

In somewhat younger Precambrian sediments there are still larger cells, fossils greater than one millimeter in diameter (with some as large as eight millimetres). They were first described in 1899 by Walcott, who discovered them in rocks from the Grand Canyon. They have since been found in nearly a dozen other rock units throughout the world. The oldest seem to be those from Utah and from Siberia, each about 950 million years old, and those from northern India, which could be even older (from 910 to 1,150 million years old).

Studies of both the morphology and the size of unicellular fossils therefore suggest that there is a break in the fossil record between 1,400 and 1,500 million years ago. Below this horizon cells with eukaryotelike traits are rare or absent; above it they become increasingly common. Moreover, the data suggest that the diversification of the eukaryotes began shortly after the cell type first appeared, apparently within the next few hundred million years. By a billion years ago there had been substantial increases in cell size, in morphological complexity and in the diversity of species. All these indicators also suggest, of course, that oxygen-dependent metabolism, which is highly developed even in the most primitive eukaryotes, had already become established by about 1.5 billion years ago.

The prokaryotes that must have held exclusive sway over the earth before the development of eukaryotic cells were less diverse in form, but they were probably more varied in metabolism and biochemistry than their eukaryotic descendants. Like modern prokaryotes, the ancient species presumably varied over a broad range in their tolerance of oxygen, all the way from complete intolerance to absolute need. In this regard one group of prokaryotes, the cyanobacteria, are of particular interest in that they were largely responsible for the development of an oxygen-rich atmosphere.

Like higher plants cyanobacteria carry out aerobic photosynthesis, a process cambrian the latter mechanism would obviously have been impossible. The lack of atmospheric oxygen would also have indirectly reduced the concentration of ammonia to very low levels. Ammonia is dissociated into nitrogen and hydrogen by ultraviolet radiation, most of which is filtered out today by a layer of ozone (O_3) high in the atmosphere; without free oxygen there would have been little ozone, and without this protective shield atmospheric ammonia would have been quickly destroyed.

It is likely that the capability for nitrogen fixation developed early in the Precambrian among primitive prokaryotic organisms and in an environment where fixed nitrogen was in short supply. The vulnerability of the nitrogenase enzymes to oxidation was of no consequence then, since the atmosphere had little oxygen. Later, as the photosynthetic activities of the cyanobacteria led to an increase in atmospheric oxygen, some nitrogen fixers adopted an anaerobic habitat and others developed heterocysts. By the time eukaryotes appeared, apparently more than half a billion years later, oxygen was abundant and fixed nitrogen (both NH_3 and NO_3^-) was probably less scarce, and so the eukaryotes never developed the enzymes needed for nitrogen fixation.

At present oxygen-releasing photosynthesis by green plants, cyanobacteria and some protists is responsible for the synthesis of most of the world's organic matter. It is not, however, the only mechanism of photosynthesis. The alternative systems are confined to a few groups of bacteria that on a global scale seem to be of minor importance today but that may have been far more significant in the geological past.

The several groups of photosynthetic bacteria differ from one another in their pigmentation, but they are alike in one important respect: unlike the photosynthesis of cyanobacteria and eukaryotes, all bacterial photosynthesis is a totally anaerobic process. Oxygen is not given off as a by-product of the reaction, and the photosynthesis cannot proceed in the presence of oxygen. Whereas oxygen appears to be required in green plants for the synthesis of chlorophyll *a*, oxygen inhibits the synthesis of bacteriochlorophylls.

The anaerobic nature of bacterial photosynthesis seems to present a paradox: photosynthetic organisms thrive where light is abundant, but such environments are also generally ones having high concentrations of oxygen, which poisons bacterial photosynthesis. These contradictory needs can be explained if it is assumed that anaerobic photosynthesis evolved among primitive bacteria early in the Precambrian, when the atmosphere was essentially anoxic. The photosynthesizers could thus have lived in matlike communities in shallow water and in full sunlight.

Somewhat later such bacteria gave rise to the first organisms capable of aerobic photosynthesis, the precursors of modern cyanobacteria. For the anaerobic photosynthetic bacteria the molecular oxygen released by this mutant strain was a toxin, and as a result the aerobic photosynthesizers were able to supplant the anaerobic ones in the upper portions of the mat communities. The anaerobic species became adapted to the lower parts of the mat, where there is less light but also a lower concentration of oxygen. Many photosynthetic bacteria occupy such habitats today.

Photosynthetic bacteria were surely not the first living organisms, but the history of life in the period that preceded their appearance is still obscure. What little information can be inferred about that early period, however, is consistent with the idea that the environment was then largely anoxic. One tentative line of evidence rests on the assumption that among organisms living today those that are simplest in structure and in biochemistry are probably the most closely related to the earliest forms of life. Those simplest organisms are bacteria of the clostridial and methanogenic types, and they are all obligate anaerobes.

There is even a basis for arguing that anoxic conditions must have prevailed during the time when life first emerged on the earth. The argument is based on the many laboratory experiments that have demonstrated the synthesis of organic compounds under conditions simulating those of the primitive planet. These synthesis are inhibited by even small concentrations of molecular oxygen. Hence it appears that life probably would not have developed at all if the early atmosphere had been oxygen-rich. It is also significant that the starting materials for such experiments often include hydrogen sulfide and carbon monoxide (CO), and that an intermediate in many of the reactions is hydrogen cyanide (HCN). All three compounds are poisonous gases, and it seems paradoxical that they should be forerunners of the earliest biochemistry. They are poisonous, however, only for aerobic forms of life: indeed, for many anaerobes hydrogen sulfide not only is harmless but also is an important metabolite.

It was argued above that oxygen must have been freely available by the time the first eukaryotic cells appeared, probably 1,400 to 1,500 million years ago. Hence the proliferation of cyanobacteria that released the oxygen must have taken place earlier in the Precambrian. How much earlier remains in question. The best available evidence bearing on the issue comes from the study of sedimentary minerals, some of which may have been influenced by the concentration of free-oxygen at the time they were deposited. In recent years a number of workers have investigated this possibility, most notably Preston E. Cloud, Jr., of the University of California at Santa Barbara and the U.S. Geological Survey.

One mineral of significance in this argument is uraninite (UO_2), which is found in several deposits that were laid down in Precambrian streambeds. In the presence of oxygen, grains of uraninite are readily oxidized (to U_3O_8) and are thereby dissolved. David E. Grandstaff of Temple University has shown that streambed deposits of the mineral probably could not have accumulated if the concentration of atmospheric oxygen was greater than about 1 per cent. Uraninite-bearing deposits of this type are found in sediments older than about two billion years but not in younger strata, suggesting that the transition in oxygen concentration may have come at about that time.

Another kind of mineral deposit, the iron-rich formations called red beds, exhibits the opposite temporal pattern: red beds are known in sedimentary sequences younger than about two billion years but not in older ones. The red beds are composed of particles coated with iron oxides (mostly the mineral hematite, Fe_2O_3), and many are thought to have formed by exposure to oxygen in the atmosphere rather than under water. It has been proposed that the oxygen may have been biologically generated. This hypothesis is consistent with several lines of evidence, but objections to it have also been raised. For example, most red beds are continental deposits rather than marine ones and are therefore susceptible to erosion; it is thus conceivable that red beds were formed earlier than two billion years ago as well as later but that the earlier beds have been destroyed. It is also possible that the oxygen in the red beds had a nonbiological origin; it may have come from the splitting of water by ultraviolet radiation. This has apparently happened on Mars to create a vast red bed across the surface of that planet, where there are only traces of free oxygen and there is no evidence of life.

Perhaps the most intriguing mineral evidence for the date of the oxygen transition comes from another kind of iron-rich deposit: the banded iron formation. These deposits include some tens of billions of tons of iron in the form of oxides embedded in a silica-rich matrix; they are the world's chief economic reserves of iron. A major fraction of them was deposited within a comparatively brief period of a few hundred million years beginning somewhat earlier than two billion years ago.

A transition in oxygen concentration could explain this major episode of iron sedimentation through the following hypothetical sequence of events. In a primitive, anoxic ocean, iron existed in the ferrous state (that is, with a valence of +2) and in

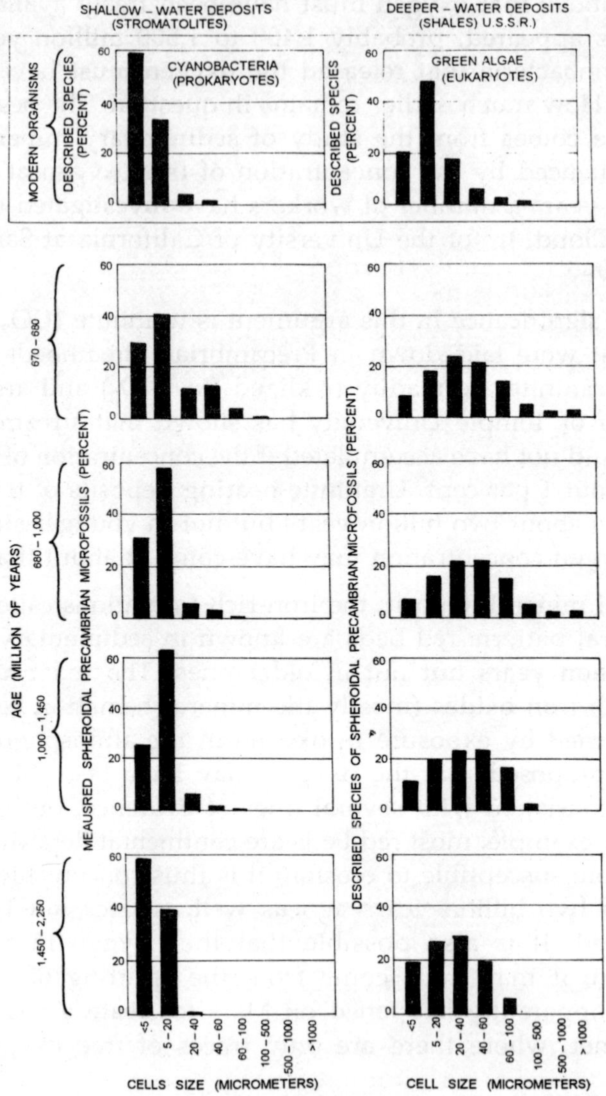

Fig. 8.7: Size of Fossil Cells provides evidence on the origin of the eukaryotes. Spheroidal microfossils of various ages were measured and assigned to eight size categories, a similar procedure was followed with spheroidal members of two groups of modern microorganisms, the prokaryotic cyanobacteria and the eukaryotic green algae. The range of sizes for the modern species overlaps, but the largest cells are observed only among the eukaryotes. The oldest fossils examined have a distribution of sizes similar to that of prokaryotes, but Precambrian rock units younger than 1,450 million years include larger cells that are probably eukaryotic, and the proportion of larger cells increases in later periods. The larger cells tend to be more abundant in shales than in stromatolitic sediments. Because shales are deposited offshore that fact would be explained if early eukaryotes were predominantly free-floating rather than mat-forming.

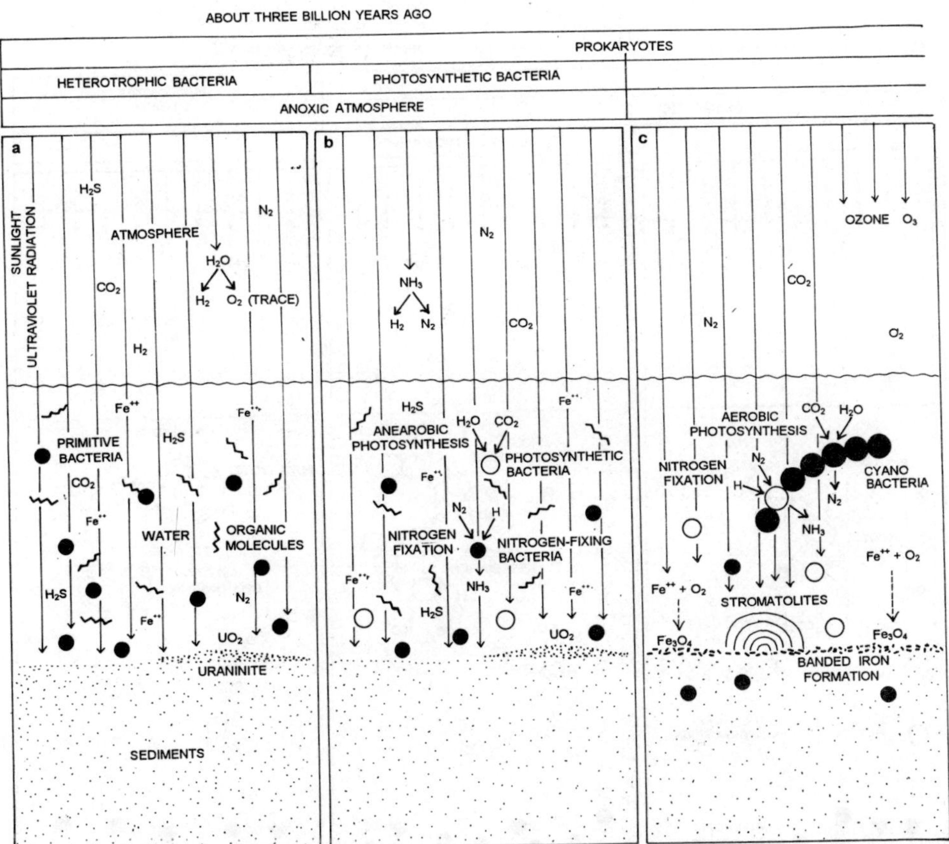

Fig. 8.8: Organism and Environment evolved in counterpoint during the Precambrian. The first living cells (a) were presumably small, spheroidal anaerobes. Only traces of oxygen were present. They survived by fermenting organic molecules formed nonbiologically in the anoxic environment. The role of such ready-made nutrients was diminished, however, when the first photosynthetic organisms evolved (b). This earliest mode of photosynthesis was entirely anaerobic. Another early development was nitrogen fixation, required in part because ultraviolet radiation that could then freely penetrate the atmosphere would have quickly destroyed any ammonia (NH_3) present. A little more than two billion years ago (c) aerobic photosynthesis began in the precursors of modern cyanobacteria. Oxygen was generated by these stromatolite-building microorganisms, but for some 100 million years little of it accumulated in the atmosphere; instead it reacted with iron dissolved in the oceans, which was then precipitated to create massive banded iron formations. Only when the oceans had been swept free of iron and similar materials (d) did the concentration of free oxygen begin to rise towards modern levels. This biologically

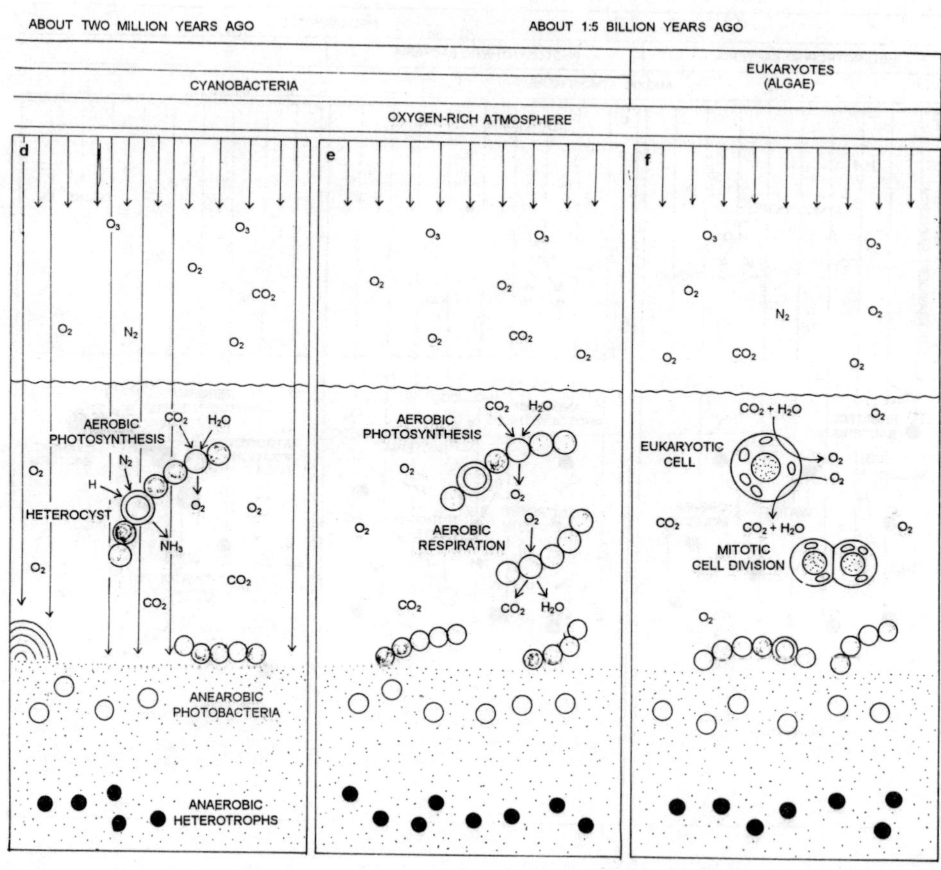

induced change in the environment had several effects on biological development. Anaerobic organisms were forced to retreat to anoxic habitats, leaving the best spaces for photosynthesis to the cyanobacteria. In a similar manner nitrogen-fixing organisms had to adopt an anaerobic way of life or develop protective heterocyst cells. Atmospheric oxygen also created a layer of ozone (O_3) that filtered out most ultraviolet radiation. Once the oxygen-rich atmosphere was fully established (e) cells evolved that not only could tolerate oxygen but also could employ it in respiration. The result was a great improvement in metabolic efficiency. Finally, about 1,450 million years ago, the first eukaryotic cells emerged (f). From the start they were adapted to a fully aerobic environment. The new modes of reproduction possible in eukaryotes in particular the advanced sexual reproduction that evolved later, led to rapid diversification of the group.

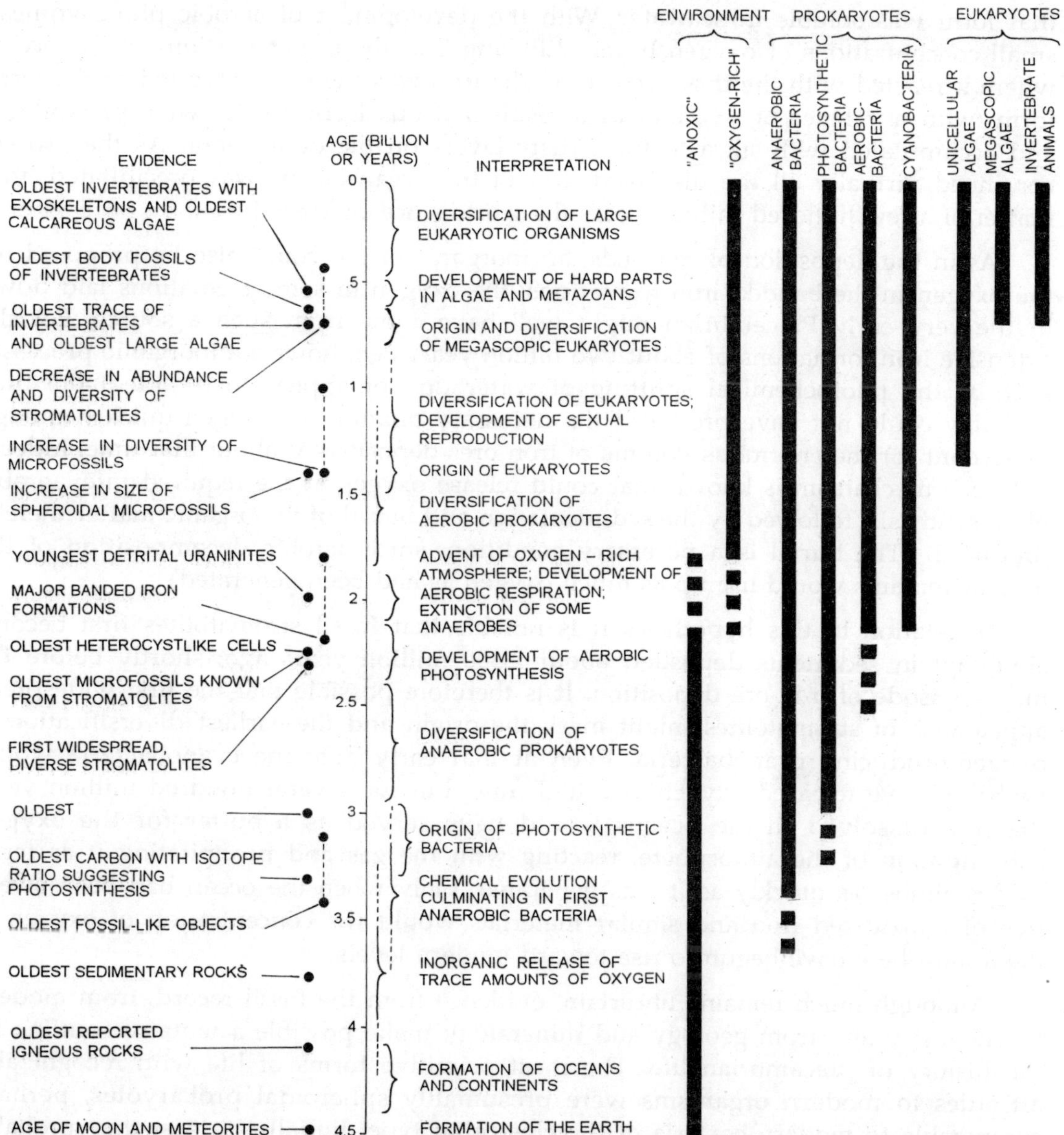

Fig. 8.9: Major Events in Precambrian evolution are presented in chronological sequence based on evidence from the fossil record, from inorganic geology and from comparative studies of the metabolism and biochemistry of modern organisms. Although the conclusions are tentative, it appears that life began more than three billion years ago (when the earth was little more than a billion years old), that the transition to an oxygen-rich atmosphere took place roughly two billion years ago and that eukaryotes appeared by 1.5 billion years ago.

that form was soluble in seawater. With the development of aerobic photosynthesis small concentrations of oxygen began diffusing into the upper portions of the ocean, where it reacted with the dissolved iron. The iron was thereby converted to the ferric form (with a valence of +3), and as a result hydrous ferric oxides were precipitated and accumulated with silica to form rusty layers on the ocean floor. As the process continued virtually all the dissolved iron in the ocean basins was precipitated: in a matter of a few hundred million years the world's oceans rusted.

As in the deposition of red beds, an inorganic origin could also be proposed for the oxygen in the banded iron formations; the oxygen in some formations laid down in the very early Precambrian might well have come from such a source. For the extensive iron formations of about two billion years ago, however, inorganic processes such as the photochemical splitting of water do not appear to be adequate; they probably could not have produced the necessary quantity of oxygen quickly enough to account for the enormous volume of iron ores deposited at about that time. Indeed, only one mechanism is known that could release oxygen at the required rate: aerobic photosynthesis, followed by the sedimentation and burial of the organic matter thereby produced. (The burial is a necessary condition, since aerobic decomposition of the organic remains would use up as much oxygen as had been generated).

In relation to this hypothesis it is notable that fossil stromatolites first become abundant in sediments deposited about 2,300 million years ago, shortly before the major episode of iron-ore deposition. It is therefore possible that the first widespread appearance of stromatolites might mark the origin and the earliest diversification of oxygen-producing cyanobacteria. Even at that early date the cyanobacteria would probably have released oxygen at a high rate, but for several hundred million years the iron dissolved in the oceans would have served as a buffer for the oxygen concentration of the atmosphere, reacting with the gas and precipitating it as ferric oxides almost as quickly as it was generated. Only when the ocean had been swept free of unoxidized iron and similar materials would the concentration of oxygen in the atmosphere have begun to rise toward modern levels.

Although much remains uncertain, evidence from the fossil record, from modern biochemistry and from geology and mineralogy make possible a tentative outline for the history of precambrian life. The most primitive forms of life with recognizable affinities to modern organisms were presumably spheroidal prokaryotes, perhaps comparable to modern bacteria of the clostridial type. Initially at least they probably derived their energy from the fermentation of materials that were organic in nature but were of nonbiological origin. These materials were synthesized in the anoxic early atmosphere and were of the type that during the age of chemical evolution had led to the development of the first cells.

The first photosynthetic organisms apparently arose earlier than about three billion years ago. They were anaerobic prokaryotes, the precursors of modern photosynthetic bacteria. Most of them probably lived in matlike communities in shallow water, and they may have been responsible for building the earliest fossil stromatolites known, which are estimated to be about three billion years old.

The rise of aerobic photosynthesis in the mid-Precambrian introduced a change in the global environment that was to influence all subsequent evolution. The resulting increase in oxygen concentration probably led to the extinction of many anaerobic organisms, and others were forced to adopt marginal habitats, such as the lower reaches of bacterial mat communities. Nitrogen-fixing organisms also retreated to anaerobic habitats or developed heterocyst cells. With little competition for those regions having optimum light the cyanobacteria were able to spread rapidly and came to dominate virtually all accessible habitats. With the development of the citric acid cycle and its more efficient extraction of energy from foodstuff, the dominance of the biological community by aerobic organisms was confirmed. When the major episode of deposition of banded iron formations ended some 1,800 million years ago, the trend toward increasing oxygen concentration became irreversible.

By the time eukaryotic cells arose 1,500 to 1,400 million years ago a stable, oxygen-rich atmosphere had long prevailed. Adaptive strategies needed by earlier organisms to cope with fluctuations in the oxygen level were unnecessary for eukaryotes, which were from the start fully aerobic. The diversity of eukaryote cell types present by about a billion years ago suggests that some form of sexual reproduction may have evolved by then. Within the next 400 million years the rapid diversification of eukaryotic organisms had led to the emergence of multicellular forms of life, some of them recognizable antecedents of modern plants and animals,

In style and in tempo evolution in the Precambrian was distinctly different from that in the later, Phanerozoic era. The Precambrian was an age in which the dominant organisms were microscopic and prokaryotic, and until near the end of the era the rate of evolutionary change was limited by the absence of advanced sexual reproduction. It was an age in which the major benchmarks in the history of life were the result of biochemical and metabolic innovations rather than of morphological changes. Above all, in the Precambrian the influence of life on the environment was at least as important as the influence of the environment on life. Indeed, the metabolism of all the plants and animals that subsequently evolved was made possible by the photosynthetic activities of primitive cyanobacteria some two billion years ago.

Chapter—9

Distribution of Animals

Zoogeography is the science which deals with the distribution of all the animals over world. First serious attempt to map out the geographical regions was made by *Sclater* (1857) who based his conclusions upon the distribution of birds. These regions are: 1. Palacartic region, 2. Ethiopian region, 3. Indian region, 4. Australian region, 5. Neotropical region, 6. Neartic region. These are, however, very apparent objections to utilizing vagrant and barrier defying creatures like birds for this purpose. Accordingly, *Mkurray* in 1866 and more in detail *Alfred Russel Wallace* in 1876, divided the surface of earth into zoological regions based chiefly on the distribution of mammals.

Many other regional classifications have been made, based on either discontinuous distribution of one particular class or upon the distribution of temperature variations and similar climatic factors. None of these is satisfactory. Therefore, *Wallace* combined Sclater's system. *Blanford* (1980), *Lydekk* (1896), *Heilprin* (1887), *Gadow* (1913), *Schmidt* (1954) etc. also made their contribution for the study of zoogrophy.

Karl P. Schmidt (1954) divided the earth into three zoogeographical realms:

1. Arctogaean,
2. Neogaean, and
3. Notogaean.

Arctogaean includes two regions the (*a*) *Holarctic* with sub-region: *Arctic, Nearctic, Caribbean* and *Palearctic.*

(*b*) *Paleotropical* with sub-regions: *Oriental, Ethiopian* and *Malagasy.*

Neogaean includes a single sub-region the *Neotropical. Notogaean* consists of 2 regions: the *Australian* having the Australian and Papuan sub-regions and the *Oceanian,* which is formed of the New Zealandian, oceanic and Antarctic sub-regions.

Wallace's classification is generally considered to be accepted system. The Indian region of *Sclater* is termed as Oriental region in the classification of *Wallace.*

Thus, six zoogeographical regions have been distinguished as follow:

1. *Palaearctic Region*

   It includes Europe, temperate Asia, North Africa and Arabia.

2. *Neartic Region*

   It includes whole of North America and Greenland.

3. *Ethiopian Region*

   It includes whole of Africa, Arabia, South of the tropic of Cancer, and Madagascar.

4. *Neotropical Region*

   It includes whole of South and Central America and West Indies.

5. *Australian Region*

   It includes Australia, New Zealand, New Guinea and neighbouring islands.

6. *Oriental Region*

   It includes India, Sri Lanka, Indochina and Malaya.

The exact boundaries of these regions are very difficult to recognize. The fauna of these regions are described only with reference to the vertebrates, although they form only 3-4% of animal kingdom.

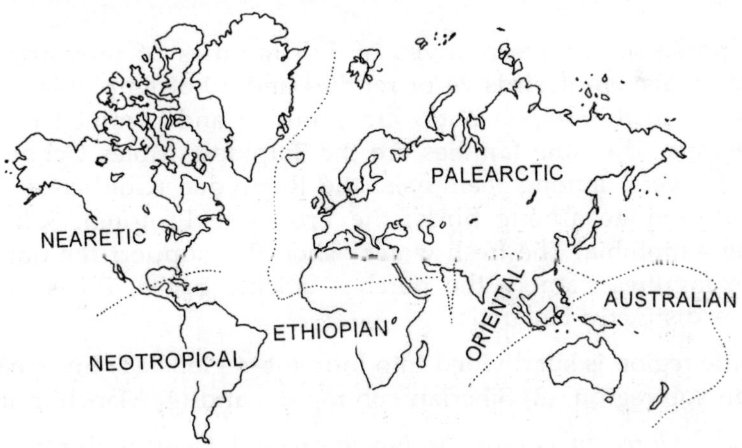

Fig. 9.1: The biogeographical regions of world.

PALAEARCTIC REGION

Extent

This, the largest of the six regions, is estimated to cover an approximate area of 14,00,000 sq. miles. It includes the whole of Europe, Iceland, the Azores, Madeira, Canary Island, Cape Verde Islands, all that portion of Africa and Arabia which lies to the north of the tropic of Cancer, Asia Minor, Persia, Afghanistan and Baluchistan and the whole of Asia north of a line which runs up to valley of the Indus, along the great Himalayan range, thence eastwards to the Nanling Mountains, south of the Yang-tse-kiang and out to sea just south of the Japanese Islands.

Climate

The climate is temperate. This region includes wet forest lands as well as dry open steppe lands. It also includes coniferous forests and a fringe of Tundra.

In the western portion there is a great in land seas which have on equalising effect on the climate. The Gulf stream materially helps to raise the temperature of Western Europe. The eastern region is much cooler. The extreme of cold are felt in North-Eastern Siberia. However, the highest summer temperature of the area is experienced in North-Western India, Arabia and Afghanistan.

The Northern part of this region is mostly low and flat, Scandinavian and the Ural mountains are the only elevated lands. In south a series of mountain ranges run across the region from west to east, including in Europe the Alps, Balkans, Caucasus etc.; in Asia—the Tian Shan, Altai, Himalayan ranges etc.

Zoological Characteristic

The fauna of this region as a whole is very similar to the Nearctic region, that is why *Heilprin* proposed a great region, *Holarctic Realm* for both Palearctic and Nearctic regions.

The region possesses representatives of 135 families of terrestrial vertebrates, namely 33 of mammals, 68 of birds 24 of reptiles and 10 of amphibia. None of these, however, is peculiar to this region but 9 are common and confined to the Palearctic and Nearctic Regions. The nine families are the Talpidae (Moles Ochotonidae (Picas) and Castoridae (Beavers) among mammals, the Regulidae (Gold-crests), Colymbidae (Divers) and Tetraomidae among birds; the Proteidae Protous), Salamanders and *Amphiuma* among Amphibia. The fresh water fishes also support the union of the two regions, for five families namely the perches, stickle backs, Pikes, Sturgeons and Polyodontidae are distinctly Holarctic.

The Palaearctic region is subdivided into four sub-regions: (1) European sub-region, (2) Mediterranean sub-region, (3) Siberian sub-region, and (4) Manchurian sub-region.

1. *European sub-region* comprises Northern and Central Europe. In this sub-region only 85 families of terrestrial vertebrates are represented and of these the reptiles and amphibia number only 6 each. Only one form of mammals are peculiar namely *Myogale* but such animals as the wolf, Hedgehog, shrew, mole and Darmouse are very characteristic as are also the tails, Pipits, Tits, Thrushes and many other birds.

2. *The Mediterranean sub-region* includes the rest of Europe, all the African and Arabian portion, Asia Mion, Persia, Afghanistan and Baluchistan. This is

the richest region possessing representatives of 120 families of terrestrial vertebrates. The Fallow-Deer, Elephant Shrew, Civet, Ichneumon, Haena, Hyrax and Porcupine are all characteristic mammals.

3. *Siberian sub-region* embraces all Northern Asia, southwards to the Himalayas. It possess representatives of 94 families of the terrestrial vertebrates. Four genera of mammals *i.e.,* the Yak, Musk, deer and a mole are almost confined to this sub-region but also ranged into the Oriental region. One of the most important members of the Siberian fauna is the freshwater seal, *Phoca ibirica* which inhabits the great fresh water lake—Baikal.

4. *The Manchurian sub-region* includes China, Mongolia, Manchurian and Corea together with the whole of Japan. It has a rich and varied fauna with representatives of 102 families of terrestrial vertebrates. At least a dozen genera of mammals including Tibetan langur, Great Panda etc. are peculiar.

NEARCTIC REGION

Extent

It includes the whole of North America and extends south as far as the middle of Mexico. It includes Greenland in the east and Aleutian islands in the west.

Climate

It has a great range of temperature. The chief physical features of the region are: (1) Large lakes and inland sea in the north eastern portion and the important ranges of high mountains in the west. (2) In the east are smaller ranges, constituting the so called Appalachian high land. (3) In the centre of this great continent is a vast extent of plain which in the north is frozen and hence barren, between latitudes 50° and 60° covered with forest, and in the south dry treeless desert.

Zoological Characteristics

The number of families of terrestrial vertebrates represented in the Nearctic Region is 120 *viz.* 26 of Mammals, 59 of Birds, 21 of Reptiles and 140 of Amphibians. The Nearctic Region possesses five peculiar families; *viz.* The Haplodontidae and Antilocapridae among Mammals; the Chamaeidae among Birds; the Aniellidae among Reptiles and Sirenidae among Amphibians.

The Nearctic differs from the palaearctic region, in the possession of several characteristic Mammals, such as Opossums (Didelphyidae) and Racoon etc., many birds such as the Blue—jays, and Turkey-buzzards etc., Reptiles such as Rattle snakes and Iguanas; Amphibia including *Axolotl, Nectures, Siren* and other large Urodeles.

There are, several peculiar genera of importance of which Scalops (web-footed Moles), Toxidae (American Badger) *Haplocerus* (Rocky mountain Goat) and Musk may be taken as examples.

Wallace has divided Nearctic Region into four sub-regions:

(1) California sub-region.

(2) Rocky Mountain sub-region.

(3) Alleghany sub-region.

(4) Canadian sub-region.

1. *California Sub-region* embraces a narrow strip of country between Sierra Nevada and the Pacific. In the North it includes island of Vancouver and the southern part of British Columbia and in the South it extends upto the head of the Gulf of California. It has 80 families of terrestrial vertebrates and of which 21 are mammals, 49 of birds, 8 of reptiles and 8 of amphibian. Three families are peculiar to this region, namely Haplodontidae, Chamaeidoe and Aniellidae. Vampires and free tailed Bats are characteristic of this region.

2. *Rocky Mountain Sub-region.* It lies immediately to the east of California and includes the whole of the dry and elevated area covered by the mountains. It is the richest portion of Nearctic region. It has 107 families of terrestrial vertebrates out of which 25 are mammals, 55 of birds, 18 of reptiles and 9 of amphibian. Although there are no peculiar families but are several characteristic genera like Prong-buck (*Antilocapra*), Rocky mountain goat (*Haplocerus*), Prairie dog (*Cynomys*), American Bison and poisonous lizard (*Heloderma*).

3. *Alleghany sub-region* comprises the United States east of the Rocky mountain sub-region and South of the Great Lakes and includes Novo Scotia. Some 99 families of terrestrial vertebrates including, mammals 8, birds 53 reptiles 16 and amphibia, 12 are represented in this sub-region. The peculiar animals are: opossum, star nosed moles, vampire bats etc. Turkeys and Sirenids (muded) are the other two families.

4. *Canadian-sub region.* All the remaining portion of North America and Greenland constitute the great Canadian sub-region. In this sub-region are to be found the representative of only 75 families including 20 of mammals, 44 of birds, 3 of birds, 3 of reptiles and 8 of amphibia. Thus it is the poorest region. But there are characteristic families including Deer, Reindeer, Elk, Bison, Sheep, Gluttons, Lemmings, the polar deer etc.

ETHIOPIAN REGION

Extent

The Ethiopian Region consists of the whole of Africa and Arabia, south of the tropic of Cancer, together with Madagascar and the small adjacent is lands. *Darlington* (1957) has not included Madagascar but has considered it as separate region on account of its distinctive fauna.

Climate

This region is mainly tropical. It is a large block of rain forests and isolated mountains. The eastern part has wide grassy plains. Its southern part is warm temperate with mixed vegetations.

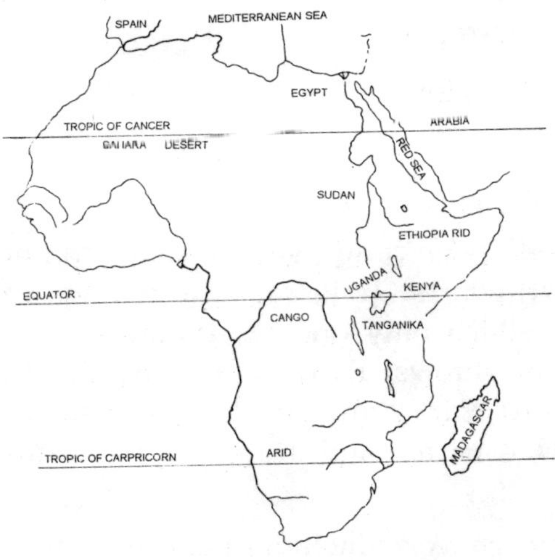

Fig. 9.2: The Ethiopian region with adjacent regions (Madagascar, Spain and Arabia).

Zoological Characteristics

The fauna of Ethiopean Region is rich, varied and well-marked. There are about 161 families of terrestrial vertebrates. Out of these 30 are peculiar.

Among mammals 12 families are peculiar which are: Chiromyidae (Aye Aye), Chrysochloridae (Golden moles), Centetidae (Tenress), Potamogalidae (Potamogale), Protelidae (Earth wolf), Battryergidae (African mole rats), Lophimyidae (created rats), Pedetidae (African Jumping Hares), Anomaluridae (African flying Squerrels), Giraffidae (Giraffes), Hippopotamidae (Hippopotamus) and Orycteropodidae (Aard-Varks).

Among birds 13 families are peculiar; Promeropidae (Promerops), Acerocharidae (Helmet Birds), Vangidae (Vangidal), Philepittidae, Musophagidae (Plankton eators), Coliidae (Colies), Serprentariidae (Secretary birds) Scopidae (Hammer-head bird) Mesoenatidae, Numididae (Guinea fowl) and Balaenicipitidae (whale head).

Among reptiles following families are peculiar: Rhachiodontidae (Egg-eating snakes), Gerrhosauridae, Zonuridae (Girdled lizard) and Uroplatidae.

Among amphibian family Dectylethridae (Clawed foads) is peculiar.

In addition to these families, a large number of important genera are confined to this region, some of them are:

Macroscelidae (Elephant—shrews), Hyaenidae (Hyaena) Elephantidae (Elephants) Rhinocerotidae (Rhinoceroses), Procavidae (Hyraces) Equidae (Horse etc.) Manidae (Pangolins); and among Birds the Ploceidae, Nectariniidae (Sunbirds) Zosteropidae (White Eyes); Indicatoridae (Honey Guides), Capitonidae (Barbets), Bucerotidae (Hornbills) and Struthionidae (Ostriches).

The Ethiopean Region is divided into four sub-regions:

1. South African sub-region.
2. East African sub-region.
3. West African sub-region.
4. Malagasy sub-region.

1. *South African sub-region* comprises the southern portion of the continent from the Cape northwards. It is represented by 133 families of terrestrial vertebrates of which only one *i.e.,* Prome ropidae is peculiar. The Golden Moles, Elephant shrews, Earth Wolf, African Mole Rats, African Jumping Hares and Aard Varks are all characteristic mammals. Among birds—Secretary birds, ostriches and in reptiles Egg-eating snakes are characteristic to this part.

2. *The East African sub-region* includes the rest of the tropical Africa and tropical Arabia. It has 145 families of terrestrial vertebrates of which only two namely Crested-Rats among mammals and Whale-head among birds are peculiar. Certain other familiar mammals, though not absolutely confined but are highly characteristic, are Giraffes, Zebras and the Rhinoceroses.

3. *The West African sub-region* includes most of the African forest region from the river Gambia eastwards to beyond Lake Chan and Southwards to embrace the water shed of the Congo. This area has no peculiar families although 134 occur in this subregion. The most characteristic mammals are the Gorilla, Chimpanzee, the Monkeys, Potamogal, the African flying squirrel and Okapi.

4. *Malagasy sub-region.* It includes Madagascar, Mauritius, Seychelles and neighbouring islands. The fauna of this sub-region is one of most interesting in the whole world. 68 families of terrestrial vertebrates found here, no fewer than 8 are absolutely confined to it. These includes two of mammals Chiromyidae (Aya Aye) and Centetidae (Trenecs); 5 of birds, *i.e.,* Aerocharidae (Helmet Bird), Vangidae, Philepittidae, Mesoenatidae and Leptosomatidae, and one of reptiles, *i.e.,* Uroplatidae.

The most characteristic feature is that all the 35 known species of sub-family Lemurinae of Family—Lemuridae are confined to this sub-region. Of birds 55 families are represented 15 families of reptiles occur here, presence of 2 genera of 9 guanidae is of special interest. Since this is an essentially New world group. Chamaelions are very characteristic of Madagascar because they are more abundant than in any other part of the world.

NEOTROPICAL REGION

Extent

It embraces South America, most of Mexico and West Indies. It is joined to the Nearctic Region by Central American isthmus and separated from other regions by sea.

Climate

Neotropical region is mainly tropical but southern South America continues into south temperate zone. From the west to the east runs the river Amazon with its hundreds of square miles of evergreen forests. In the west the long range of the Andes has high mountain forests, plateau land and gentle slopes.

Zoological Characteristics

155 families of terrestrial vertebrates found here and not less than 39 families are absolutely confined to it.

Among mammals 10 families are peculiar: Cebidae (American Monkeys), Callitrichidae (Marmosets), Solenodontidae (Solenodonts), Dasyproctidae (Agouties), Chinchillidae (Chinchillas), Caviidae (Cavies), Dinomydidae (Dinomys), Bradypodidae (Sloths), Calnolestidae (Selvas) and Myrmecophagidae (Ant-eaters).

Among birds 23 families are peculiar *viz.* Coerebidae (Honey Creepers), Phytotomidae (Plant Cutters), Pipridae (Manakins), Oxyrhamphidae, Dendrocolaptidae, Conopophagidae, Formicarridae, Pteroptochidae, Galbuldidae, Bucconidae, Rhamphastidae Tinamidae, Steatornithidae, Momotidae, Todidae, Palamediedae, Psophiidae, Aramidae, Cariamidae, Eurypygidae, Thinocorythidae, Opisthoconidae and Rheidae.

Among reptiles two families are peculiar *viz.* Xenosauridae and Dermatemydidae.

Among amphibians, 4 families are peculiar; Dendrobratidae, Dendrophryniscidae. Hemiphractidae and Amphignathodontidae.

Neotropical Region is divided into four sub-regions:

1. Chillian sub-region.
2. Brazalian sub-region.
3. Mexican sub-region.
4. Antilean sub-region.

1. *Chillian sub-region*. It includes western coast of South America and embraces the summits of Andes of Peru and Bolivia. This region does not contain any peculiar family but characteristic forms are chinchillas, Llamas, Rhears etc.
2. *Brazilian sub-region* includes all the rest of South America terminating Northwards at the Isthmus of Panama. This is the richest of the Neotropical sub-regions, containing representatives of 133 families of which 27 are of mammals, 71 birds, 23 reptiles and 12 amphibia. This sub-region is the great home of nearly all the arboreal vertebrates of South American such as American monkeys, Vampire bats, American porcupines, sloths and opossums.

There are in addition many other animals as Spiny-mice, Cavies, American Tapirs etc. found in the forest. Among reptiles and amphibians, the tree snakes, coral snakes, iguanas, alligators, Solid-chested tree frogs, typical tree frogs etc. are common.

3. *Mexican sub-region* constitutes all the Neotropical country North of the Isthmus of Panama. It possesses representatives of 127 families of terrestrial vertebrates which include 24 of mammals, 67 of birds, 26 of reptiles and 10 of amphibia.
4. *The Artillian sub-region* includes all the West Indies except Tobago and Trinidad. Since the area is wholly made up of islands, most of which are small, it is hardly surprising to find that the number of families of terrestrial vertebrates is much less. Some 76 families, including 7 mammals, 47 birds, 16 reptiles and 6 amphibia are represented. There is a remarkable absence of mammals in this region as in other insular sub-region for there are no Primates, Carnivores, Ungulates or Edentata. The only rodents are the spiny mice. The birds of the following important families are quite absent *i.e.*, Plant cutters Manakins, American creepers, Ant thrushes etc.

The negative characteristics of this region are also very remarkable. Except in Central America and West Indies, there are no insectivores, civets, oxen, sheep and antelopes.

AUSTRALIAN REGION

Extent

The Australian Region includes the whole of Australia New Zealand, Newguinea, the Molluca and other neighbouring islands, and practically the whole of islands in the Pacific ocean. Its boundary on the west is a line drawn between the islands of Bali and Lombok (Wallace's line), thence to the east of Celebes or the Philippine Islands. The line then goes due eastwards along the tropic of Cancer to include the Sandwich Islands, thence curves round to the south of New Zealand and Auckland Islands, Tasmania and Australia.

Climate

The northern part of the Region, north Australia and New Guinea, lies with in the tropics with high summer temperature and much of the area is covered by rain forest. The interior of the Australian continent is also hot, but dry, while further south the climate becomes mainly temperate.

Zoological Characteristics

There are 134 families of terrestrial vertebrates. About 30 families are peculiar.

Among mammals 8 families are peculiar *viz.*, Macropodiae (Kangaroos), Phalangeridae (Phalangers), Phascolomydae (Wombats), Peramelidae (Bandicoots), Notoryctidae (Marsupial), Dasyuridae (Dasyures), Echidnidae (*Echidna*) and Ornithorhynchidae (*Ornithorhynchus*).

Among birds 17 families are peculiar namely: Paradiscidae (Birds of Paradise), ptilonorthynchidae (Bower Birds), Meliphagidae (Honey Eaters), Drepanididae (Drepanis), Artichornithidae (Scrub Birds), Xenicidae Menuridae (Nestor Parrots), Loriidae (Loreis), Cyclopsittacidae Stringopidae (Owl-Parrots), Phinochaetidae (Kagu), Gonudiae (Crowned Pigeons), Didunculidae(Tooth-billed Pigeon), Apterygidae (Kiwis), Dromaeidae (Emus), and Casuariidae (Cassowaries).

Among reptiles 3 families are peculiar: Pygopodidae (Scalefooted Lizards), Hatteriidae (Turtra), and Carettchelydidae (Fly-river turtle).

Among amphibians 2 families are peculiar. Ceratobatrachidae and Genyophrynidae.

The Australian Region is also characterized by the absence of certain important groups. Thus, only few mammals are there except marsupials and monotremes. The other orders being represented by some bats and small rodents. Apes and monkeys, insectivores, carnivores, ungulates and edentates are entirely absent. Among birds the finches, buntings and wood peckers are absent.

The Australian Region has been divided into four sub-regions:

1. Australo-Malayan sub-region.
2. Australian sub-region.
3. Polynesian sub-region.
4. New Zealand sub-region.

1. *Austro-Malayan sub-region* comprises all the islands of the Malya Archipelago not included in the Oriental Region together with New Guniea, the Moluccas and the Solomon Islands. In this sub-region, there are 113 families of terrestrial vertebrates of which 4 are peculiar. These are crowned-pigeon and Fly-River Turtle, both confined to New Guinea and two little known Amphibian families Ceratobatrachidae and Genyophrynidae from the Solomon Islands and Sudest Island respectively. Since most of this area consists of small islands, a large number of peculiar genera and species exist.

In New Guinea, three marsupial genera are peculiar. The Babirussa is another remarkable and peculiar member of the Austro-Malayan fauna occurring in Celebes, Buru and the Sulu Islands. Among birds, Birds of Paradise, Bower Birds, Honey-Eaters, White eyes, Frog mouth, Lories and Megapodes attain their highest development in this area; while Honey-peckers, Cuckoo-Shrikes, Fly-catchers, Pitatas and Fruit-pigeons are very numerously represented. Lastly in New Guinea and the neighbouring islands, the amphibian families Narrow-mouthed Toads and Tree frogs are characteristic.

2. *Australian Sub-region* consists of the whole of Australia and Tasmania. In the sub-region 98 families of terrestrial vertebrates are represented, namely 15 mammals, 67 birds, 13 of reptiles and 3 of amphibians. Of these about half a dozen, which include wombats, marsupial mole. Duck-Bills, Scrub-Birds, Lyre-birds and Emus are confined to this region. This area is notable as being the great home of marsupials, for out of 41 known genera, 34 are represented and 24 confined absolutely to it. The avifauna too is highly peculiar; though only 3 families are confined to this subregion yet the proportion and peculiar species is larger than in any other sub-region in any part of the world.

The characteristic but not peculiar animals are Kangaroos, bandicoots, Thylacine, Bower-birds, Honey-eaters, Creepers, Swallow-shrikes, Cockatoos, Bustard-Quails, Cobras. Scale-footed lizards, Varanus, Side-necked Tortoise. The Cobras form about 2/3 of the all snakes found in Australia and all of them are poisonous so that the absence of Vipers and Rattle snakes is more than compensated. Lastly, the entire absence of all tailed amphibians is noteworthy.

3. *Polynesian sub-region.* The rest of the Islands as far north as the Tropic of Cancer and including the Sandwich Islands are embraced in the Plynesian sub-region. This is made up entirely of small islands, and as such the absence of certain forms is to be regarded as more characteristic. The families of terrestrial vertebrates is only 53, of these 3 are of bats and 37 of birds, whose occurrence in these remote islands is probably due to their superior powers of dispersal. Of reptiles, 9 families occur and of amphibians only 2. The other two families of mammals include the Muridae (Mice) and the Cervidae (Deer).

The families peculiar to the Plynesian are the Drepanididae confined to Sandwich Islands; Kagu only found in New Caledonia and tooth-billed pigeons confined to Samoa.

4. *New Zealand sub-region.* In includes New Zealand, Norfolk island, Auckland, Campbell and Macquaire islands. These are 34 families of terrestrial vertebrates. Among mammals there are 3 families, family of Noctiliomidae (free-tailed bats) Vespertilinoidae (typical bats) and Muradiae (Mice rats).

Among birds 5 families namely the Xeicides, Nestorids (Nestor parrots), Stringopids) (Owl-parrots), Apterygids (Kiwis) and Hatterids are confined to this sub-region.

Among reptiles 13 families are peculiar. Snakes are absent. The amphibian is only represented by one frog, *Liopelma*.

The peculiar animal known as Tuatra (*Sphenodon punctatus*) belonging to family Rhynchocephalidae (class—Reptilia). It is commonly known as *living fossil*.

ORIENTAL REGION

Extent

This Region includes those portions of continental Asia which are not comprised in the Palaearctic and Ethiopian Region together with the Malaya Archipelago as for east including Bali, Borneo, the Philippine islands and Formosa.

Climate

The Indian sub-region is in its northern portion, chiefly composed of plain and desert, more particularly in the watersheds of great rivers Indus and Ganges. Its fauna as a whole shows a desert affinity to that of the Ethiopian Region; while the desert in the north-west is debatable ground, and may be regarded as a transitional tract between the Oriental and Palaearctic Regions. The southern portion of India is more luxurient than the northern one, and is largely covered with tropical forest, with a series of elevated tracts culminating in Western and Eastern Ghats. Ceylon, the Indo-Chinese sub-region and most of the Malayan islands are almost entirely covered with tropical forests of the most luxurient character and posses varied and extremely rich fauna.

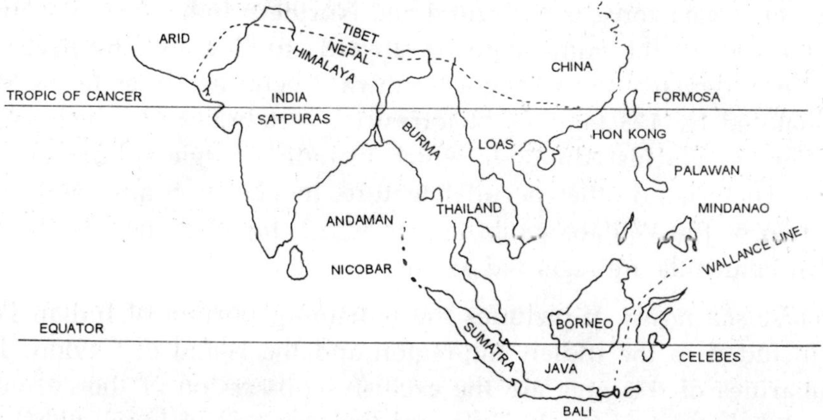

Fig. 9.3: The Oriental region (solid lines) and adjacent land (broken lines).

Zoological Characteristics

The terrestrial vertebrates are represented with 153 families of which 10 are peculiar.

Among mammals 4 families are represented by Hyalobatidae (Gibbons), Tarsiidae (Tarsiers), Galeopithecidae (Flying lemures) and Tupaiidae (Tree-shrews).

Birds are with only one family—*i.e.*, Eurylaemidae (Broad Bills).

Among reptiles 5 families are peculiar they are Elachistodontidae, Uropeltidae (Shield tails), Lanthanotidae, Gavialidae (*Gavialis*) and Platysternide (Big Head Tortoise).

In addition to the above, several well known species and genera are quite characteristic. Among Mammals the Orangutan (*Simia satyrus*) the Macaque monkey (*Macacus*), the Tiger (*Felis tigris*) and Indian elephant (*Elphas maximus*), the Malayan Tapir (*Tapirus indicus*), and three out of the five known species of Rhinoceros are nearly or quite confined to the Region, while families Tragulidae (Chevrotains) and Manidae (Pangolin) are very characteristic. Although only one family of Birds is given above as peculiar, yet many other have their monopolies in the Oriental Region. They are Starlings drongos, Orioles, Honey Peckers, Bulbuls, Pittas and many others. Reptiles and Amphibians are very well-represented in this region for besides the five peculiar families above enumerated, are following the characteristic Illysiidae (Cylinder Snakes). Xenopeltidae, Acrochordidae (Wart Snakes) and pitvipers etc.

Wallace divided Oriental region into four sub-regions:

(1) Indian sub-region

(2) Ceylonese sub-region

(3) Indo-Chinese sub-region

(4) Indo-Malayan sub-region

1. *Indian sub-region* consists of Central and Northern India, from the river Indus and the foot of the Himalayan southwards to Goa and the River Krishna, the line of demarcation taking a southward bend nearly as far as Mysore. It is inhabited by 123 families of terrestrial vertebrates of which only one is peculiar *i.e.,* Elachistodontidae which include a single species of Colubrine Snake. There is no other peculiar feature. Indeed by Scalter and others, the first two of the Wallace's sub-regions which together include the whole of Indian Peninsula are regarded as one.

2. *Ceylonese sub region.* It includes the remaining portion of Indian Peninsula, not included in the Indian sub-region and the island of Ceylon. The main peculiarities of this area are the exclusive possession of the curious family of snakes known as Shield-Tails and the presence of Loris which does not occur in Northern India, though recorded from Eastern Burma. The genere Platacanthomys (Spiny-Rat) is also characteristic, only occuring elsewhere in Cochin China. On the whole, 122 families of terrestrial vertebrates are represented and are identical to the fauna of Indian sub-region.

3. *Indo-Chinese sub region.* It includes China, South to the Palaeartic region, Burma, Thailand and the islands of Hainan, Formosa and Andamans. It contains as many as 138 families.

Among mammals three genera *i.e.*, Takin, Hapalomys and Panda are confined to this sub-region. It also contains Gibbons, Flying lemurs, Java Rhinoceros, Malyan Tapir etc.

Among birds, Wrens are common. One family of tortoise *i.e.*, Platysternidae is peculiar. Among amphibians Salamander and Disc-tongued frogs occurs.

4. *Indo-Malayan sub-region*. Includes the Malayan Peninsula and all those islands of the Malaya Archipelago which fall with Oriental Region. In this region occur 132 families of terrestrial vertebrates of which one, the Lanthanotidae is peculiar and another *i.e.*, Tarsidae is practically so. In addition, several genera of mammals are peculiar. Among these may be mentioned Orang-Utan, Proboscis-monkey, Heringale, Cynogale, Malayan Bedger and some small Rodents. The green Tupaia (Tree-shrews) is well-represented and characteristic while the Gibbons, Flying Lemurs, two species of Rhinoceros, the Malayan Tapirs and the Broad-bill are peculiarly Oriental forms which this region shows with the Indo-Chinese. The typical Australian Cockatoos are represented by a single species in the Philippines while the Megapodes another essentially Australian group occur in the Philippines, Borneo and the Nicobar Islands.

ZOOGEOGRAPHY OF INDIAN SUB-REGION

Blanford (1901) in *Phil. Trans. Royal Soc. London* discussed the geographical distribution of animals on the earth in British India. In his discussion he included India Pakistan, Kashmir, Gilgit, Ladakh, Nepal, Bhutan, Sikkim, others Cis-Himalayan States, Assam, countries between Assam and Burma such as Garo, Khasi and Naga Hills, Manipur, Burma with Karenni, Tenasserim and the Mergui Archipelago, Andaman and Nicobar Islands. *B. Prasad* (1942) added Laccadive and Malolive islands.

Sub-regions of India

Owing to the heterogeny nature of the faunas of the different parts further divisions of Indian sub-region is not an easy task. Its division has been attempted by a number of scientists. A brief history of such attempts is as follows:

Jordan (1862) divided it on the basis of birds; *Gunther* (1864) on the basis of reptiles; *Blanford* (1876) on molluscs; *Wallace* (1876) considered the distribution of all the groups of animals. *Blanford* (1888) on mammals, *M. Smith* (1931) based his findings of reptiles and amphibians; *B. Prasad* (1921) sub-divided Indian sub-region into 5 division. *Mahendra* (1942) divided Indian sub-region into 10 sub-division on the basis of distribution of reptiles and amphibians. These are:

1. The arid and semi-arid province of North India.
2. The Western Himalayas.
3. Southern Burmese province.

4. Trans-Gangatic province.
5. Gangatic plain and adjacent part as far as South as 20° latitude.
6. South India below 20° latitude excluding Travencore.
7. Travencore province.
8. Ceylon.
9. Andaman Islands.
10 Nicobar Islands.

1. *The arid and semi-arid province of North India.* It consists of W. Pakistan, Punjab, Western Rajasthan and Cutch. The fauna is allied to that of the countries to the west of present W. Pakistan and also almost all general and species are characteristic.

Among lizards are *Teratoscincus, Stenodactylus, Alsophylax, Agamura, Pristrus, Eremias* etc. Among snakes are *Leptotyphlops Contia, Lytorhynchus* etc., are present. Many species are endemic. Very few turtles occur but none are endemic or peculiar to it.

Amphibians are rare the only genera found being the cosmopolitan *Rana* and *Bufo.*

2. *Western Himalayas.* It embraces Kashmir, Simla, Kumaun and Garhwal Districts, adjacent mountainous regions of Western Tibet, Alpine and Punjab etc. This zone is poor in amphibians and reptiles. These are certain endemic species *e.g. Gymnodactilus fasciolatas, G. lawderonus, Japalura major, J. Kumaonesis* and *Phrynocephalus theobaldi* and *Leiolopiṣma ladacense.*

3. *The Trans-gangetic province.* It consists of the northern part of Bihar and undivided Bengal, the whole of Assam, Burma north of 20° latitude Nepal, Himalayas to the east of Nepal, Sikkim and Bhutan. The herpetological fauna is both rich and characteristic. Amongst the endemic genera are *Ptyctoaemus, Mictopholis, Oriocalotes, Stolizteaia, Tylotriton* etc.

4. *Southern Burmese Province* comprises southern Burmese province including Burma approximately south of 20° latitude. It includes certain characteristic snake species and some endemic amphibians.

Among snakes are *Gymnodactylus Oldham, G. Consobinoides, Riopa anguina, Bungarus flaviceps, Doliophis intestinalis, Coluber melanurus.* Among turtles are *Platysternus megacephalum, Triony Cartilagineus, Geoemyda grandis* etc.

Among amphibians *Kalobula macrodactyla, Glyphoglossus molossus, Cellulla gnttulata* are characteristic of this region.

5. *Gangatic plain and adjacent Regions as far South as 20° lat.* The southern boundary is very vague and approximate. This subdivision is characterised by the extreme poverty of fauna and has no endemic species. There is complete absence of house gecko. The Indian gharial, *Gavialis* ranges as much as in the Indus system as in Ganges, Brahmputra and Mahanadi.

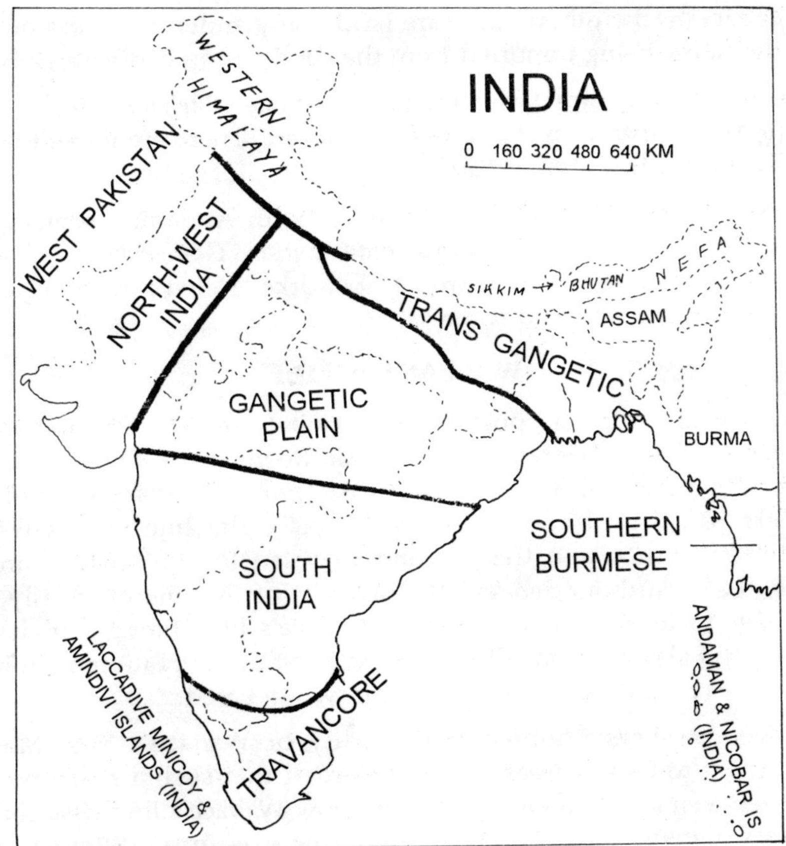

Fig. 9.4: The Indian sub-region.

6. *South India below 20° lag. excluding the Travancore Province.* This province is funistically characteristic and has *Riopa lineala, Barkudia insularis, Gomnodactylus dekkanenois, Hemidactylus subtriedrus* and *Hemidactylus reticulatus.*

Only a few wide spread species of fresh water turtles are found.

7. *The Travancore.* Consisting of the hilly country south of Lat. 12° or 13° on the west, and south of the Coleroon river in the east. The distinctness of this division is well-recognised, the fauna being extremely rich in endemic forms. The number of characteristic genera and spp. is too large.

8. *Cylon.* It forms a distinctly zoogeographic province, with clear affinities to the Travancore sub-region.

9. *Andaman Islands.* Besides having a few endemic forms, which differentiate the fauna of this region as separate from others, and Andaman are faunistically allied to the Burmese sub-region on one hand, and to the

Nicobars on the other. There are hardly any endemic genera owing evidently to the fauna being imported from the continent in comparatively recent times.

The endemic species are *Gymnodactylus rubidus, Phelsuma andamanesis, Calotes andamanisis, Typhlope andamansis, Boiga andamanesis* etc. There are no amphibians (except 2 spp. of *Rana*) and no fresh water turtles.

10. *Nicobar Islands.* It's fauna has affinities with Andaman Islands. It has fauna like *Calotes mystaceus Mabuya andamanesis, Gonicephalus suberistatus.* The endemic species are few in number like *Mabuya rugifera, Glotes jubatuo Leiolopisma macrotis* and some other.

WALLACE'S LINE

The Wallace's line is the supposed boundary between the Oriental and Australian faunas. The faunas of the Oriental and Australian Regions are extra-ordinarily different. What happens when they meet in the Indo-Australian Archipelago has long fascinated Zoogeographers. As first drawn by *Wallace* (1860), the line ran between Bali and Lombom, between Borneo and Celebes and between the Philippines and Sangi and Taland islands. Later authors debated the position of the line especially the northern part of it. *Huxley,* for example, who named Wallace's line, thought it should split the Philippine, putting Palawan in the Orient and the rest of the Philippine in the Australian region. Still later even the existence of the line was debated.

Later zoogeographers doubted the validity of Wallace's line. Majority of the scientists thought it to be imaginary, some regarded that such a sharp boundary could not be drawn between any two faunal regions. Now Wallace's line is no longer accepted as the boundary between the two regions, as there is great difference between the faunas of Bali and Lombok, that are separated by only about 32 km. The rich fauna of Borneo and comparatively poor faunas of Celebes and the Philippines cannot be denied. The difference is regarded due to impoverished young fauna in one area and an older richer fauna in the other.

WEBER'S LINE

In 1904 Weber drew another line between the Oriental and Australian regions. It demarcates fauna to the west as predominentally Oriental and to the east Australian. However, one problem its that if a line is thus drawn to separate 50 per cent Oriental and 50 per cent Australian elements for one group of animals, it may not be the same for another group. Majority of the zoogeographers do not recognize Weber's line and prefer a broad belt as a separate zone. This is due to the difference in mobility of different kinds of animals. A further proof of these differences is that the flora of New Guinea is mainly Oriental while its fauna is strictly Australian.

Dickenson and others (1928) gave the term `Wallacea' to separate eastern boundary of the Oriental region at or near Wallace's line and western boundary of the Australian Region just west of New Guinea. In that the intervening area may be regarded as a separate subtraction-transition area. *Darlington* (1957) also recognized that this is an appropriate and useful term. Wallacea bounds the Oriental and Australian regions.

BATHYMETRIC DISTRIBUTION

The Bathymetric distribution concerns itself with the vertical range of organisms in space, i.e., from the highest Alpine peak to the abyssal depth of the sea. One would find a series of contrasting conditions which of necessity profoundly effect the organism. Of these the primary conditions are:

(1) It determines the method of breathing. With a few exceptions, if the animals are terrestrial and breath in air they can not live in the water and those who are aquatic animals and breathe in water cannot survive outside of water.

(2) The presence or absence of light not only modifies the animal directly but indirectly through the effect upon the food supply for assimilative plants, which form the ultimate nourishment of all animate nature,cannot exist where light is wholly absent.

(3) The third primary condition is the presence or absence of a substratum, without which the organisms must be self supporting, either buoyant or able to swim. This condition therefore determines the bionomic group to which the animal belongs, whether plankton or nekton on the one hand, or benthos on the other.

Like fresh water ponds and lakes marine environment has been classified into two zones by *Hedgepath* (1957). (1) *Pelagic Zone.* It includes the entire water mass lying above the ocean floor. Thus the animals which live here are called *Pelagic.* They may be *plankton* (microscopic) or *nekton* (large swimming animals). (2) *Benthic zone.* It includes all the sea bottom, thus the animals living in or on the sea bottom are called *benthonic.*

The secondary conditions limiting bathymetric distribution are whether the water is fresh or salt and the increase of pressure with the depth.

ORGANIC REALMS

Hence from the bathymetric as well as from the geographic stand point three organic realms may be recognised. The bathymetric divisions are:

(1) Geobiotic or terrestial

(2) Limnobiotic or fresh water inhabiting (lakes and rivers)

(3) Halobiotic or marine or salt-water inhabiting (the sea).

1. Geobiotic or Terrestrial Realm

It extends from high tide mark along the shore of continents and islands to the summit of the highest elevation. It ranges in altitude, therefore, from lowlands to the

Alpine peak and each division—low land up land, prairie, high plain or mountain range has its own fauna and flora, governed by many factors but in part at least by altitude. The so called timber line, the limit of tree growth is governed by altitude, although vary in different regions due to climate and latitude.

One additional terrestrial subrealm of Geobiotic realm is the *eryptozoic* or *subterranean caves*. The dryness and absence of light is very harmful to the plants which are the food suppliers to the animals. Examples of permanent cave dwellers are *Peromyscus leucopus* (white-footed mouse) having bulging eyes, and ling tectile whiskers and ears; *Typhlotriton spelaeus* with normal eyes in larval stage but in adult the eyes degenerate; *Gronias nigrilabris* (cave cat fish) is partially blind. Internal parasites and wood boring larvae of insects also dwell in dark environment and are consequently modified.

2. Limnobiotic or Freshwater Realm

The terrestrial waters such as ponds, lakes and rivers contain very limited fauna due to freshness of water and continuous flowing character. This is a condition to which the great majority of invertebrates with their planktonic larvae can not adapt themselves. Certain lakes and relic seas are the only bodies of fresh water of sufficient depth to have deep sea characteristic but we do not find such profound modifications in lacustrine forms as in those which live in the deep sea bottom. For example biolumieniscence is characteristic of deep sea nocturnal animals but none are found in the deep lakes, although in each instance they might be very useful in the struggle for existence.

3. Halobiotic or Marine Realm

It is a very important realm, for here we find all the contrasting characteristics abundantly developed. From the ages during which the ocean has existed, it has given sufficient time to the inhabitants for evolution.

The halobiotic realm is divided into four subrealms as follows:

(a) Strand

(b) Flat sea or shallow sea

(c) Pelagic

(d) Abyssal.

1. Strand

The strand or tidal zone is the transitional area between the main and terrestrial realms, for here the inhabitants are left bare twice daily by the receding tide and have to endure drying, either by means of closable shells or other devices or burrow down into the moist sand, or must be able to breathe both the air and the water. The tidal Zone is of course of very variable extent, owing to the differing height of tides and range from a width of a few feet to several and in the rare cases to many miles. The tides, when run into a constantly narrowing area, may grow to a height of 60 ft at a time.

2. *Flat Sea*

The flat sea of *Neritic province* is the ocean upto a depth of 200 meters above the continental shelf below low tide mark. The continental shelf is formed by the action of storm waves, which are continually cutting back the shore line nd depositing the debris, together with other land waste, upto the sea especially at its outer edge. This margin is 200 meters and marks the extreme limit of wave action.

This province is very important t biologically for it has both light and its attendant vegetation and a substratum where upon benthonic organisms may dwell. All of these factors, light, plant food, movement of water, warmth, and isolation, make this area a variable hot bed for evolution.

The shore fauna is certainly the most representative of all faunas.

Foraminifera with beautiful shells of lime slowly gliding on the ponds of sea-wind. Calcareous sponges like little vases and more irregularly moving sponges, hydroed zoophytes, sea-anemones and corals often like beds of flowers; unsegmented worms such as the living films which glide on the sea weeds and the nemertines or ribbon worms provided with a remarkable protrusible proboscis, the higher ringed worms or annelids like *Nereis, Aphrodite* etc., the starfishes creeping up the rocks, the brittle stars, the sea urchins and the sluggish sea cucumbers. The beautiful colonies of most animals or Bryozoa: myriads of crustaceans such as water fleas, beach fleas, sand hoppers, shrimps, hermit crabs, shore crabs etc. Bivalves innumerable such as cockles and mussels, oysters and razor fish, many gastropods, cuttle fishes; a large representation of ascidians, the lancelets, true shark fishes like sandribs, poisonous sea snakes, numerous shore birds and an occasional mammals like otter and seal—on the whole a more representative fauna than anywhere else.

The flat sea is divisible into (a) A *pelagic zone* and (b) The *Littoral zone*. The pelagic zone is characterized by the absence of substratum while the littoral zone is the substratum of sea bed.

3. *Pelagic Realm*

The pelagic realm embraces all the open seas down to the depth to which effective sunlight penetrates. It is characterised physically by the absence of a substratum. In the upper portion there is variable temperature and frequent and violent wave action, while in its lower strata the movement of the waters and the temperature are greatly reduced.

Assimilating plant life, which forms the ultimate food supply of all animals is dependent upon the presence of red, orange, and yellow rays which virtually restricts it to the upper 200 m. of oceanic water.

Due to absence of substratum no *benthonic* form is found in this realm. Only either the *planktonic* or *nektonic* forms exist here.

4. *Abyssal Realm*

Beneath the limits of the continental shelf and pelagic realm, the abyssal realm is found which extends from below 200 meters. Two region abyssal can be distinguished: the region between 200-600 meters may be referred as *abyssal pelagic* (where no substratum is there so all the animals swim or float); and below 600 meters as *abyssal benthos* (the substratum is present).

Characteristics of Abyssal Realm

These are as follow:

(i) *Absence of Light*. Light may exist in the upper transitional strata but it lacks the rays which are essential for assimilating plants, hence non exists. The animals, therefore, are all carnivorous or feed upon dead organic matter. Below the transitional zone, there is darkness, except for bioluminescence is profound.

(ii) *Quiescence*. There is no movement of water except the sluggish ocean currents of the greater depth, the progress of which is very slow.

(iii) *Cold*. Below a certain depth, the waters of all the oceans of world are nearly to the freezing point of fresh water. The main temperature of Atlantic at the surfaces 68°F, at 00 fathoms 37°F, at 1000 fathoms 35.6°F.

(iv) *Pressure*. The pressure of abyssal water is enormously increasing directly with the depth. The ratio of increase being about one ton to a square inch of surface for every 1000 fathoms of depth.

Thus the abyssal realm constitutes a simple biotic environment of vast extent but of comparatively uniform and changeless character and hence not conducive to rapid evolution and change. None of the deep sea creature is old geologically speaking for while from 25 to 35 paleozoic genera are known in the purest shallow seas none of the animals which people the deep sea is older than the Mesozoic. They seem to be all migrants from shallow waters which have become adapted to the deep sea conditions but there is in no instance of evolution of a new race of animals exclusively restricted to the abyssal realm.

Vertical Distribution

Life exists at all the depths. *Murray* estimated the number of forms as follow:

Down to	200	meters about	4200	species
"	200	"	600	"
"	4000	"	400	"
"	5000	"	150	"

Among the invertebrates, the sponges form an important element in the deep sea fauna, and coelentrates such as the coals, hydroids and their allies, while not so numerous as the sponges, are also well represented. Echinoderms, among which are the brittle stars and stalked crinoids are present; and holothurians or sea cucumbers are also found. The modern stalked crinoids are rarely found beyond a depth of 2000 fathoms, although free-swimming species, *Bathymetra abyssicola* has been obained at 2900 fathoms.

Bryozoa range to 3000 fathoms. Brachiopods have been found from 2900 and probably so much deeper. Of the molluses, the Pelecypoda have a very great veritcal distribution, *Mytillus* sp. being from the shore to 3000 fathoms. In the greater depth the shells of bivalves are exceedingly delicate, being sometimes quite transparent. The majority of gastropods are shallow-water forms, although a number of them are found at depth from 1000 to 2000 fathoms. The shells and their ornamentation are more delicate than in their shoal water relatives. Cephalopods inhabit waters of moderate depths only.

Of the Anthropods, we have branacles from 3000 fathoms, ostracods from 2000, decapods such as crabs, shrimps, lobsters ranging down to 2,500, though true crabs are largely shallow-water forms.

Of the *Vertebrates* elasmobranches, the true shark have been taken at depths from 345 to 400 fathoms and the rays down to 608 fathoms. None show deep sea characteristics, except perhaps the luminous shark, *Spinax niger,* which ranges from 500 to more than 1500 fathoms. The chimaeroids or silver sharks, on the other hand, often with huge eyes and long body and tail, are distinctly abyssal. *Chimaera affinis* ranges from 200 to 1300 fathoms, *Harriotta* from 700 to 1000 and *Chimaera monstrosa* down to 1000.

The Escociformes group of teleost fishes have a few deep sea members such as *Cetomimus,* with a huge mouth and very small eyes.

GEOGRAPHICAL DISTRIBUTION

Life occurs in almost all the habitats ranging from high mountains to the deepest sea bottom. One would find a series of contrasting conditions which of necessity profoundly affect the organism.

The distribution of animals and plants may be of following two types:

1. *Distribution in Space*

It is of two types:

(*i*) *Geographical distribution.* It deals with horizontal or surficial distribution of animals on land and in water in different continents and on islands. The study of geographical distribution is called *Zoogeography.*

(ii) *Bathymetric distribution.* It deals with the vertical or altitudinal distribution of animals on land and in water. The bathymetric distribution is further divided into:

(a) *Geobiotic* or *Terrestrial*

(b) *Limnobiotic* or *fresh water inhibiting*

(c) *Holobiotic* or *marine water inhibiting.*

2. Distribution in Time

It is *geological distribution* or the *durational distribution* of animals from the very beginning of the life on the earth upto recent time.

Pattern of Distribution

The distribution of life over the surface of earth is world wide. There is no such place where the living organisms are not found. In spite of that the distribution of animals and plants is not uniform for example, a particular species of animals is present in one area but entirely absent from other. Three patterns have been observed regarding distribution of animals.

1. *Cosmopolitan Distribution*
2. *Bipolar Distribution*
3. *Discontinuous Distribution.*

1. Cosmopolitan Distribution

The animals which are found all over the world are said to have cosmopolitan distribution. *Mytilus, Artemia* (Shrimp), bats, rats, hawks, cockroaches etc., are world wide in distribution. These animals have a wide range of adaptations to a wide variety of environmental conditions.

There are two types of cosmopolitan animals (a) *Eurytopic,*when they have wide range and uniform distribution and (b) *Steno-topic,*when they have restricted range of distribution.

2. Bipolar Distribution

In bipolar distribution the animals are only confined to the polar regions i.e., only in the waters of arctica and antarctica, A few examples are *Lampta, Grammaria, Sardina* etc. (Phylum Coelenterata); *Limacina* (Mollusca); *Oncorhynchus,*(Pacific Salmon) and *Lamma cornubica* (shark).

3., Discontinuous Distribution

Some of the present day animals which are closely related are found distributed in widely separated areas. Some examples will illustrate the discontinuous distribution. Marsupials are found in Australia, and South America. In Australia there are several kinds of marsupials like Kangaroos, Wombets, Koalas and Bandicoots. Only the

Opossum is found in the eastern states of America and a few marsupials in the South America.

Another example is that of Tigers. The Tigers are found only in India and neighbouring regions like Nepal. Lions are only found in India and Africa. Elephants are found in the forests of India, Nepal and Africa. Two species of Alligators are found one in South-Eastern United States and the other in China. The lung-fishes (Dipnoi) are represented by only three living genera, *Neocera-todus* in Australia, *Protopterus* in Africa and *Lepidosiren* in South America.

Another example of discontinuous distribution is the Camel. They are present in parts of Africa, Asia and South America. The Llama found in Andes. Different species of Tapirs are found in Malaya, and South America and Islands of Jawa and Sumatra.

Many more examples are there for the discontinuous distribution. It can be explained if we understand the geological data and animal evolution.

From the Palaeozoic to the Mesozoic era, there were only two major land masses Gondwana and Laurasia and these were in contact at times. Gondwana Centered around the South Pole while Laurasia overlapped the equator and extended well into the northern hemisphere. During the Cretaceous, these masses fragmented to form the present continents and these have since drifted apart very slowly toward their present positions. Gondwana gave rise to Africa, South America, Australia, Antarctica, Arabia and India. Laurasia broke up into North America and Eurasia.

A species originates in a definite more or less restricted area and then tends to spread in all directions occupying suitable habitates. When an impassable barrier is encountered, the migration of the species stops unless it is removed, by geologic or climatic changes. Sometimes *bridges* are established and the animals are able to cross the barriers. Three type of bridges have been recognised (1) A *corridor bridge* is a broadly continuous connection as exists between Europe and Asia (2) A *filter bridge* is more temporary in duration as is the Beringstrait which connects Asia with North America (3) A *Sweepstakes bridge* is the one which depends upon accidental transportation on floating ice, log of wood clinging mud of the birds claw or even on board the ships.

It is in connection with this type of history of the earth and animals that the problem of widely discontinuous distribution must be understood. Marsupials, Ratitae birds and lung fishes had a world wide distribution to begin with. Later on by subsedence of land and continental drift, they became isolated. Where the competition with superior animals was severe, they have been eliminated, in other places they survive till the present day. The barrier of deep sea between Bali and Lombock has existed since Palaeozoic times and it has never been bridged. This is the reason for the distinctness of fauna in the two islands.

DISPERSAL

All the animals are capable of multiplying in large numbers. If all the progenies of a particular species in allowed to live in the same area, the result will be shortage of food, and the struggle for existence. Moreover, there is a population pressure due to which the individuals migrate to expand to boundaries of their range or, sometimes, are perished. The geographical range of the species is increased due to dispersal. *Darlington* (1957) defined dispersal as, "The total geographical movement of groups of animals and the movement of individuals (eggs, youngs and adults) especially their outward scattering."

Means of Dispersal

There are several agencies which help in the dispersal of animals. These are:

1. By wind

Some animals are dispersed either by the power of flight or by being carried passively by means of wind. Several small insects are blown by air to a few hundred metres from their place of birth. Sometimes, by air, they are also taken away to higher levels. Air-dispersal is found in insects, birds and bats.

2. By hurricanes

Small living animals are transported by cyclonic storms of great velocity about 120 km. an hour. Some American birds have reached in this way the coasts of Britain and France.

3. By water

Some aquatic animals and larvae migrate from one place to another with water currents. Rafts of trees or floating debris also carry the animals, such as many species of arthropods to pass down large rivers and sometimes are seen far at sea. Small arctic animals move on to the ice.

4. By land

Active movements of the species are plausible by land. For finding new pastures, active movements of animals are well-known. Land-bridges are used as passages of migration.

5. By human agency

Several animals reach new localities by accidental or international transportation.

6. By hosts

By this means some parasitic and commensal animals are transported to various new localities.

7. Migration

It is also an important, factor for the dispersal of animals. There are two types of migration (*i*) permanent and (*ii*) seasonal migration.

Thus the distribution of animals into different regions is only possible due to the dispersal. But there are a large number of barriers to these dispersals.

BARRIERS TO DISPERSAL

Barriers may be defined as the factors which hinder in the normal distribution of animals. Following types of barriers are there:

1. Typographical barriers.
2. Climatic barriers.
3. Vegetative barriers.
4. Large bodies of water as barriers.
5. Impurity and lack of salinity of sea water.

1. Typographical Barriers

High and extensive mountain ranges act as barriers in the distribution of many terrestrial animals. They are effective if they are parallel with the equator as in Asia and Europe. Here we find a remarkable difference between the species occupying the northern and those occupying southern slopes. This is true of the Great Himalayan Range which is covered by snow. On the south of it, there are hot, moist plains of India having tropical fauna which resembles largely with that of Africa. In the north, the climatic conditions are changed so that the animals of this region resemble largely with that of Europe.

In the New World, the mountain ranges run North to South and their influence upon the distribution is very less. *Mayr* found that there are distinct altitudinal races including lowland, midmountain, and alpine races.

2. Climatic Barriers

Climate seems to limit the range of several animals. But it is believed that, in many cases, it is not the climate itself so much as the change of vegetation consequent on climate which produces the effect.

Degree of heat is very important in limiting animal distribution. However, the distribution of tiger and elephant shows that this does not very much alter their distributions. For instance, tiger has its normal home in the hot districts of India but the Indian archipelago is also found in colder climates such as in the elevated regions of the Caucasus and the Altai chain and in the Himalayas perhaps on snow and also in the cold plains of Manchuria. Elephants likewise do not suffer from cold, provided a sufficient amount of water is available.

The influence of temperature is much more marked in the cold-blooded animals. Thus the amphibians and reptiles are tropical and temperate in their distribution, rapidly diminishing in numbers towards the poles.

Lack of moisture also controls the distribution and the effect is more marked in the extreme cases where it produces the desert conditions where only forms adapted to withstand drought can survive, for others it forms insuperable barrier. The most notable desert barrier is Sahara, due to presence of this desert the fauna of North Africa and S. Africa are totally distinct. Similarly, the Kalahari desert is responsible for the distinct fauna of Central Africa from that of cape of Good Hope.

The increase of moisture may produce swampy conditions which may make the area impassable for creatures not adapted to them. Increased humidity also have marked effect in the vegetation and also in the spread of insect-life.

3. *Vegetation Barriers*

The influence of vegetation on dispersal is both direct as well as indirect. Its direct effect is on arboreal animals, which can not cross the regions where forests do not occur. In the same way, the larger terrestrial animals can not penetrate through the dense forest of tropics. For example, several species of elephants i.e., *Elephas* and *Mastodon* of N. America did not succeed in crossing the southern region of Mexican plateau to the South.

The vegetation also effects the distribution indirectly as it forms the important source of food. For example the primates live in tropical forests as they feed on nuts, fruits and blossoms. The animals with short-crowned browsing teeth would needtrees and shrubbery for the food while long-crowned grazing teeth need extensive pasturage of harsh-grasses. many insects and catterpillars feed upon certain plants and if these plants are absent, those insects and catterpillar will not exist. Similarly some plants are dependent upon certain insects for pollination.

4. *Large Bodies of Water*

Larger bodies of water such as giant river systems and oceans act as most effective barriers in the dispersal of terrestrial vertebrates like amphibians, reptiles, mammals and flightless birds. Fresh water fishes like carps, catfishes etc., are unable to migrate through the large bodies of salt water.

To the modern amphibia, salt water constitutes a most effective barrier as the common salt is poisonous to them, even 1 per cent solution preventing the development of their larvae. Amphibia are, therefore, almost completely absent from oceanic islands. Moreover, the tailed amphibians such as sirens, newts, salamanders etc. are restricted to northern hemisphere for the southern land masses, Australia, Africa and S. America are isolated by oceans and are incapable of crossing through small migratory roots. The anurans have a wider range of distribution while the burrowing caecilians are confined to S. America and Africa.

For crocodiles and marine turtles, the seas do not act as barrier. Giant tortoises are confined to certain oceanic islands e.g., Galapagos islands of the western Indian ocean and totally absent on the mainland of S. America, Africa and Eurasia.

The serpents, though many of which are good swimmers, but are incapable of passing large bodies of salt water with the exception of some sea snakes. Adult lizards are also incapable of passing an oceanic barrier, though the eggs in some cases could be transported to the islands.

With the birds, it is only the flightless forms such as the ostrich, rhea, apteryx etc. which are incapable of oceanic migration for even small birds are carried far on favouring winds.

Among mammals, except for whales and seals, bodies of water of a greater expanse than 20 to 50 miles are impassable when not frozen.

Similarly land masses form barriers to the spread of sea-life.

5. *Impurity and lack of salinity of Sea-water*

Impurity and salinity of water also act as barriers. Sea-water forms a barrier for sponges, branchiopods, cephalopods, asteroids and echinoids.

Moody (1962) described two types of barriers:

(*i*) Physical barriers include bodies of water, dry land, high mountain ranges, deserts, open plains, forests as well as climatic factors.

(*ii*) Biological barriers include absence of food suitable to species in question, presence of competitors for the same food supply or nesting sites, or presence of predator animals.

However, a species is limited in distribution by the sum-total of external influences. The range of equilibrium is finally governed by *Liebig's Law of Minimum*, which is the most valuable attribute. Animals having adaptability can extend their ranges to regions, which offer conditions of life differing from those in the centre of dispersal. Lack of adaptability prevents dispersal.

Chapter—10
Evolution by Dobzyansky

Macromolecules and Evolution

The diversity of life is staggering. About 400,000 species of plants and 1,500,000 species of animals have been described and named, but the census is far from complete. The diversity of the living world is apparent not only in the large number of species, but also in their heterogeneity. Organisms are extremely diverse in size, way of life, and habitat, as well as in structure and form.

Yet, despite their prodigious diversity, organisms share much in common. Oxygen, hydrogen, and carbon are common chemical elements in all organisms, together they account for about 98.5 per cent by weight of any living being. Four kinds of macromolecules—proteins, carbohydrates, lipids, and nucleic acids—are the basic molecular constituents of all living processes. The genetic information of all organisms, from bacteria to man, is encoded in the double-helical structure of DNA. The processes of transcription and translation and the genetic code are essentially uniform throughout all life.

Certain similarities are shared by some, not all organisms. These similarities can be used to classify, i.e., to characterize, some groups of organisms and distinguish them from others. Centuries ago biologists noted that living things can be arranged in a hierarchic fashion, that some organisms (and groups of organisms) resemble each other more than they resemble other organisms (or groups). The basic process responsible for the hierarchy of similarities among living things is of course evolution—some organisms resemble each other more than others because they are more closely related by lines of descent.

No modern classification of life entirely ignores evolution. The three prevailing systems of classification, are known as phenetic, cladistic, and evolutionary. In this chapter we shall discuss the methods used to infer relationships of evolutionary descent (phylogeny) on the basis of resemblance and dissimilarity.

Phylogenetic relationships are ascertained on the basis of several complementary sources of evidence. In some cases the fossil record provides definitive evidence of the phylogenetic relationships among groups of organisms. However, the fossil record is far from complete, and is often seriously deficient. Second, comparative studies of living forms also provide information about phylogeny. Comparative anatomy is the branch of science that in the past has contributed the most information, although additional knowledge came from comparative embryology, cytology, ethology, biogeography, and other biological disciplines. In recent years the comparative study of informational macromolecules (proteins and nucleic acids) has become a powerful tool for the study of phylogeny.

Morphological similarities among organisms were probably always recognized by men. Among the Greeks, *Aristotle* (384-322 B.C.) and later his followers and those of *Plato*, particularly *Porphyry* (234-305 A.D.), classfied organisms (as well as inanimate objects) on the basis of similarities. The Aristotelian system of classification was further developed by some medieval Scholastics, notably *Albertus Magnus* (1193-1280) and *Thomas Aquinas* (1225-1274). The modern foundations of taxonomy were laid in the eighteenth century by *Linnaeus* (1707-1778), and by the lesser-known botanist *Adanson* (1727-1806). *Cuvier* (1769-1832), another great taxonomist, suggested that similarities of form are related to similarities of function. *Lamarck* (1744-1829) dedicated much of his work to the systematic classification of organisms, and proposed that their similarities are due to ancestral relationships.

The modern theory of evolution, originating from *Darwin* (1859), provides a causal explanation of the similarities among living beings. Organisms evolve by a process of descent with modification. Changes, and therefore differences, gradually accumulate over the generations. The more recent the last common ancestor of a group of organisms, the less their differentiation; similarities of form and function reflect phylogenetic propinquity. For this reason, phylogenetic affinities can be inferred on the basis of relative similarity.

A distinction is necessary between resemblances due to propinquity of descent and resemblances only due to similarity of function. *Homology* is correspondence of features in different organisms due to inheritance from a common ancestor. The forelimbs of humans, dogs, whales and chickens are homologous; the skeletons of these limbs are all constructed of bones arranged according to the same pattern because they were inherited from an ancestor with similarly arranged forelimbs. *Analogy* is correspondence of features due to similarity of function but not related to common descent. The wings of birds and flies are analogous. Their wings are not modified versions of a structure present in a common ancestor, but rather have developed independently as adaptations to a common function, flying. The similarities between the wings of bats are partially homologous and partially analogous. The skeletal structure is homologous because of common descent from the forelimb of a reptilian ancestor, but the modifications for flying are different and independently evolved, and in this respect are analogous.

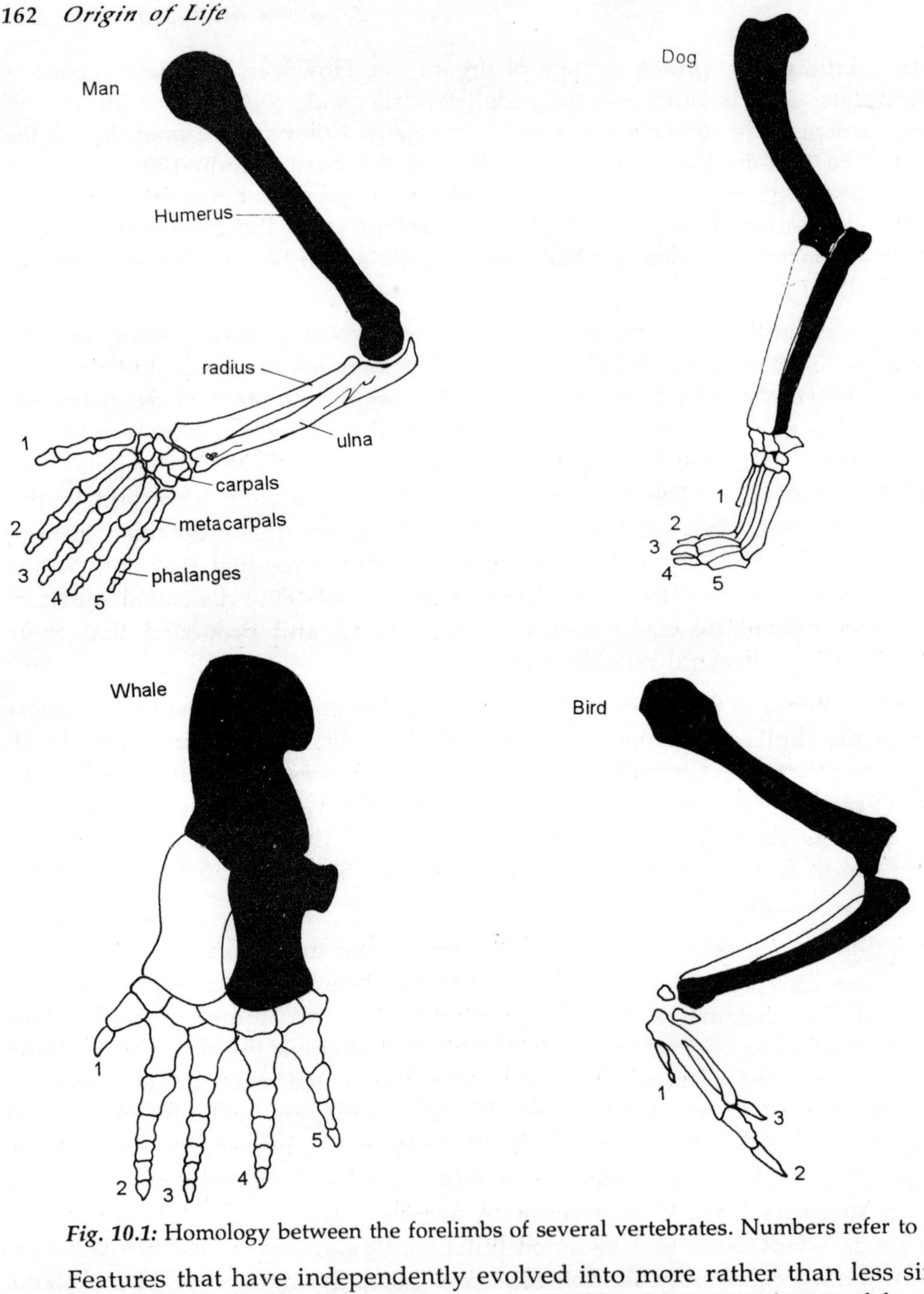

Fig. 10.1: Homology between the forelimbs of several vertebrates. Numbers refer to digits.

Features that have independently evolved into more rather than less similar ones are said to be *convergent*. Convergence often occurs due to similarity of function, such as the evolution of wings in the ancestors of birds, bats, and flies. The shark (fish) and

Bird

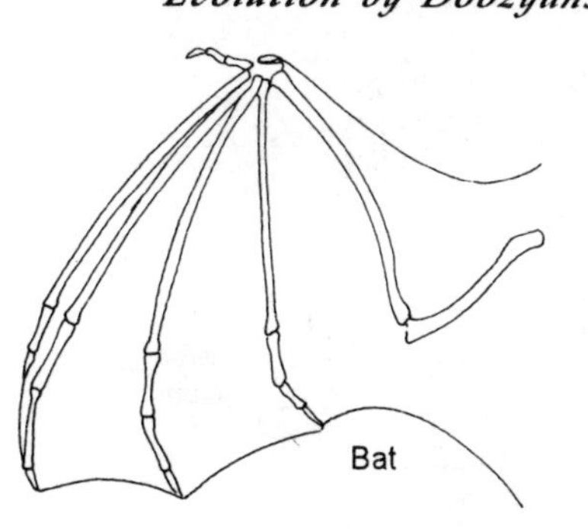

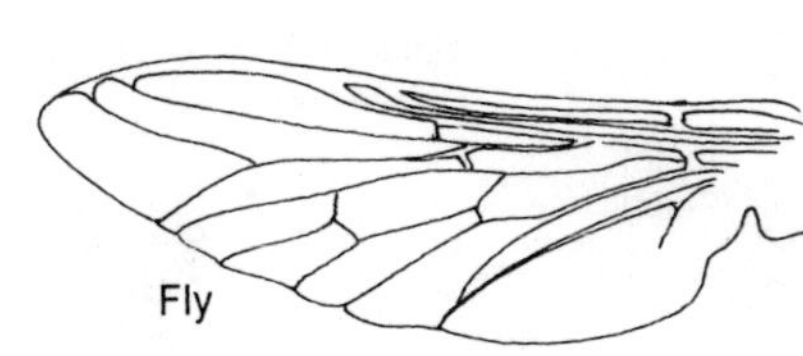

Fig. 10.2: Analogy between the wings of birds, bats, and flies. The wings of flies and birds (or bats) are analogous; they have evolved independently in these organisms as adaptations to flying. The similarities between the wings of bats and birds are partially homologous and partially analogous. The skeletal structures are homologous owing to common descent from the forelimb of a reptilian ancestor, but the modifications for flying are analogous, having evolved independently.

the dolphin (mammal) exhibit similarities in external morphology due to convergence, since they have evolved independently as adaptations to aquatic life.

Taxonomists speak also of *parallel* evolution. Parallelism often involves similarities due to both homology and analogy. According to *Simpson* (1961), "parallelism is the development of similar characters separately in two or more lineages of common ancestry and on the basis of, or channeled by, characteristics of that ancestry." Parallelism and convergence are not always clearly distinguishable. Strictly speaking, convergent evolution occurs when descendants resemble each other more than their ancestors did with respect to some feature. Parallel evolution implies that two or more lineages have changed in similar ways, so that the evolved descendants are as similar to each other as their ancestors were. The evolution of marsupials in Australia paralleled the evolution of placental mammals in other parts of the world. There are Australian marsupials resembling true wolves, cats, mice, squirrels, moles, sloths, and anteaters. The evolution of these placental mammals and of the corresponding marsupials proceeded independently but in parallel lines due to their adaptation to similar ways of life. Some resemblances between a true anteater (*Myrmecophaga*) and a marsupial anteater (*Myrmecobius*) are due to homology—both are mammals. Others are due to analogy—both feed on ants.

Parallel and convergent evolution are also common in plants. New World cacti and African euphorbias are similar in overall appearance although they belong to separate families. Cacti and cactus-like euphobias are succulent, spiny, water-strong,

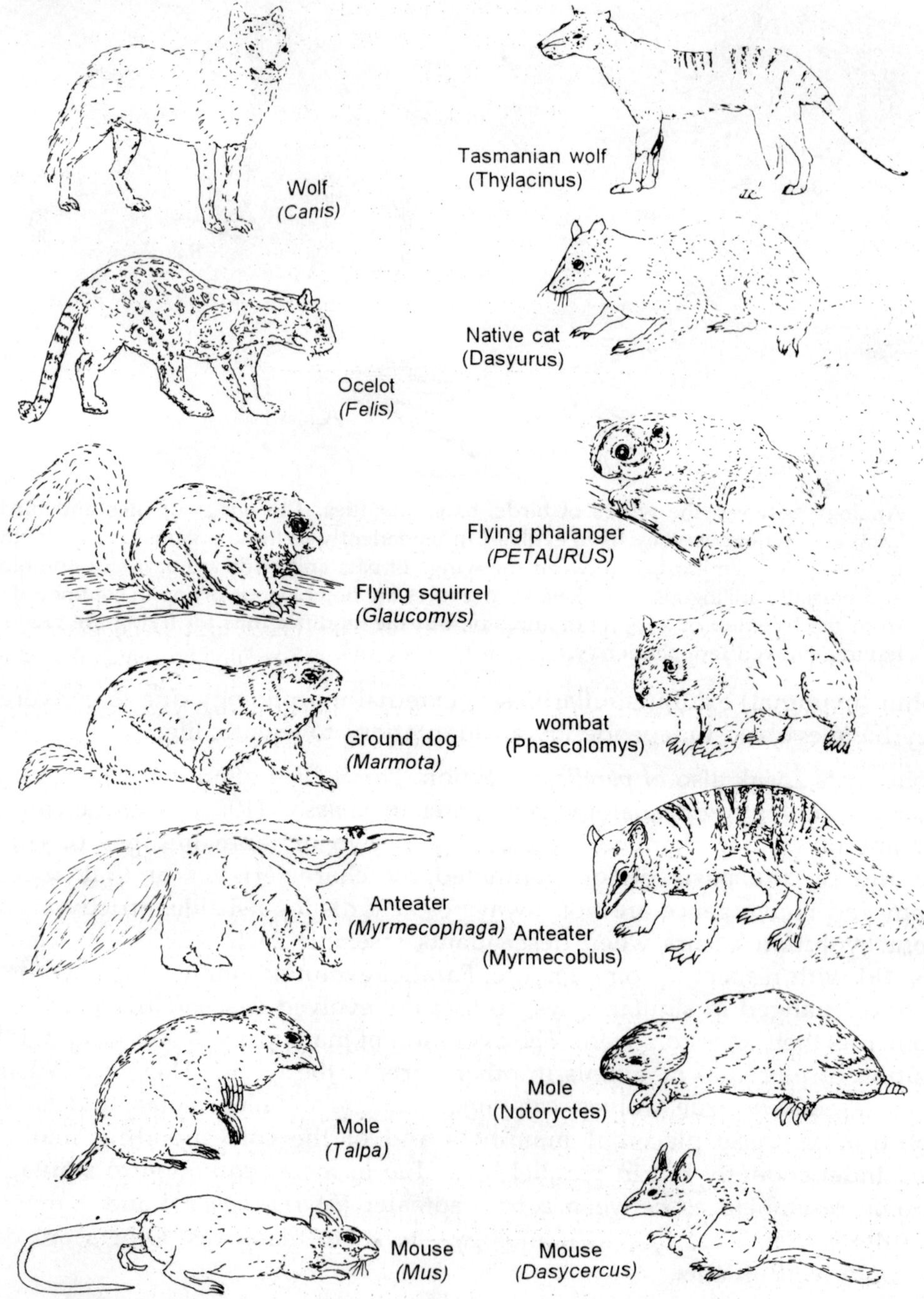

Fig. 10.3: Parallel evolution of placental and Australian marsupial mammals. [After Simpson and Beck, 1965.]

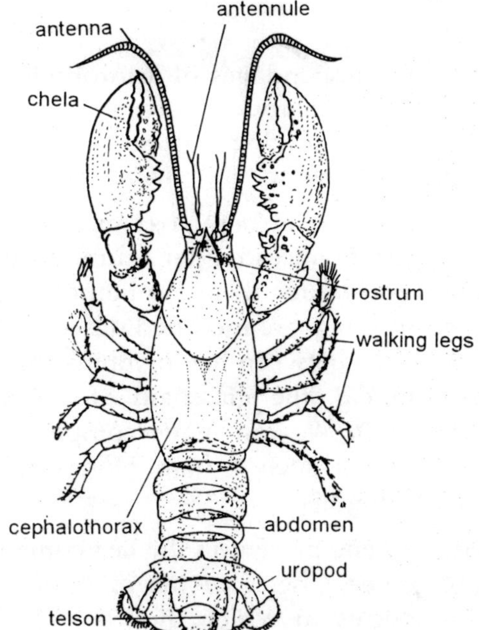

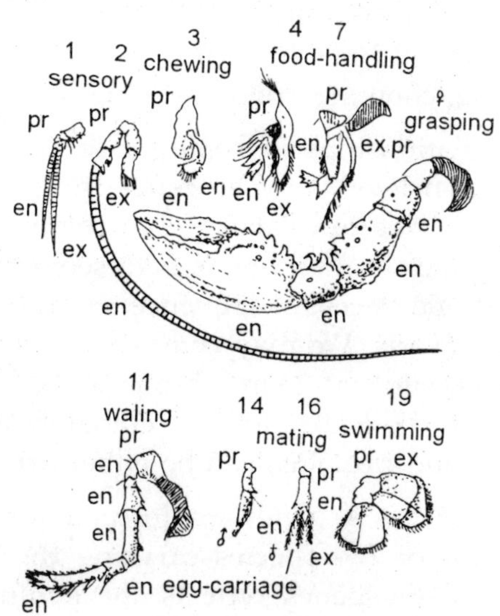

Fig. 10.4: Serial homology in the crayfish. The appendages are all modified versions of a basic pattern. This pattern has a basal structure, called the *protopodite* (*pr*), which may consist of more than one segment. There is also a two-fingered structure. The "finger" nearest the midline of the body is called the *endopodite* (*en*), and the lateral one is called the *exopodite* (*ex*); each consists of one or several segments. The appendages have derived through modifications, repetitions, and occasional loss of the elemetns of the basic pattern.

flowering plants adapted to the arid conditions of the desert. Their cactus-like morphologies have evolved independently as responses to similar environmental challenges. Cacti and cactus-like euphorbias are instances of convergent evolution.

Parallel and convergent evolution are particularly common in the reproductive structures of flowering plants. Because fossil remains of these structures are rare, it is often difficult to ascertain whether convergent or parallel evolution rather than homology is involved. A characteristically difficult situation occurs in the group of families having bell-shaped or tubular flowers with joined petals. Heaths (*Erica*), primroses (*Primula*), phloxes, mints (Labiateae), figworts (Scrophulariaceae), gentians, bluebells (*Campanula*), honeysuckles (*Lonicera*), sunflowers (Compositae), and others were all at one time grouped together in the superorder Sympetalae. Although some of these flowers may be strictly homologous, botanists are now certain that others, perhaps the majority, represent parallelisms or convergences.

Evaluation of Homologies

So far we have discussed homology between different organisms. A different kind of homology may be recognized, namely between repetitive or serial structures of the same organism. This has been called *serial homology*. Serial homology exists, for example, between the arms and legs of man, or among the seven cervical vertebrae of mammals, or among the branches or leaves of a tree. The jointed appendages of arthropods are elaborate examples of serial homology. Nineteen pairs of appendages exist in the crayfish, all built according to the same basic pattern but serving diverse functions—sensing, chewing, food-handling, walking, mating, egg-carrying, and swimming.

Relationships in some sense akin to those between serial homologs exist at the molecular level between genes and proteins resulting from ancestral gene duplications. The α, β, γ, and δ hemoglobin chains and the genes coding for them are homologous, similarities in their amino acid sequences occur because they are modified descendants of a single ancestral sequence. Homologous comparisons between hemoglobins can be of two kinds. We may compare the same hemoglobin, e.g., the alfa chain, of different species, such as man, chimpanzee, dog, and snake. Or we can make comparisons between the α, β, γ and δ hemoglobins of a single species, such as man. Homologous genes (and proteins) can be either orthologous or paralogous.

Orthologous genes are descendants of an ancestral gene present in the last common ancestor of the species carrying the orthologous genes. Therefore the evolution of orthologous genes reflects the evolution of the species in which they are found. Comparisons among orthologous genes serve to establish species phylogenies. For example, comparison of the α hemoglobins of different mammals leads to inferences about mammalian phylogeny. *Paralogous* genes are descendants of a duplicated ancestral gene. The evolution of paralogous genes reflects divergence that has accumulated since the genes duplicated. Homologies between paralogous genes of the same species, e.g., the α, β, γ and δ hemoglobins in man, serve to establish gene phylogenies, i.e., the evolutionary history of duplicated genes within a given lineage. For example, the duplication leading to the α and β hemoglobins of man occurred further in the past than the duplication giving rise to the β and δ hemoglobins.

In establishing gene phylogenies comparisons should be made between paralogous genes (or proteins) of the *same* species. For example, myoglobin and the α and β hemoglobin chains of man should be compared with each other, and so on. Data otbained from different species should corroborate each other, since the phylogeny of duplicated genes is the same in all species carrying the same set. Differences between paralogous genes in different species (say the α hemoglobin in man and the β hemoglobin of dogs) reflect the phylogeny of the duplicated genes as well as the evolution of the species.

Whether or not the similarities between features of different organism are due to homology cannot always be decided unambiguously, but may be a matter of probability. To determine phylogenetic relationships we must be able to distinguish similarities that are homologous from those that are analogous or simply accidental. Moreover, degrees of homology must be quantified in some way in order to determine the propinquity of common descent among species. Difficulties also arise in determining the degree of homology among various organisms. Consider the five forelimbs shown in Figure 9.1. Are the homologies greater between man and the dog than between man and the whale; or between man and the whale than between man and the chicken? The fossil record may sometimes provide the appropriate information, but as we have noted the fossil record is far from complete. The evidence from comparative studies of living forms and paleontological evidence must be examined together.

The comparative studies of proteins and nucleic acids presented later in this chapter provide quantitative estiamtes of the degree of homology between organisms. Methods used to distinguish homology from accidental similarity are discussed in this section. In general, two complementary criteria help to distinguish homology from other similarities (Simpson, 1961). (1) With respect to any given feature (a limb, organ, behavioural pattern, protein, etc.) *minuteness of resemblane* suggests homology. (2) The greater the *number of similar features* between any two organisms, the more likely it is that homology, and therefore common ancestry, is responsible for the similarities of any one feature.

The detailed correspondence between the parts of the forelimbs of man, dogs, horses, chickens, and whales, indicate that they are homologous. In all these forelimbs there are scapulas, humeri, radii, ulnas, carpals, metacarpals, and phalanges. The number of digits is five in man, dogs, and some whales, three in chickens, and only one in horses. But the correspondence in the structure and position of the forelimb bones is too precise to be accounted for by anything but common ancestry. Comparisons of other features, for example of the rest of the skeleton, show that many other traits are quite similar in the five animals. By comparing multiple features, the relative degree of similarity among all five animals can be established. For example, it becomes apparent that man, dogs, horses, and whales are more similar to each other than any of them is to the chicken.

Inger (1958) has shown that simple streamling and oral suckers have evolved independently in separate groups of tadpoles living in fast streams. Nevertheless, characteristic details of the oral discs permit identification of the phylogenetic groups to which the tadpoles belong. The homologies are more numerous and detailed among tadpoles of the same phylogenetic group than among tadpoles belonging to different groups but living in torrents. Examination of the detailed structure of convergent features usually makes it possible to detect analogy because resemblance rarely extends to the fine details of complex traits.

As a rule, similarities due to analogy, and in general to parallel or convergent evolution, lack the detailed correspondence of parts observed in cases of homology. Moreover, the number of similar features is limited. The Tasmanian world (*Thylacinus*) and the true wolf (*Canis*) resemble each other, but their similarities are related to a particular habit of predatory adaptation. For characteristics not related to that adaptation, *Canis* exhibits greater and more detailed similarities with the cat (*Felis*) or the squirrel (*Glaucomys*) than with the Tasmanian wolf. Bats, birds, and flies have evolved wings as adaptations to flying, but the correspondence in the morphology of the wings is far from precise. Moreover, bats, birds, and flies are quite different in many other features.

Assume that a group of organisms, say mammals, have been studied with the purpose of deciding their most likely phylogeny. A great variety of homologous features have been compared. How do we evaluate all this information, which might be conflicting in some cases? Traditionally, comparative anatomists and taxonomists have found from experience that within a group of organisms some features are more reliable indicators of phylogeny than others. That is, not all homologous features are weighted equally. An element of subjectivity is thus introduced; experts do not always agree on what traits should be given greater weight. To circumvent this problem, some taxonomists have suggested that as many characters as possible be studied and all be given equal weight. The data are then processed and "objective" taxonomic classifications are otbained. These procedures and the theory behind them are known as "phenetics" or "numerical taxonomy". The methods of numerical taxonomy yield "dendrograms"—organisms are connectede in a branching arrangement in which organisms or groups of them are connected at nodes. Whether dendrograms should be interpreted as phylogenies depends on whether it can be assumed that the traits under consideration have evolved at approximately constant rates.

Numerical taxonomy, cladistics, and evolutionary taxonomy differ not only in their methods of classification, but also in the theories behind the methods. Numerical taxonomy has been severely criticized by some leading taxonomists and evolutionists, e.g., Simpson (1961) and Mayr (1969). Many taxonomists and comparative anatomists prefer to give different weight to different traits on the basis of accumulated knowledge and their experience rather than rely on the impersonal manipulations of data made by a computer. However that may be for anatomical, behavioural, and certain other traits, there is little doubt that the methods of numerical taxonomy are appropriate to evaluate the data otbained from the study of a single protein or DNA sequence.

Determination of phylogenetic relationships on the basis of anatomical, embryological, behavioural, and ecological knowledge requires intensive study of the biology of the organisms. There are specialized treaties in which such studies are presented, and we shall not review them here. In the following sections we shall, however, summarize certain methods that can be applied without much knowledge of the biology of the organisms under study. One method is based on the linear arrangement of genes in polytene chromosomes; its application is limited to Diptera. The other methods are based on the direct study of DNA and proteins. For the most part these latter methods have been developed only recently. They are powerful methods to ascertain phylogenetic relationships. Needless to say, decisions concerning phylogeny should be made on the basis of all available information, including biochemical, chromosomal, anatomical, embryological, ethological, ecological, biogeographical, and paleontological knowledge.

Chromosome Phylogenies

Changes in the number and shape of chromosomes have occurred in evolution through chromosomal fusions, fissions, translocations, inversions, duplications, and deletions. Number, size, and shape of chromosomes have been used to estalish phylogenetic relationships among related species. The giant polytene chromosomes of Diptera exhibit sequences of individually recognizable bands that permit detection of inversions and other chromosomal rearrangements. Ancestral relationships can sometimes be inferred from chromosomal rearrangements, particularly those involving *overlapping* inversions.

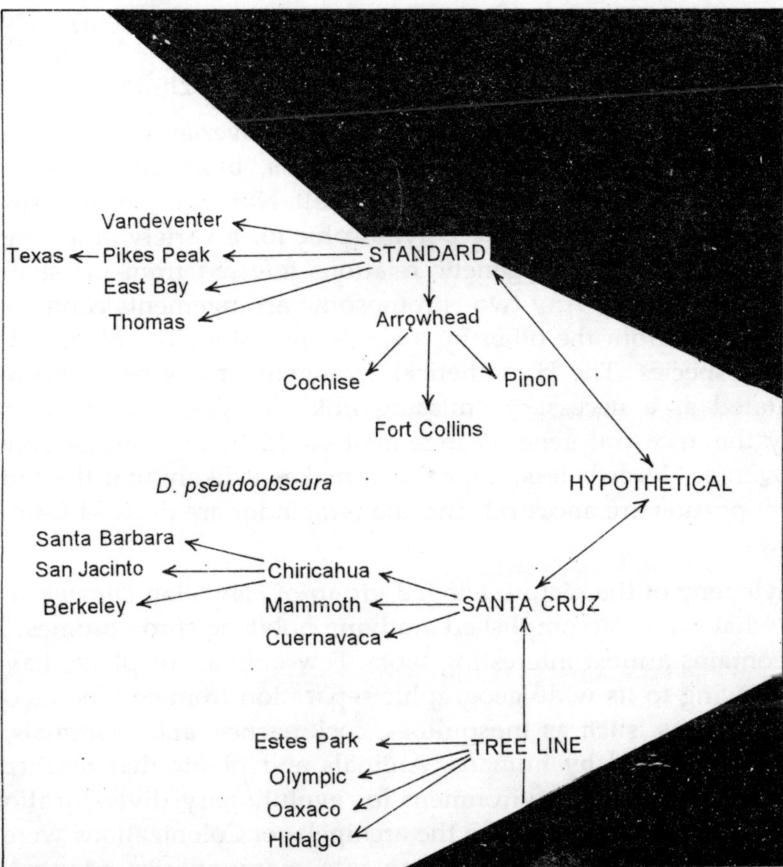

Fig. 10.5: Phylogeny of the gene arrangements in third chromosomes of *Drosophila pseudoobscura*, *D. persimilis*, and *D. miranda*. The phylogenetic relationships are inferred from the sequence of bands in polytene chromosomes of the salivary glands. The Standard cromosomal arrangement is found in both *D. pseudoobscura* and *D. persimilis;* all other sequences exist in only one species. The Hypothetical arrangement has never been found, but is postulated as a "missing link", perhaps extinct, between Standard and Santa Cruz. The pattern suggests that the ancestral arrangement is likely to be one of the four: Standard, Hypothetical, Santa Cruz, or Tree Line.

Overlapping inversions are chromosomal inversions that include part of previously inverted chromosome segments. Suppose that the sequences of bands in three different homologous chromosomes are ABCDEFGH, AEDCBFGH, and AEDFBCGH. The second sequence can arise from the first, or the first from the second, by a single inversion involving the segment BCDE. Similarly, the third sequence can arise from the second, or vice versa, by a single inversion of the segment CBF. However, the third sequence cannot arise from the first, nor the first from the third, by a single inversion. The second sequence represents a necessary intermediate step between the first and the third. If no other information is available, there is no way of knowing which one of the three sequences may have been the ancestral one, but only three phylogenetic relationships are possible: 1→ 2 → 3 or 3 → 2 → 1, or 1← 2 ← 3. Any phylogenies involving the direct transitions 1 → 3 or 3 → 1 can be excluded.

Phylogenies have been established by means of overlapping invesions in organisms with polytene chromosomes, such as *Drosophila*, black flies (Simuliidae), midges (Chironomidae), and mosquitoes (Culicidae). Natural populations of *Drosophila* pssudoobscura and *D. persimilis* are polymorphic for a variety of arrangements in their third chromosomes. The phylogenetic relations inferred from the sequences of bands are shown in Figure 10.5. Any two chromosome arrangements connected by an arrow can be derived one from the other by a single inversion. The Standard arrangement is found in both species. The Hypothetical arrangement has never been found but has been postulated as a necessary "missing link" between Standard and Santa Cruz. Conceivably the ancestral gene arrangement could be any one or some other one as yet undiscovered. Nevertheless, the pattern makes it likely that the four arrangements in the central portion are ancestral, and the remainder are derived from them as shown by the arrows.

The phylogeny of the picture-winged group of Hawaiian *Drosophila* is a remarkable example of what can be accomplished studying polytene chromosomes. The archipelago of Hawaii contains a most interesting biota. Few animals or plants have colonized the archipelago, owing to its wide geographic separation from continents or islands. Some groups of organisms, such as mosquitoes, cockroaches, and mammals, never reached Hawaii until introduced by humans. Animals and plants that reached the Hawaiian islands found a propitious environment for evolutionary diversification; much of the native flora and fauna is endemic to the archipelago. Colonizations were often followed by "adaptive radiations"—diversification into many species adapted to the various conditions existing in separate islands.

The total area of the archipelago is some 16,000 square kilometers, about half the size of Belgium or less than one-twenieth the size of California. Yet Hawaii contains around 500 endemic species of drosophilids, nearly one-third of the number known for the entire world. Greater morphological and ecological diversity exists in Hawaiian *Drosophila* than in the rest of the world. A phylogeny of 101 species based on chromosomal arrangements. The chromosomes do not provide information about the direction of ancestor-descendant relationships. Any one of the species in the diagram could be the ancestor from which all others derived through gradual accumulation of overlapping inversions. However, *D. primaeva,* is probably the most similar to the ancestral species from which the species have derived through the steps indicated in the diagram. The chromosomal arrangements of *D. primaeva* show several homologies with continental species of the *D. repleta* group, from which Hawaiian *Drosophila* are thought to be derived.

Table 10.1: Endemic flora and fauna in the Hawaiian archipelago. (After Carlquist, 1970, and other sources.)

	Genera		*Species subspecies, varieties*	
	Total number	*Percent endemic*	*Total number*	*Percent endemic*
Ferns	37	8%	168	65%
Flowering plants	216	13	1729	94
Land molluscs	37	51	1064	99+
Insects	377	53	3750	99+
Drosophilids*	4	75	510	100
Birds	41	39	71	99

*Species introduced by man are not included.

Comparisons of the chromosomal arrangements found in different islands of the Hawaiian archipelago have helped elucidate the pattern of colonizing migrations between islands. A minimum of 22 inter-island migrations are postulated to account for the evolution of the picture-winged species. For the most part, colonizations have occurred from geological older to younger islands. Hawaii, the most recently formed island, has received colonizers from all other large islands, but it has not been the origin of any colonizing migrant. Kauai, the oldest of the larger islands, has been the source of colonizations to all other large islands, but has received only one successful colonization, from Maui.

Methods of "Hybridizing" DNA

Genetic information is encoded in the nucleotide sequence of DNA molecules. The degree of genetic differentiation between two species may be measured by the proportion of nucleotide pairs that are different in their DNA. This can be accomplished by making hybrid molecules between single strands of DNA from different organisms. Doty and his colleagues discovered that the two strands of DNA could be dissociated and reassociated in vitro, and succeeded in obtaining hybrid molecules made from strands coming from two different species of bacteria or viruses. It was soon realized that these techniques could be used to measure DNA differentiation between species.

Two different measures of DNA differentiation can be obtained by hybridization: (1) the extent of the hybridization reaction, *i.e.,* the fraction of the DNA of two species

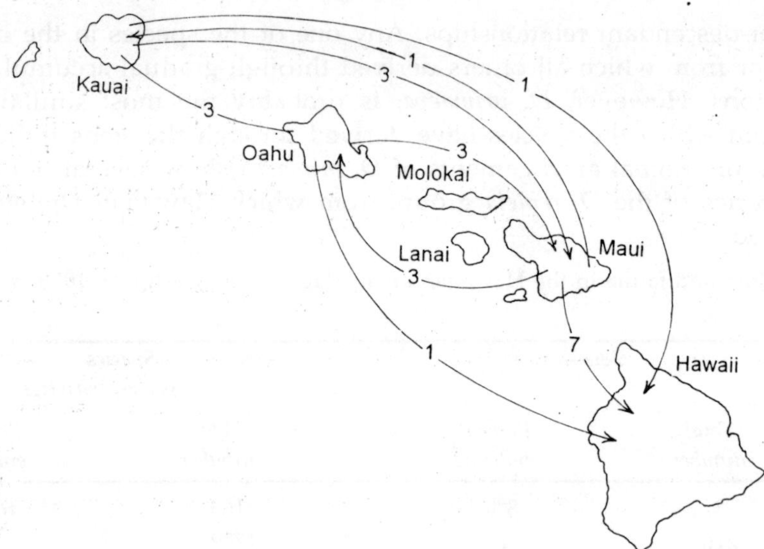

Fig. 10.6: Minimum number of colonizations postulated to account for the evolution of the picture-winged *Drosophila* species of Hawaii. The arrows indicate the direction of migration; the numbers in each arrow indicate the minimum number of separate colonizations postulated in each direction.

that forms hybrid molecules; (2) the proportion of nucleotide pairs that are complementary in the hybrid molecules. Single strands of DNA may reassociate into double-stranded molecules even though not all nucleotide pairs are complementary. The proportion of noncomplementary nucleotides is estimated by observing the temperature at which the two strands of the hybrid DNA molecule separate. A brief description follows of the methodology used to determine the degree of homology between DNA of different species.

Double-stranded DNA can be denatured or "melted" into single strands by heating it to about 100°C. The heat breaks the hydrogen bonds between the two complementary strands. If the solution is rapidly cooled, the strands remain separated. One common technique to measure the proportion of DNA sequences that are homologous in two species starts by trapping the single-stranded polynucleotide segments of one species, A, in a homogeneous matrix such as agar or nitrocellulose membrane filters. The agar or filter is sheared into small pieces containing single-stranded DNA fragments.

The filter-or agar-bound DNA of species A is then incubated at temperatures around 60°C in a solution containing single-stranded DNA segments of the same species, A, and a second species, B. The unbound DNA of species A and B in solution has been previously denatured and sheared into small segments about 500 nucleotides in length. The free single-stranded DNA of species A is labelled with a suitable radioactive isotope, such as tritium, ^{3}H, or phosphorous, ^{32}P. The solution contains a small constant amount of the radioactive DNA of species A, but the amount of DNA of species B is varied in separate experiments. The solution is incubated for several hours to permit association between the free and the abound DNA. The remaining free DNA is then washed out. The amounts of free DNA of species A and B that have formed duplexes with the bound DNA can then be determined, since the DNA of species A is radioactive.

The results of a typical experiment are shown in Figure 10.7. Denatured DNA from *Drosophila melanogaster* was immobilized in a nitrocellulose membrane filter, later cut into circular pieces 7 mm in diameter. The filters holding DNA were incubated in a solution containing 1 μg ($=10^{-6}$ grams) of ^{3}H-labeled DNA from D. *melanogaster,* and variable amounts of DNA from some other species. In the control experiments the two DNAs in solution are from *D. melanogaster*. As the amount of unlabeled DNA increases, the amount of radioactive DNA that pairs with the filter-bound DNA decreases and gradually approaches zero.

When the solution contains 1 μg of labeled DNA from *D. melanogaster* and variable amounts of DNA from *D. simulans,* the results are different . As the amount of *D. simulans* DNA from *D. melanogaster* pairing with the filter-bound DNA gradually decreases to about 20 per cent of the total paired DNA. But when the amount of *D. simulans* DNA in solution is increased beyond 100 mg, the proportion of labeled D. *melanogaster* pairing with the bound DNA is not reduced below 20 per cent. This implies that about 20 per cent of the filter-bound DNA of D. *melanogaster* has no complementary sequences in *D. simulans*. When the species tested is D. *funebris,* the proportion of labelled D. *melanogaster* DNA forming duplexes never decreases below 80 per cent. Therefore, about 80 per cent of D. *melanogaster* DNA has no homologous sequences in D. *funebris*. Armadillo and D. *melanogaster* DNA apparently have no homologous sequences.

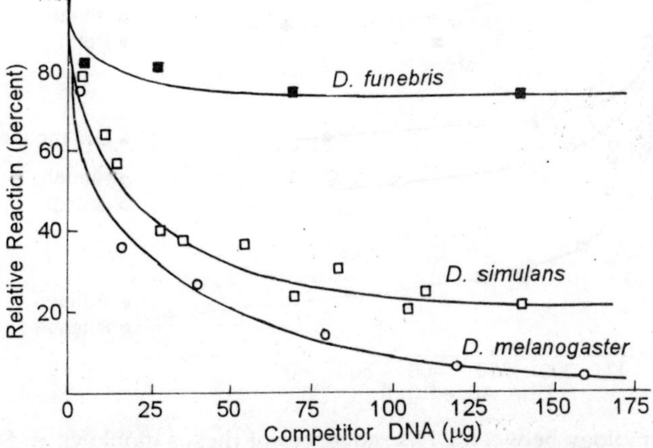

Fig. 10.7: Homology between the DNA sequences of *Drosophila melanogaster* and those of two other species. The experiment is carried out by hybridizing 1 μg of filter-bound DNA from D. *melanogaster* with 1 μg of radioactively labeled DNA from D. *melanogaster* and varying amounts of DNA from other species. All DNA's are first separated into single strands by heat denaturation.

The degrees of homology between DNA from rheus monkeys and other primate species are shown in Figure 10.7; 130 micrograms of agar-bound rhesus DNA were incubated with 0.5 μg of ^{32}P-labeled rhesus DNA and variable amounts of DNA from various species in turn. All the rhesus DNA molecules have homologous sequences in baboons; the baboon and rhesus DNA fragments pair equally efficiently with the agar-bound DNA from rhesus. The baboon and rhesus monkey cannot be differentiated by this technique. About 30 per cent of rhesus DNA sequences have no homologs in either chimpanzee or human DNA; about 50 per cent have no homologs in galago DNA, and so on.

Table 10.2 shows the results of an experiment using 0.5g of agar-bound human DNA incubated with 0.5 μg of ^{14}C-labeled human DNA and variable amounts of DNA from other species. Human and chimpanzee DNA are not distinguishable by this procedure. Man and chickens have 10 per cent homologous sequences. The proportion of DNA from other species that has homologous sequences with human DNA ranges between 10 and 100 per cent.

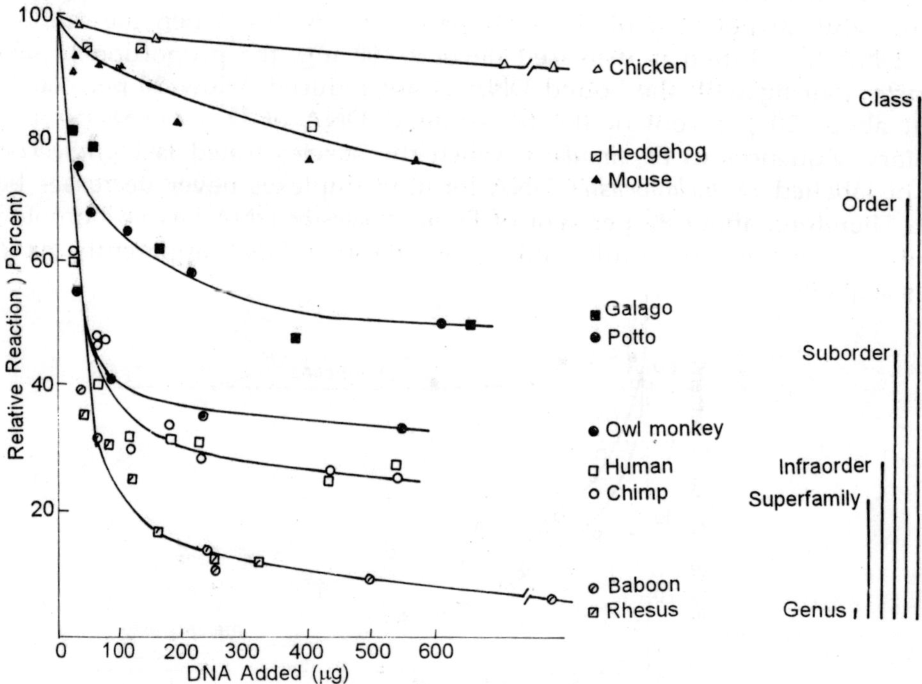

Fig. 10.8: Homology between DNA sequences of rhesus monkeys and those of various other species. The techniques used are basically similar to those in Figure 10.8. The amount of filter-bound DNA from rhesus monkey was 130 μg. The DNA in solution included half a microgram of labeled rhesus DNA and varying amounts of DNA from other species. The proportions of nonhomologous DNA are indicated by the bars at right, where the taxonomic separation between rhesus monkeys and the other species is indicated.

Table 10.2: Competition between 200 μg (or more) of DNA of various species and 0.5 μg of ^{14}C-labeled human DNA for binding with 0.5 g of agar-bound human DNA

	Degree of taxonomic differentiation	*Percent inhibition of human-human DNA binding*
Man		100%
Chimpanzee	Family	100
Gibbon	Family	94
Rhesus monkey	Superfamily	88
Capuchin monkey	Superfamily	83
Tarsier	Suborder	65
Slow loris	Suborder	58
Galago	Suborder	58
Lemur	Suborder	47
Tree shrew	Suborder	28
Mouse	Order	21
Hedgehog	Order	19
Chicken	Class	10

In the experiments described in the previous paragraphs the proportion of sequences forming duplexes with bound DNA depends on the temperature of incubation, the length of the reaction, and other experimental conditions. Moreover, the sequences forming duplexes need not be complementary for every nucleotide. Sequences form whenever a major fraction of the nucleotides are complementary; some noncomplementary nucleotides may be interspersed between homologous segments. The proportion of noncomplementary nucleotides in interspecific DNA duplexes can be determined by the rate at which the DNA strands separate at increasing temperatures.

The techniques used to determine the proportion of noncomplementary nucleotides in hybrid DNA duplexes are conceptually simple. DNA duplexes are formed using bound DNA as described above, or simply placing two types of single-stranded DNA in free solution for several hours to allow hybrid double strands to form. The hybrid DNA is then heated by increasing the temperature at a rate of 1°C every few minutes. The duplex DNA gradually dissociates into single strands that are collected at regular intervals. The proportion of duplex DNA dissociated at each temperature is then plotted. The proportion of noncomplementary nucleotides is determined by comparing the dissociation curves of hybrid DNA and control DNA duplexes, the latter formed by re-association of DNA strands from a single organism. The critical parameter called thermal stability (TS) is the temperature at which 50 per cent of the duplex DNA has dissociated. The difference (ΔTS) between the TS of hybrid DNA and the control is known to be approximately directly proportional to the proportion of unpaired

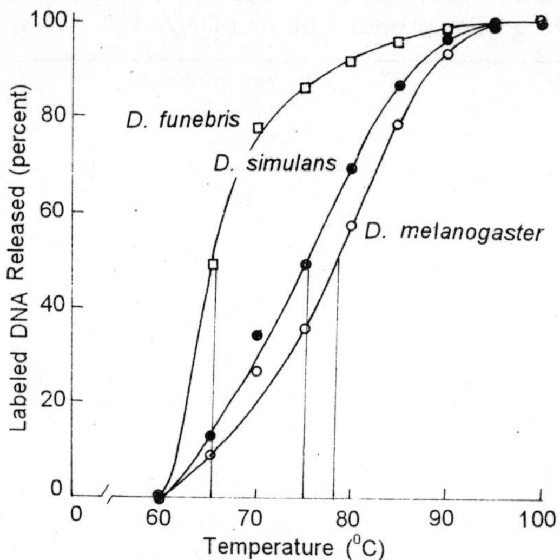

Fig. 10.9: Thermal stability profiles of DNA duplexes having one strand from *Drosophila melanogaster* and the other from the species indicated. The critical parameter called thermal stability (TS) is the temperature at which 50 per cent of the duplex DNA has dissociated. The difference (DTS) between the TS of hybrid DNA and that of nonhybrid DNA (D. *melanogaster* with D. *melanogaster*) corresponds approximately to the percentage of mismatched nucleotide pairs in the hybrid DNA duplex. The TS for the nonhybrid duplex DNA is 78°C, for the D.*melanogaster*-D. *simulans* duplex DNA 75°C, and for the D.*melanogaster—D. fundebris duplex* DNA 65°C. Thus, the proportion of nucleotide pairs different from D. *melanogaster* is three per cent for D. *simulans* and 13 per cent for D. *funebris*.

nucleotides in the hybrid DNA. For some time it was thought that each degree centigrade in ΔTS corresponded to 1.5 per cent mismatched nucleotide pairs. More recent experiments indicate that the correspondence may be 1°CΔTS= 0.65 per cent mismatched nucleotides. These estimates are, however, subject to substantial experimental errors. For the time being, one may assume that the correspondence is approximately 1°CΔTS=1 per cent mismatched nucleotides.

DNA Phylogenies

The thermal stability profiles shown in Figure 10.9 were obtained using duplexes between ^{3}H-labeled DNA from *Drosophila melanogaster* and DNA of other species bound to nitrocellulose membrane filters as described above (see Figure 10.8). The TS for homologous duplexes of *D. melanogaster* DNA is 78°C. The TS for *D. melanogaster*—D. *simulans* hybrid DNA is 75°C and ΔTS=3°C. For D. *melanogaster-D.* funebris hybrid DNA, TS=65°C and ΔTS–13°C (Laird and McCarthy, 1968). Therefore, the hybrid DNA's of D. *melanogaster* and D. *simulans* are noncomplementary in about 3 per cent of their nucleotides, while those of *D. melanogaster* and D. *funebris* have about 13 per cent mismatched nucleotides.

A phylogenetic tree of several genera related to the house sparrow (*Passer domesticus*) is shown in figure 10.10. Thermal stabilities were determined using first the slate-coloured junco (*Junco hyemalis*) and then the hermit thrush (*Catharus guttatus*) as references for DNA binding. The data from both sets of comparisons were combined to construct the phylogeny. The numbers in the tree branches are the differences in thermal stability (ΔTS), and thus represent approximately the per cent of nucleotide replacements that have occurred in the evolution of each branch.

Thermal-stability profiles can be used to determine rates of nucleotide change per unit time. This requires that the time of divergence of the phyletic lines be known from paleontological or other data. Table 9.3 gives estimates of the nucleotide differences between various primats including man. The corresponding phylogenetic tree and the per cent of nucleotide changes occurring in each branch are depicted in Figure 10.12. It is apparent in the last two columns of Table 9.3 that the rate of nucleotide change per unit time is not the same in the various lineages, even if we allow for some experimental error. For example, the nucleotide changes per year between man and galagos (5.2) or between the green monkey and galagos (5.0) are about three times as large as the nucleotide changes between man and chimpanzees (1.6) or between man and the gibbon (1.8). The rate of nucleotide change is more nearly the same in all lineages if it is calculated in terms of the generations elapsed rather than the number of years. The generation time is longer in large animals, such as man and the chumpanzee, than in small animals, such as the galago or the green monkey.

Up to the present time DNA-hybridization studies have provided limited information on phylogenetic relationships. The techniques are laborious and expensive, due in part to the need to label the DNA of one species with a radioactive isotope. Difficulties also arise because the results obtained are very sensitive to the experimental conditions used. Inconsistent results are not rare and are sometimes observed in a

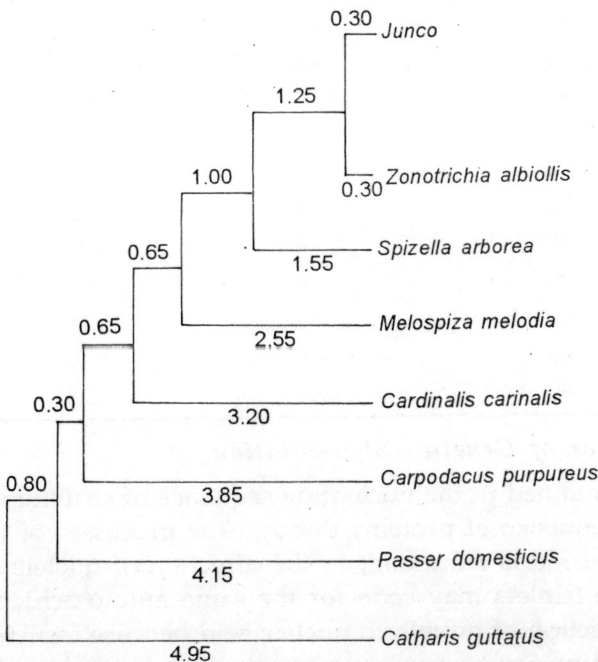

Fig. 10.10: Phylogeny of several species of birds related to the house sparrow (*Passer domesticus*), based on thermal-stability profiles of DNA hybrid duplexes. The numbers on the branches are estimated changes in the percentage of nucleotide-pair substitutions that have occurred in evolution.

single set of experiments. *Hoyer* and colleagues (1972) measured ΔTS between man and orangutan. They first used labeled human DNA and unabeled orangutan DNA, and then labeled orangutan DNA and unlabeled human DNA. ΔTS was 2.9°C in the first experiment, but only 1.9°C in the second experiment, although indentical procedures were used in both cases.

Most recently, techniques have been developed to obtain the nucleotide sequence of the messenger RNA coded by specific genes. Although extremely laborious at present, these techniques will provide precise measures of the differentiation of specific genes in separate species. They may become an important phylogenetic tool of the future.

Table 10.3: Rates of nucleotide change in the evolution of primate DNA.

DNAs compared	*Millions of years since divergence*	*Percent nucleotide changes*	*Percent changes per million years**	*Nucleotide changes per year*
Man and:				
Man	0	0%	-	-
Chimpanzee	15	2.4	0.08%	1.6
Gibbon	30	5.3	0.09	1.8
Green monkey	45	9.5	0.11	2.1
Capuchin	65	15.8	0.12	2.4
Galago	80	42.0	0.26	5.2
Green monkey and:				
Green monkey	0	0	-	-
Rhesus	15	3.5	0.12	2.4
Man	45	9.6	0.11	2.2
Chimpanzee	45	9.6	0.11	2.2
Gibbon	45	9.6	0.11	2.2
Capuchin	65	16.5	0.12	2.4
Galago	80	42.0	0.25	5.0

Electrophoretic Measures of Genetic Differentiation

The information contained in the nucleotide sequence of structural genes is passed on to the amino acid sequence of proteins through the processes of transcription and translation. The genetic code is redundant in the direction of nucleic acid to protein—two or more nucleotide triplets may code for the same amino acid. The genetic code is ambiguous in the direction of protein to nucleic acid because individual amino acids may be coded for by any one of two or more nucleotide triplets. If the nucleotide sequence of a structural gene is known, the amino acid sequence (called the *primary structure*) of the coded protein is unambiguously determined. Knowledge of the primary structure of a protein does not allow unambiguous determination of the nucleotide

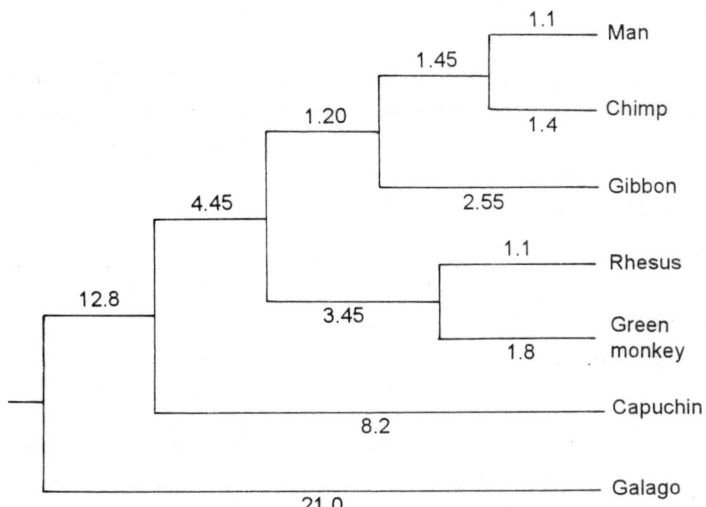

Fig. 10.11: Phylogeny of primate species, based on thermal-stability profiles of DNA hybrid duplexes. The numbers on the branches are estimated changes in the percentage of nucleotide-pair substitutions that have occurred in evolution.

sequence coding for it. Proteins different in their primary structures are encoded by different genes, although proteins with identical primary structures may or may not be coded by identical genes.

As already explained, proteins differing in their net electric charge can be separated by the techniques of gel electrophoresis. Not all amino acid changes in primary structure can be detected by gel electrophoresis, only those resulting in charge differences, and occasionally those changing the conformation (*secondary and tertiary structures*) of the proteins. For this reason, and also owing to the redundancy of the genetic code, not all allelic differences can be detected by electrophoresis of the corresponding proteins, but proteins differentiated by gel electrophoresis are always products of different genes. Thus, electrophoretic techniques underestimate the amount of genetic variation. With this important reservation, protein electrophoresis provides the means of estimating the amount of variation in natural populations. The rationale is that a number of genes can be selected for study without knowing a priori whether or not the genes are variable, or how variable they are. Therefore, a moderate sample of gene loci may be assumed to represent an unbiased or random sample of the genome of the organisms studied. Estimates of genetic variation obtained from the study of a few loci can be extrapolated to the whole genome.

The same rationale justifies the use of electrophoresis to estimate the degree of genetic differentiation between different populations. A moderate number of proteins is studied in two populations. the proteins are chosen without previous knowledge of whether or not they are different in the two populations. The average degree of genetic differentiation observed in such a random sample of proteins estimates the amount of genetic differentiation between the two populations over the whole genome. Gel electrophoresis is therefore a fairly simple method of measuring genetic differentiation between species. When several species are studied a dendrogram can be constructed that reflects the relative degrees of genetic differentiation among the species. Interpreting the dendrogram as a phylogeny requires the assumption that the amount of genetic differentiation averaged over a number of gene loci is approximately proportional to the time since their last common ancestor.

In recent years gel electrophoresis has been used to measure genetic differentiation among closely related species in various groups of organisms, from annual plants, through insects and other invertebrates, to fishes, reptiles, amphibians and mammals. These studies have provided much relevant, and in some cases critical, information about phylogenetic relationships.

The data obtained by electrophoresis are gene and genotypic frequencies for electrophoretically detecrable alleles. The gene and genotypic frequencies observed in two different species are transformed into measures of genetic distance using any one of a variety of statistical methods. Two statistics widely used are "genetic identity," I, and "genetic distance," D, calculated as follows. Let A and B be two different populations, and 1 a given locus. The normalized probability that two alleles, one from each population, are identical is

$$l_1 = \frac{\Sigma a_1 b_1}{\sqrt{\Sigma a_i^2 \Sigma_i^2}},$$

where a_i and b_i are the frequencies of the *i*th allele in populations A and B, respectively. I*i* equals one when two populations have the same alleles in identical frequencies, and zero when two populations have no alleles in common. When several gene loci are considered, the *mean genetic identity* is defined

$$I = \frac{J_{AB}}{\sqrt{J_A J_B}}$$

where J_{AB}, J_A and J_B are the arithmetic means, over all gene loci sampled, of $\Sigma a_i b_i$, Σa_i^2, and Σb_i^2 respectively. The value of *I* may range from zero (complete genetic differentiation) to one (complete genetic identity).

The *mean genetic distance,* D, between two populations is defined as

$$D = -\log_e l.$$

D can range in value from zero (no genetic differences) to infinity. If it is assumed that nucleotide substitutions within a gene occur independently from each other, and that the number of nucleotide substitutions per locus follows a Poisson distribution, D may be interpreted as a measure of the average number of electrophoretically detectable nucleotide substitutions that have accumulated in the evolution of two populations since they separated from a common ancestral one. The statistics *l* and D were used to measue the amount of genetic differentiation during the speciation process.

Electrophoretic Phylogenies

Assume that a set of loci have been studied by electrophoresis in each of a number of related species. The D values for all comparisons between pairs of species in the group can be used to construct a dendrogram that reflects genetic affinities among the species. Various mathematical methods may be employed to construct dendrograms based on measures of genetic distance. Different methods do not always yield identical dendrograms, although the results are generally fairly similar. Some of the numerical methods used to cluster species proceed, in essence, as follows. The two most similar species are first connected together as a cluster; the species most similar to the previous two is then added, and so on. To add a species to an existing cluster, one of two different criteria may be used: (1) the species to be added may be the one that is most similar to the *last* previously added species, or (2) one may choose the species that is the most similar, on the average, to *all* the species already clustered. There are methods for constructing dendrograms that do not require one to assume that evolutionary rates are homogeneous throughout a group of species. The latter methods are therefore preferable whenever dendrograms are intended to represent phylogenies.

The North American minnows are a group of fishes of the family Cyprinidae that consists of about 250 species. Genetic distances among nine species (each one classified in a different monotypic genus) of minnows from California are given in Table 10.4. The distances are based on electrophoretic studies of 24 gene loci. Two species, *hesperoleucus* and *Lavinia*, are genetically very similar, D=0.055. The two most different species are *Pogonichthys* and *Notemigonus*, D=1.118. Figure 10.12 is a phylogenetic tree of the nine species based on their genetic distances. The dendrogram indicates probable

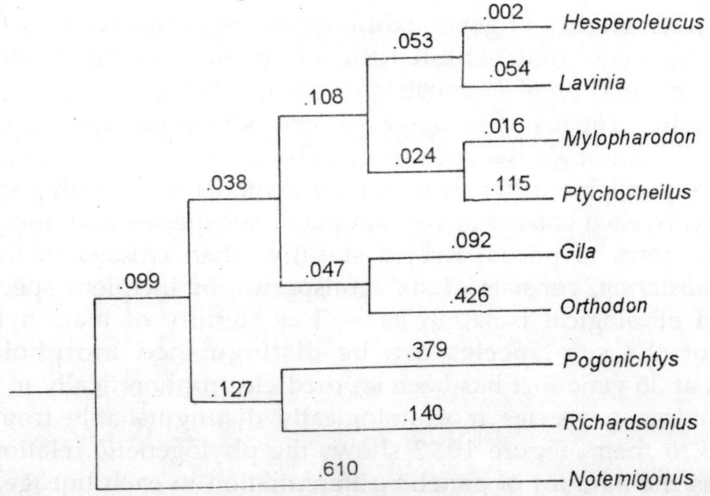

Fig. 9.12: Phylogeny of nine species of minnow fishes (each classified in a differcent monotypic genus), based on electrophoretic differnces at 24 gene loci coding for enzymes and other proteins. The numbers on the branches are estimated nucleotide substitutions (detectable by electrophoresis) per locus that have taken place in evolution.

phylogenetic relationships, as well as the amount of genetic change that occurred in each branch. The lineage going to *Notemigonus* experienced 0.610 electrophoretically detectable nucleotide substitutions per locus since it separated from the last ancestor common with the other eight species. The lineage leading to these latter species had 0.099 substitutions per locus before it split into two lineages, on giving rise to *Pogonichthys* and *Richardsoninus,* the other to the remaining six species, and so on. The tree shown in figure is based on the matrix of genetic distances given in Table 10.4. If we had electrophoretic data for additional species, the resulting tree would be not only more detailed, but also the relationships among the nine species in figure might be somewhat changed.

Table 10.4: Genetic distances (above diagonal) and similarities (below diagonal) between nine species of California minnows.

	1	2	3	4	5	6	7	8	9
1. *Hesperoleucus*	—	0.055	0.095	0.194	0.518	0.705	0.432	0.251	0.901
2. *Lavinia*	0.946	—	0.147	0.216	0.616	0.746	0.519	0.354	0.919
3. *Mylopharodom*	0.909	0.863	—	0.131	0.546	0.600	0.453	0.174	0.790
4. *Ptychocheilus*	0.824	0.806	0.877	—	0.541	0.600	0.526	0.333	0.989
5. *Orthodom*	0.596	0.540	0.579	0.582	—	1.079	0.776	0.518	1.094
6. *Pogonichthys*	0.494	0.474	0.549	0.549	0.340	—	0.519	0.679	1.118
7. *Richardsnius*	0.649	0.595	0.636	0.591	0.460	0.595	—	0.443	0.976
8. *Gila*	0.778	0.701	0.840	0.717	0.596	0.507	0.642	—	0.884
9. *Notemigonus*	0.406	0.399	0.454	0.372	0.335	0.327	0.377	0.413	—

Electrophoretic studies of genetic differences are particularly useful for establishing phylogenies of species that exhibit little or no morphological differentiation. The *D.willistoni* group of *Drosophila* consists of several closely related species endemic to the New World tropics. Six species are siblings, morphologically nearly indistinguishable, although the species of individual males can be identified by slight but diagnostically reliable differences in their genitalia. Two sibling species, D. *willistoni* and D. *equinoxialis,* each consist of two allopatric subspecies with incipient reproductive isolation in the form of partial hybrid sterility when crossed in the laboratory. One species, D. *paulistorum,* consists of six semispecies, or incipient species, with more or less developed ethological isolation as well as sterility of male hybrids. Neither the subspecies nor the semispecies can be distinguished morphologically. Genetic differentiation at 36 gene loci has been studied electrophoretically in six sibling species and in D. *nebulosa,* a species morphologically distinguishable from the siblings but closely related to them. Figure 10.13 shows the phylogenetic relationships among the taxa, as well as the amount of genetic differentiation in each lineage, estimated on the basis of the matrix of genetic distances. In this figure the vertical distances between neighboring taxa are approximately proportional to their genetic distances. The data in the matrix and in the figure also permit estimation of the amount of genetic differentiation at various stages of the speciation process.

To determine the phylogeny of a group, the results of electrophoretic studies should be incorporated with any other relevant evidence. Different studies provide complementary evidence, and are likely to agree on the whole. The taxa of the D. *willistoni* group have been intensively studied with respect to their reproductive affinities, chromosomal gene arrangements, sexual behavior, geographic distribution, ecology, and statistical differences in morphological parameters. The relationships shown in this figure agree very well with those in figure 10.13 based on electrophoretic data. One advantage of the genetic data obtained by electrophoresis is that the degree of differentiation between taxa or groups of taxa can be quantified.

The techniques of gel electrophoresis have been extensively used in recent years to study phylogenetic relationshps among related species in a large variety of animal and plant groups. This methodology is useful only when the taxa are genetically not

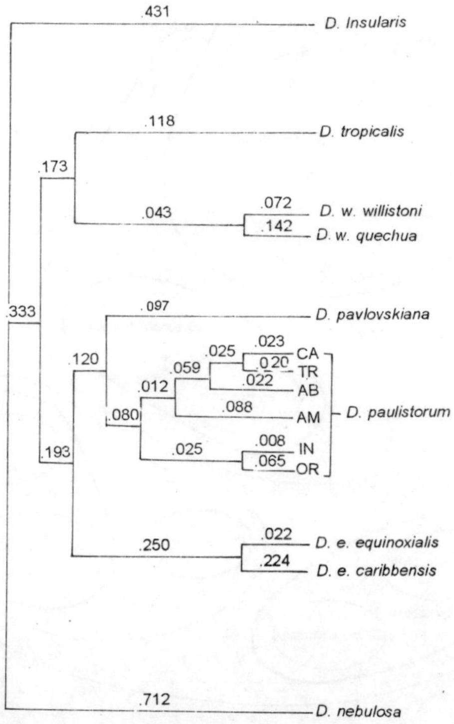

Fig. 10.13: Phylogeny of species related to *Drosophila willistoni,* based on electrophoretic differences at 36 gene loci coding for enzymes. The numbers on the branches are estimated nucleotide substitutions (detectable by electrophoresis) per locus that have taken place in evolution. The vertical distances between neighboring taxa are roughly proportional to their genetic differentiation. There are seven species in the phylogenetic tree. Two species, *D. Willistoni* and *D. equinoxialis,* consist of two sub-species each D. *paulistorum* represents a complex of six semispecies or incipient species, called Centroamerican (CA), Transitional (TR), Andean-Brazilian (AB), *Amazonian* (AM), Interior (IN), Orinocan (OR).

very different, *i.e.,* when species of the same genus or closely related genera are compared. In electrophoretic gels two forms of a protein are judged to be the same if they have identical migrations. Whenever two proteins migrate at different rates we know that they are structurally different but we do not know whether they differ in one, two, or several amino acid replacements. The average number of amino acid replacements per protein is estimated from the proportion of proteins that are identical in two species, by assuming that the number of replacements follows a Poisson distribution. That is, estimates of genetic distance between taxa are based on the zero

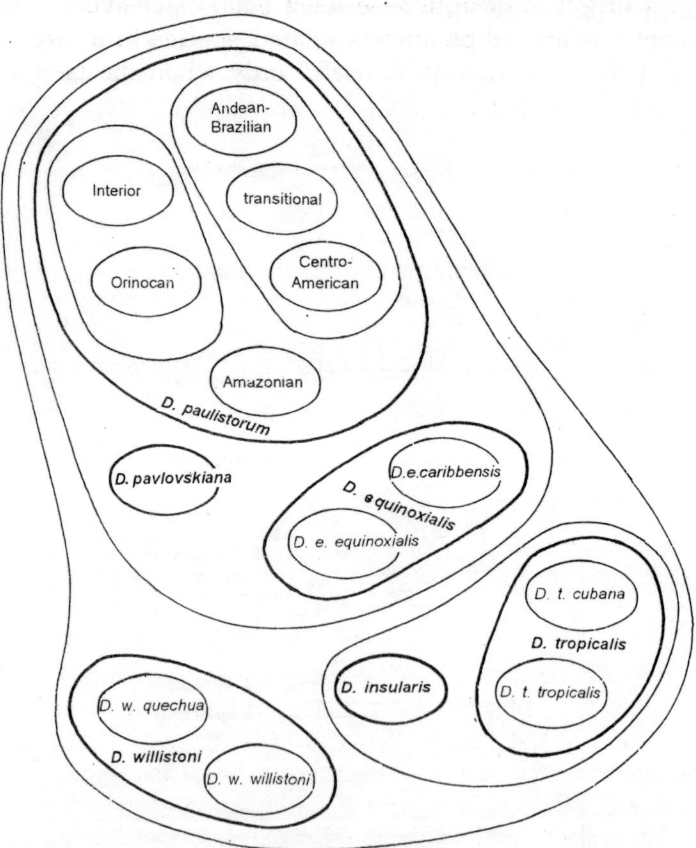

Fig. 10.14: Phylogenetic relationships of the sibling species of the *Drosophila willistoni* group, based on the study of reproductive affinities, chromosomal differences, sexual behaviour, geographic distribution, ecology, and morphometric differences. The diagram represents a cross-sectional view of phylogenetic branches . The phylogenetic relationships are very similar to those shown in figure 10.13 although they are based on different evidence. Note that this diagram includes two subspecies of D. *tropicalis*, but does not include D. *nebulosa.*

class of the Poisson distribution. If two taxa are electrophoretically different in every protein, the zero class is empty and no estimates can be made of the average number of amino acid substitutions per protein. If the proportion of proteins with identical electrophoretic mobility (the zero class in the Poisson distribution) is very small, estimates of genetic distance are unreliable because they are potentially subject to large errors. Moreover, when genetically very different taxa are compared it is likely that proteins with ostensibly identical electrophoretic mobilities may in fact differ in amino acid composition, since not all amino acid replacements are detectable by electrophoretic techniques. Other biochemical methods described below make possible determination of phylogenetic relationships among taxa not closely related.

Immunological Techniques

Estimates of the degree of similarity between proteins can be obtained by immunological techniques, such as immunoelectrophoresis, immunodiffusion, quantitative precipitation complement fixation, and turbidimetry. In outline, the immunological comparison of proteins is performed as follows. A protein, say albumin, is purified from an animal, say man. The purified protein is injected into a mammal, such as a rabbit. The rabbit develops an immunological reaction and produces antibodies against the foreign protein (antigen). The antibodies produced by the immunized rabbit will thereafter react not only against the specific antigen used, but also against other related proteins. The greater the similarity between the protein used to immunize the rabbit and the protein tested, the greater the extent of the immunological reaction. The degrees of dissimilarity between the albumin of man (or whatever protein used in the original immunization) and albumins from different species are expressed as "immunological distances."

An efficient and sensitive immunological method that requires only small amounts of purified protein is *microcomplement fixation*. "Complement" is a series of sequentially acting chemical substances found in vertebrate serum. When complement is added to antigens and antibodies under appropriate experimental conditions, it becomes fixed within the three-dimensional lattice of the antibody-antigen complexes. The amount of antigen-antibody reaction is measured by determining the amount of complement fixed in the reaction. Suitably prepared ("sensitized") red blood cells are added, and any complement not fixed by the antibody-antigen complexes is available to lyse the cells. The amount of lysed cells is determined spectrophotometrically. The number of lysed red blood cells is proportional to the amount of free (unfixed) complement. The greater the amount of lysed cells the lesser the extent of the antigen-antibody reaction. Immunological distances measured by microcomplement fixation are approximately proportional to the number of amino acid differences of related proteins.

Microcomplement fixation of various proteins has been used to determine immunological distances in a variety of organisms, including mammals, birds, and amphibians. Table 10.5 shows the immunological distances between man, apes, and Old World monkeys calculated from data by Sarich and Wilson (1967). Antibodies were prepared independently against albumin obtained from man (*Homo sapiens*), chimpanzees (*Pan troglodytes*), and gibbons (*Hylobates lar*), These were reacted against albumins obtained from seven species of apes and six species of Old World monkeys (cercopithecoidea: *Macaca mulatta, Papio papio, Cercocebus galeritus, Cercopithecus aethiops, Colobus polykomos,* and *Presbytis entellus*). The albumins of the two species of chimpanzee appear identical. The tests with antiserum prepared against man show, that albumins from the African apes (chimpanzee and gorilla) are more similar to human albumin than the albumins from the Asiatic apes (orangutan, siamang, and gibbon) albumins from the Old World monkeys are most different. The antiserum to chimpanzee produced similar results; chimpanzee albumin is more similar to the albumins of man and gorillas than to those of the Asian apes, while the Old World monkeys are most different. The antiserum to gibbon albumin indicates that the albumins of gibbon and siamang are very similar; it also indicates that orangutan albumin is not much more different from albumins of the other Asian apes than the albumins of the African apes.

Table 10.5: Immunological distances between albumins of various Old World primates.

Species tested	*Antiserum to*		
	Homo	*Pan*	*Hylobates*
Homo sapiens (man)	0	3.7	11.1
Pan troglodytes (chimpanzee)	5.7	0	14.6
Pan paniscus (pygmy chimpanzee)	5.7	0	14.6
Gorilla gorilla (gorilla)	3.7	6.8	11.7
Pongo pygmacus (orangutan)	8.6	9.3	11.1
Symphalangus syndactylus (siamang)	11.4	9.7	2.9
Hylobates lar (gibbon)	10.7	9.7	0
Old World monkeys (average of six species)	38.6	34.6	36.0

Lysozyme is anenzyme present in most animal species as well as in many plants and microorganisms. The immunological distances between lysozyme of man or the baboon and a variety of primates are shown in Table 10.6. Compared with human lysozyme, the lysozyme of chimpanzees appears to be identical (Immunological Distance=0), that of orangutans very similar (I.D.=1), but that of gorillas quite different (I.D.=32). These results are not fully consistent with those obtained with albumin, since the albumin of gorillas is more similar to that of man than to that of orangutans. Another anomaly in Table 10.6 is that the Old World monkeys appear more closely related to the New World monkeys than to man and the apes (see the last column in the table). Yet, there is ample evidence indicating that the phyletic line leading to the New World monkeys separated from the lineage leading to the Old World monkeys before that latter separated from the lineage leading to man and the apes. One more inconsistency in Table 10.6 involves the immunological distance between man and the baboon; when anti-human serum is used the I.D. is 126, but when anti-baboon serum is used it is 66.

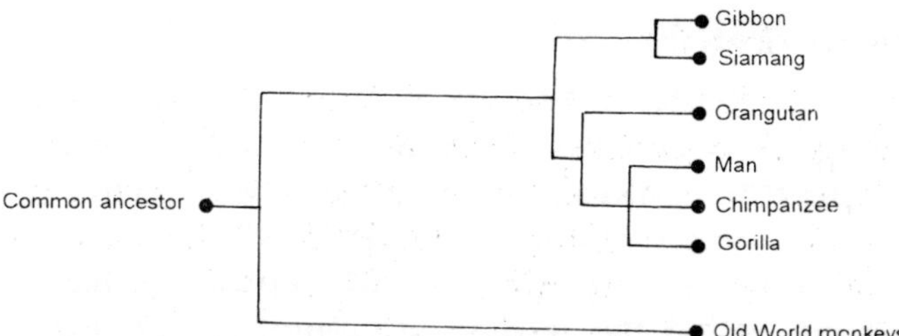

Fig. 10.15: Phylogeny of man, apes, and Old World monkeys, based on immunological differences between their albumin proteins. Man, the chimpanzee, and the gorilla appear more closely related to each other than either one of them is to the orangutan.

Table 10.6: Immunological distances between the lysozymes of man or baboon and those of various primates. The lysozyme was obtained from milk.

Species tested	*Anti-lysozyme to*	
	Man	*Baboon*
Homo sapiens (man)	0	66
Pan troglodytes (chimpanzee)	0	67
Pongo pygmaeus (orangutan)	1	76
Gorilla gorilla (gorilla)	32	38
Old World monkeys:		
Cercopithecus aethoips (green monkey)	93	6
Cercopithecus talapoin (talapoin)	114	3
Macaca mulatta (rhesus)	122	1
Macaca speciosa (stump-tailed macaque)	124	2
Macaca fascicularis (crab-eating macaque)	130	2
Macaca radiata (bonnet macaque)	131	2
Papio cynocephalus (baboon)	127	0
New World monkeys:		
Saimiri sciureus (squirrel monkey)	127	36
Saguinus oedipus (tamarin)	134	37
Callithrix jacchus (marmoset)	137	36

These inconsistencies do not invalidate the use of microcomplement fixation or other immunological methods for evolutionary studies. rather, the inference to be drawn is simply that phylogenies should be based on all available evidence, not just on one single trait. Similar inconsistencies occur in other kinds of studies. For example, if gross external morphology alone is considered, dolphins and seals might appear more closely related to some fishes than to terrestrial mammals. In order to establish phylogenetic relationships, data obtained from the immunological study of a given protein should be combined with immunological studies of other proteins, with other biochemical evidence, and with morphological, behavioural, and any other relevant information.

Amino Acid Sequences of Proteins

Proteins are polymer molecules made up of amino acids. The same 20 kinds of amino acids are found in all organisms. Proteins may consist of one, two, or more polypeptides. Each polypeptide is coded by one gene. The sequence of amino acids in a polypeptide is called its primary structure. Polypeptide chains have a helical structure, called the secondary structure of the protein. Polypeptide helices fold into three-dimensional configurations, called the tertiary structure. Under normal physiological conditions the various amino acids of a polypeptide interact through hydrogen bonds (H—H), disulfide linkages (S-S), and other kinds of bonds to produce a stable configuration. That is, the secondary and tertiary structures of a polypeptide are determined by its primary structure. Proteins consisting of more than one polypeptide have a *quaternary structure,* which refers to the three-dimensional topology of the asssociated polypeptides. Hemoglobin A, the most common form of hemoglobin found in human adults, consists of four polypeptides, two alpha and two beta chains. Many other proteins consist of two or more polypeptides.

The primary structure of a polypeptide is determined by the nucleotide sequence in the DNA coding for it. The number of differences between two homologous polypeptides or proteins reflects the number of differences in the corresponding genes. Genetic differentiation between species, and therefore probable phylogenetic relationships, can be inferred from the degree of differentiation in the primary structure of proteins.

The common procedure for establishing the amino acid sequence of polypeptide requires, first breaking the protein into small fragments of peptides, and then determining the amino acid sequence in each peptide. The polypeptide is broken into fragments with enzymes that hydrolyze the bonds between contiguous amino acids at specific sites. The resulting peptides are separated from each other using such procedures as column chromatography and two-dimensional paper chromatography. The amino acid sequence in each peptide is asscertained usually with the help of an apparatus known as a "protein sequencer" or "sequinator."

The procedure just described is carried out at least twice for each polypeptide to be sequenced, using two enzymes that hydrolyze the protein at different points. Trypsin and chymotrypsin are the most commonly used enzymes. Trypsin hydrolyzes the bonds between the carboxyl group of either lysine or arginine and the amino group of the contiguous amino acid. Chymotrypsin hydrolyzes polypeptides at the carboxyl ends of either tryptophan, phenylalanine, or asparagine. Two different sets of peptides are obtained with trypsin and chymotrypsin. The complete sequence of the protein is determined from the overlaps between the amino acid sequences of the two sets of peptides.

The first protein to be sequenced was insulin, which consists of 51 amino acids. During the early 1950s it was shown that the amino acid sequences of the insulins of cattle, pigs, sheep, horses, and sperm whales are identical except for replacements in three consecutive amino acids. The procedures to establish the primary sequence of proteins are extremely laborious. Yet more than 500 sequences or partial sequences are presently known through the efforts of scores of investigators working in many laboratories. Many additional protein sequences are determined every year.

Evolutionary changes in proteins may involve not only amino acid replacements but also additions and/or deletions of amino acids. Homologous proteins may therefore differ not only in the sequence of amino acids but also in the length of the polypeptides. The alpha and beta chains of human hemoglobin are homologous—they are coded by genes that have arisen by ancestral duplication. Yet the alpha chain consists of 141 amino acids, the beta chain of 146 amino acids. In order to determine the degree of similarity between homologous proteins, the possible occurrence of additions and deletions must be taken into account. The similarities between the alpha and beta hemoglobin chains are maximized if they are aligned over 148 positions. Gaps are assumed to exist at positions 2, 48, and 56 to 60 in the alpha chain, and at positions 19 and 20 in the beta chain. Each gap may be due to either a deletion in one chain or an addition to the other.

The occurrence of additions and deletions of amino acids in homologous proteins raises a problem. A typical protein consists of about 100 or more amino acids. Each of the 20 common amino acids is likely to occur several times in any given protein. If we had the freedom to place gaps anywhere, any two proteins would be likely to have identical amino acids at a number of sites whether or not the proteins were homologous. Several methods have been devised to decide whether similarities in amino acid composition between two proteins are due to accidental coincidence or to homology.

One method makes all possible comparisons between fragments (spans) of two proteins, and uses statistical procedures to determine whether or not the amount of similarity is greater than that expected by chance alone (Fitch, 1966). The known sequence of each protein is divided into all possible spans of 30 consecutive amino acids. (The spans should be neither too long nor too short, and 30 amino acids have been found to be a convenient number). One such span would include all amino acids from position 1 to 30, a second from 2 to 31, a third from 3 to 32, and so on. A protein with x amino acids has x-29 such spans of length 30. For example, 141-29=112, and 146-29=117 spans of 30 consecutive amino acids are possible for the alpha and beta hemoglobin chains, respectively. Comparisons are made, with the aid of an electronic computer, between every span of one protein and every span of the other protein. The number of sites occupied by identical amino acids in both proteins is determined for all comparisons. If the overall number of similarities is greater than expected by chance alone, the proteins are considered homologous.

The degree of similarity between presumed homologous proteins is often so large that homology can be inferred without carrying out any statistical tests. Cytochrome *c* is a protein involved in cell respiration, and in higher animals and plants is found in the mitochondria. The amino acid sequences of the cytochromes *c* of 20 diverse organisms. The number of amino acids ranges from 103 in the bullfrog and the tuna to 112 in wheat; vertebrates generally have 104 amino acids. Homology is evident when all cytochromes *c* are aligned . The cytochromes *c* of man and the rhesus monkey are different only in position 66, where man has isoleucine but the rhesus monkey has threonine. Even evolutionary distant organisms share in common many amino acids, *e.g.*, man and baker's yeast have identical amino acids in 64 positions. All 20 sequences have the same amino acids in 20 of the 112 sites.

The degree of differentiation, or distance, between any two homologous proteins of known amino acid sequence can be measured in two ways. First, one may simply count the number of amino acid differences between two given sequences. For example, the cytochromes *c* of man and the rhesus monkey differ by one amino acid (at site 66), those of man and the horse by 11 amino acids (at sites 19, 20, 23, 54, 55, 58, 66, 68, 91, 97, and 100), and so on.

A second procedure, which yields more information than the first, makes use of the genetic code. The replacement of one amino acid by another may require, as a minimum, one, two, or three nucleotide substitutions in the corresponding DNA triplet. For example, in position 66 the cytochrome *c* of man has isoleucine, that of the rhesus monkey has threonine; the minimum number of nucleotide differences between the codons for these two amino acids is one. One the other hand, in position 20 man has methionine and the horse has glutamine; the corresponding codons differ by no less than two nucleotides. The minimum number of nucleotide differences is thus determined for each amino acid replacement between two proteins, and all such differences are added. This yields the minimum number of nucleotide changes that must have occurred in the separate evolutions of the two genes coding for the proteins compared. Table 10.7 gives the minimum numbers of nucleotide differences among the genes coding for the cytochromes *c* of 20 organisms. (Seventeen organisms are the same in Figure 10.17 and Table 10.7; the figure also includes gray whales, bullfrogs, and wheat, while the table includes donkeys, pigeons, and the yeast *Saccharomyces*.) The minimum number of nucleotide differences between the genes coding for two homologous proteins is sometimes called their *mutation distance*. The mutation distance between two genes is often greater than the number of amino acid differences between the corresponding proteins. Moreover, the actual number of nucleotide differences between two genes may be greater than their mutation distance, since the latter measures only the *minimum* number of nucleotide differences.

Whether the distance between two proteins is measured by the number of amino acid differences or by the number of nucleotide substitutions, a problem remains. The number of changes that have occurred in evolution may be greater than the number of present differences, since several intermediate conditions may have existed. For example, at a given position in a codon the changes might have been A→G→C→G. At present there would be only one difference, whereas three replacements would have occurred over time. Amino acid differences and mutation distances often underestimate the number of changes that have actually happened in evolution. The greater the time since two homologous genes or proteins had the last common ancestor, the greater on the average the error of estimation. However, methods exist to correct for this type of error.

Table 9.7: Minimum numbers of nucleotide differences in the genes coding for cytochromes c in 20 organisms. The nucleotie differences are inferred from the amino acid sequences of the cytochromes c.

	1	2	3	4	5	6	7	8	9	10	11	12	13	14	15	16	17	18	19
1. Man																			
2. Monkey	1																		
3. Dog	13	12																	
4. Horse	17	16	10																
5. Donkey	16	15	8	1															
6. Pig	13	12	4	5	4														
7. Rabbit	12	11	6	11	10	6													
8. Kangaroo	12	13	7	11	12	7	7												
9. Duck	17	16	12	16	15	13	10	14											
10. Pigeon	16	15	12	16	15	13	8	14	3										
11. Chicken	18	17	14	16	15	13	11	15	3	4									
12. Penguin	18	17	14	17	16	14	11	13	3	4	2								
13. Turtle	19	18	13	16	15	13	11	14	7	8	8	8							
14. Rattlesnake	20	21	30	32	31	30	25	30	24	24	28	28	30						
15. Tuna	31	32	29	27	26	25	26	27	26	27	26	27	27	38					
16. Screwworm fly	33	32	24	24	25	26	23	26	25	26	26	28	30	40	34				
17. Moth	36	35	28	33	32	31	29	31	29	30	31	30	33	41	41	16			
18. Neurospora	63	62	64	64	64	64	62	66	61	59	61	62	65	61	72	58	59		
19. Saccharomyces	56	57	61	60	59	59	59	58	62	62	62	61	64	61	66	63	60	57	
20. Candida	66	65	66	68	67	67	67	68	66	66	66	65	67	69	69	65	61	61	41

Protein Phylogenies

Matrices of genetic distances provide the data for constructing dendrograms using the clustering procedures mentioned earlier in this chapter, or other methods specifically designed to deal with primary structures of proteins. Overall, the relationships correspond fairly well with the phylogeny of the organisms as determined from the fossil record and other sources. There are disagreements, however. Chickens appear more closely related to penguins than to ducks and pigeons; the turtle, a reptile, appears more closely related to birds than to the rattlesnake; men and monkeys diverge from the other mammals before the marsupial kangaroo separates from some placental mammals. In spite of these erroneous relationships, it is remarkable that the study of a single protein yields a fairly accurate representation of the phylogeny of 20 organisms as diverse as those in the figure.

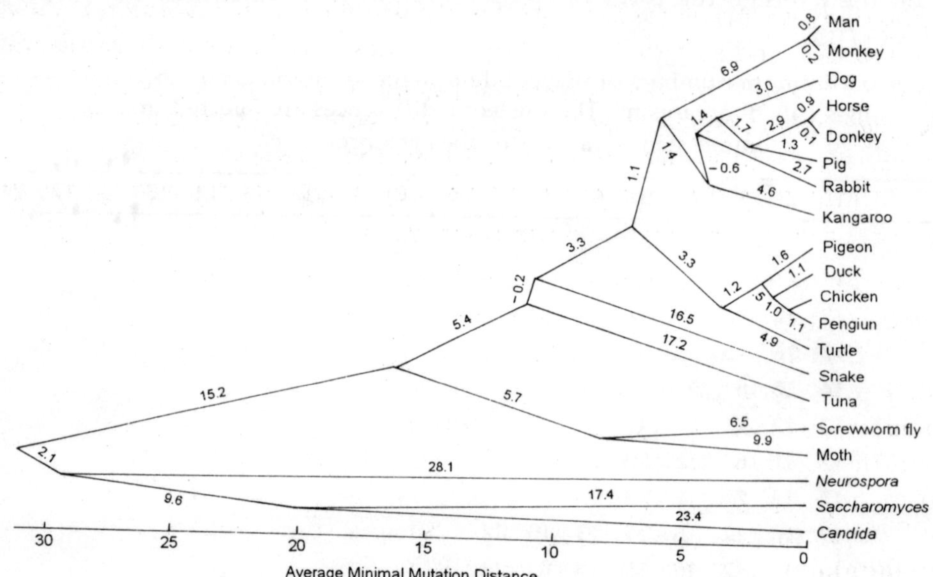

Fig. 10.16: Phylogeny of 20 organisms based on the cytochrome *c* matrix in Table 9.7. The phylogeny agrees on the whole fairly well with phylogenetic relationships inferred from the fossil record and other sources. This good agreement is remarkable since it is based on the study of a single protein and encompasses organisms ranging from yeast, through insects, fish, reptiles, ampnibians, birds, and mammals, to man. The numbers on the branches are the estimated number of nucleotide substitutions that have taken place in evolution.

Unfortunately, protein-sequencing is a very laborious procedure, and the number of known sequences is as yet limited. As more and more proteins are sequenced in more and more organisms, our knowledge of phylogeny will improve considerably. Once again, however, we must bear in mind that the information obtained from the study of protein sequences should be used in conjunction with all other kinds of information relevant to phylogeny.

Cytochromes *c* are slowly evolving proteins. Organisms as different as man, the silkworm moth, and baker's yeast have in common a large proportion of amino acids in their cytochromes *c*. The evolutionary conservatism of these cytochromes makes it possible to establish the degree of ancestral propinquity among organisms only remotely related. This same conservatism, however, makes cytochrome *c* useless for determining phylogenetic relationships among closely related organisms, since these may have cytochrome *c*, sequences that are completely or nearly identical. The primary structure of cytochrome *c* is identical in man and chimpanzee, which diverged 10 to 15 million years ago; it differs by only one amino acid replacement between man and the rheus monkey, whose most recent common ancestor lived 40 to 50 million years ago.

Fortunately, different proteins evolve at different rates. Phylogenetic relationships among closely related organisms may be inferred by studying the primary structures of rapidly evolving proteins, such as carbonic anhydrases and fibrinopeptides in mammals. Carbonic anhydrases are proteins physiologically important in the reversible hydration of CO_2 and in certain secretory processes. Two kinds of carbonic anhydrases, CA I and CA II, exist in all mammals including marsupials, while only one is found in fishes, birds, and invertebrates. The two carbonic anhydrases are the result of a gene duplication that must have occurred after the divergence of mammals and birds (some 300 million years ago), but before the divergence between marsupials and placental mammals (about 120 million years ago).

Mammal carbonic anhydrases consist of about 260 amino acids. The amino acids at 115 of the 260 positions in CA I have been identified in various primates including man. For these 115 amino acids the minimum numbers of nucleotide substitutions between man and each of five other species are as follows: chimpanzees, One; orangutans, four; vervet monkeys, six; rhesus monkeys, six; baboons, seven. The overall configuration of the dendrogram conforms well with the known phylogeny of these animals.

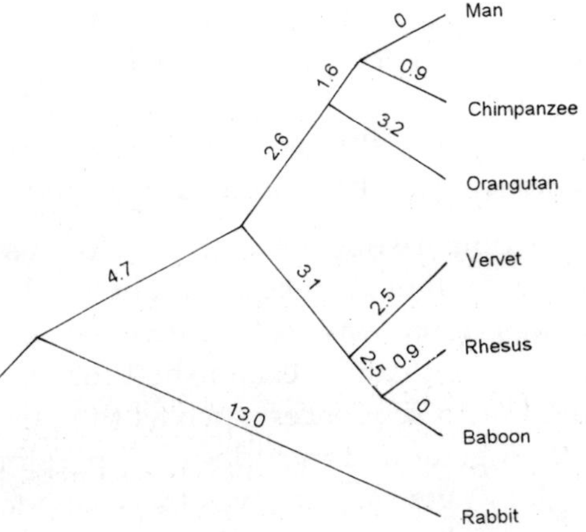

Fig. 10.17: Phylogeny of various primates, based on the sequence of 115 amino acids in carbonic anhydrase I. The numbers of the branches are the estimated numbers of nucleotide substitutions that have occurred in evolution. (After Tashian *et al.*, 1976.)

Fibrinopeptides are polypeptide segments cleaved from fibrinogen when blood clots. The fibrinopeptides are another example of fast-evolving molecules. They have been used to study the phylogeny of closely related organisms, such as the primates and the artiodactyls or cloven-hoffed mammals.

Phylogenetic relationships among species are determined by comparing the primary structures of orthologous, not paralogous, proteins (see above). Orthologous genes and proteins start their separate evolutions at the time when the corresponding species have their last common ancestor. Paralogous genes evolve independently of each other from the time when the gene duplication takes place. Erroneous inferences concerning phylogenies of species can be made if genes or proteins thought to be orthologous are instead paralogous. Assume, for instance that comparisons are made between the carbonic anhydrase CA II from man, CA II from rhesus monkeys, and CA I from chimpanzees. Among the 115 amino acids compared there are 85 differences between man CA II and chimpanzee CA I but only two between man CA II and rhesus CA II. If the three proteins were erroneously thought to be orthologous, man would appear much more closely related to the rhesus monkey than to the chimpanzee. The differences between man CA II and chimpanzee CA I have accumulated not since the time of divergence between the ancestors of man and chimpanzee, but since the duplication of the CA I and CA II genes about 200 million years ago. Grossly erroneous inferences would also be made in a phylogeny of mammals based on the amino acid sequence of the alpha hemoglobin chain of some species, the beta chain of some other species, and the delta chain of still others.

Some inconsistencies occasionally found in protein phylogenies may be due to comparisons between paralogous rather than orthologous proteins. For example, immunological studies of lysozymes show ducks and chickens closely related to each other, but very far removed from geese. This unreasonable relationship could be accounted for if the lysozymes studied in ducks and chickens were orthologous, but the lysozyme studied in geese were paralogous to the others. Recent evidence indicates that such may have indeed been the case.

Comparisons between paralogous genes or proteins serve to determine phylogenies of *genes* rather than species. The phylogeny of the globin genes produced by the various duplications was determined by comparisons between paralogous genes. Similar phylogenies can be established for other paralogous genes coding for proteins of known amino acid sequences, such as the carbonic anhydrases. It is in fact possible to construct a phylogeny that includes both paralogous and orthologous genes. Depicts a phylogeny of globin genes that gives the number of nucleotide replacements between orthologous genes of different species as well as between duplicated genes.

The Neutrality Theory of Protein Evolution

The biochemical techniques described in this chapter and the methods of comparactive anatomy lead to inferences of phylogenetic relationships based on resemblance. Similarities due to homology have to be distinguished from those due to analogy, and convergent and parallel evolution must be taken into account. For similarities due to common ancestry, the assumption is made that degrees of similarity reflect degrees of phylogenetic propinquity. On the whole, this is a reasonable assumption, since evolution is a gradual process of change. However, differences in rates of evolutionary change among lineages may be a source of error. Assume that a species S_1, diverged from the common ancestor of two other species, S_2 and S_3, before the latter diverged from each other. Assume also that a certain trait, say a protein, has evolved at a much faster rate in the lineage leading to S_3 than in the other two lineages. It might be the case that the primary structure of the protein would be more similar between S_1 and S_2 than between S_2 and S_3. The phylogeny inferred from the study of the protein might then be erroneous. Again, it is for that reason that all relevant sources of evidence must be taken into account in the construction of phylogenies.

The hypothesis has been recently advanced, however, that rates of amino acid replacements in proteins and nucleotide substitutions in DNA may be approximately constant because the vast majority of such changes are selectively neutral. New alleles appear in a population by mutation. If alternative alleles do not affect the fitness of their carriers, changes in allelic frequencies from generation to generation will be affected only by the random process of sampling—by genetic drift. Rates of allelic substitution would be "stochastically" constant, *i.e.*, would occur with a constant probability for a given protein. That probability can be shown to be simply the mutation rate for neutral alleles.

The neutrality theory of protein evolution has been championed by, King and Jukes (1969), Kimura and Ohta (1971), and others. These authors admit that the evolution of morphological, behavioural, and ecological traits is largely governed by natural selection. They propose however, that the evolution of most proteins, and of the genes coding for them, is for the most part due to chance. The neutrality theory of protein evolution assumes that, for any gene, a large proportion of all possible mutants are harmful to their carriers; these mutants are eliminated or kept at very low frequencies by natural selection. A large fraction of mutations, however, are assumed to be adaptively equivalent. Since these mutants do not affect the fitness of their carriers, they are not subject to natural selection. According to the neutrality theory, evolution at the molecular level consists for the most part of the gradual replacement of one amino acid sequence for another functionally equivalent to the first. The theory assumes that although favourable mutations occur, they are so rare that they have little effect on the overall evolutionary rate of amino acid substitutions.

Neutral alleles are not defined as adaptively identical in the mathematical sense. Operationally, neutral alleles are those whose differential contributions to fitness are so small that their frequencies change more due to the accidents of sampling through generations than to natural selection. The conditions for this can be simply stated. Assume that we have two alleles, A_1 and A_2 and that the adaptive value of one relative to the other is 1:1 +*s* (where *s* may be a positive or negative number). The two alleles are effectively neutral if, and only if

$$|N_e s| << 1, \qquad \ldots 9.1$$

where N_e is the effective size of the popultion or, approximately, the number of breeding individuals.

It is clear from the inequality that whether or not two alleles are adaptively equivalent depends not only on their differential effect on fitness, *s*, but also on the effective size of the population. Assume, for example, that *s*=0.001. In a population with 100 breeding individuals, $|N_e s| = 100 \times 0.001 = 0.1$. Selection would have little effect; the allelic frequencies would change from generation to generation mostly by random drift. Assume now that the same two alleles are present in a population of effective size 10,000. Then $|N_e s| = 10{,}000 \times 0.001 = 10$. Changes in allelic frequencies would be for the most part by selection. The allele with the highest adaptive value would gradually replace the other. In a population with $N_e \geq 10{,}000$ individuals, two allelles will be effectively neutral only if $s < 0.0001$.

We want now to find out the rate of allelic substitution, *k*, per unit time in the course of evolution. Time units can be years, generations, or multiples thereof. In a panmictic (random mating) population with N diploid individuals

$$k=2Nmx, \qquad \ldots 9.2$$

where *m* is the mutation rate of a gene per gamete per unit time (time is measured in the same units as for *k*), and *x* is the probability of ultimate fixation of an individual mutant. The derivation of Equation 9.2 is straightforward—there are 2N*m* mutants per unit time, each with a probability *x* of becoming fixed.

Let us now consider the probability of fixation, x_n, of newly arisen neutral mutant. Since there are 2N genes in the population, and all are assumed to have equal probability of becoming fixed, the probability of fixation of a newly arisen neutral mutant is simply

$$x_n = \frac{1}{2N} \qquad \ldots 9.3$$

Replacing the value of x_n in Equation 9.2, we obtain the rate of allelic substitution, k_n, for neutral allelels as

$$k_n = 2Nm\left(\frac{1}{2N}\right) = m. \qquad \ldots 9.4$$

That is, the rate of substitution of neutral alleles is precisely the rate at which neutral alleles arise by mutation, independently of the size of the population and any other parameters. This is not only a remarkably simple result, but also one with momentous implications if it indeed applies to protein evolution. For a given protein, the rate of evolutionary change would occur with a constant probability if we assume that the mutation rate remains constant. Moreover, the evolutionary rate of allelic substitutions would provide an estimate of the rate of neutral mutations in the corresponding gene. For a neutral allele that eventually becomes fixed, the average number of generations, t_n, that elapse from its appearance by mutation until it becomes fixed can be shown to be $t_n = 4N_e$. Since the value of t_n is a function only of N_e, it is the same for all gene loci within a given population, independent of their mutation rates and other variables. In a population of effective size 10,000, any newly arisen mutant that eventually becomes fixed takes, on the average, about 40,000 generation until fixation.

We now consider the evolutionary rate of allelic substitution for alleles subject to natural selection. Assume that a newly arisen mutant, A_2, is advantageous relative to the preexisting allele, A_1 so that the fitness for A_1A_1 is 1, for A_1A_2 1+*s*, and for A_2A_2 1+2*s*. If *s* is a positive number much smaller than one, the probability of fixation, x_s, of the new allele is approximately

$$x_s = 2N_e s/N. \qquad ...9.5$$

The probability of fixation of a new mutant as a function of the product $N_e s$ is shown in Figure 10.18. (When the value of $N_e s$ is close to zero, the probability of fixation becomes very nearly $1/2N$, as shown earlier). Substituting Equation 9.5 into Equation 9.2 we obtain the rate of allelic substitution, k_s, for a selective mutant without dominat effects:

$$K_s = 2Nm(2N_e s/N) = 4N_e sm. \qquad ...9.6$$

The rate of substitution (k_x) of selective alleles depends on the effective size of the population, the selection coefficient, and the mutation rate. The greater the selective advantage of an allele, the faster the rate at which it will replace the disadvantageous allele. As pointed out above, selection has a significant effect whenever $|N_e s|>1$. In such a case, $|4N_e sm|$ will be greater than m. In general, then, the rate of evolution is greater for selectively advantageous alleles than for neutral alleles.

The conclusions just reached need to be qualified, however. The value of s is unlikely to remain constant over long periods of evolutionary time. If s changes from positive to negative at different times, as might be the case, the rate of fixation could be considerably smaller than that given by Equation 9.6. This equation applies only to a simple case of selection with no dominance (*i.e.*, the heterozygote has intermediate fitness between the two homozygotes). Other modes of selection exist. With heterozygote superiority or frequency-dependent selection, more or less stable polymorphisms may persist for long periods of time. Indeed, Equation 9.6 has little operational applicability.

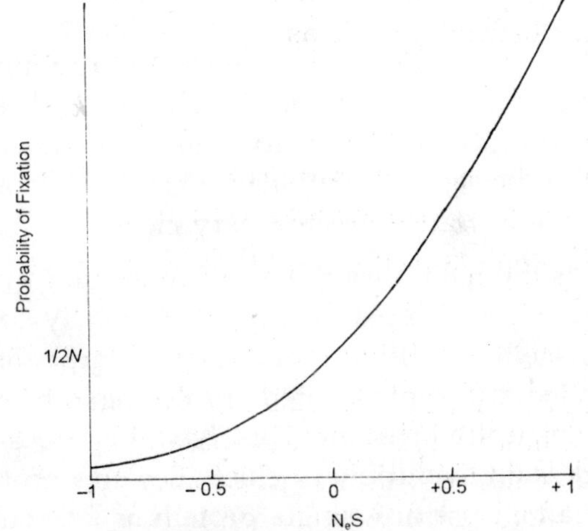

Fig. 10.18: Probability of fixation of a mutant gene as a function of the product of the effective population size (N_e) times the selection coefficient (s) of the gene. When $s=0$, the probability of fixation is $1/2N$ and depends exclusively on genetic drift. When the product $N_e s$ is greater than 1 or smaller than -1, the probability of fixation (or elimination) of the mutant allele depends largely on selection, with drift having a relatively small effect.

The Molecular Clock of Evolution

If the neutrality theory of protein evolution were correct for a large number of proteins, the implications for the study of phylogeny would be momentous. Equation 9.4 shows that the expected rate of allelic substitution for neutral alleles is simply the rate at which neutral alleles arise by mutation. For a given gene, mutation rates are likely to remain fairly constant over long periods of evolutionary time.

Allelic substitutions would therefore be expected to occur at a constant rate. Protein and gene evolution would serve as evolutionary clocks. First, the degree of protein differentiation among species would be a measure of their phylogenetic relatedness. Second, the actual "chronological" time of the various phylogenetic events could be estimated. Assume that we have a phylogeny based on cytochrome *c*. If the rate of evolution of cytochrome *c* were constant through time, the number of nucleotide substitutions that have occurred in each one of the "legs" of the phylogenetic tree would be directly proportional to the time elapsed. If the actual geological time of any one of the events in the phylogenetic tree were known from some outside source (such as the paleontological record), the times of all other events could be determined by a simple proportion. That is, once it is "calibrated" by reference to one single event, the molecular clock can be used to measure the time of occurrence of all other events in a phylogeny.

The molecular clock implied by Equation 9.4 is of course not a "metronomic clock," like timepieces in ordinary life that measure exact time with theoretically perfect regularity. The neutrality theory predicts, instead, that molecular evolution is a "stochastic clock," like radioactive decay. The *expected* rate of change is constant, although some variation occurs. Over fairly long periods of time a "stochastic clock" is nevertheless quite accurate. Moreover, it should be emphasized that each gene or protein would be a separate clock, providing an independent estimate of phylogenetic events and their time of occurrence. Each gene or protein would "tick" at a different rate (the mutation rate to neutral alleles, m, of the gene), but all of them would be timing the same evolutionary events. The joint results of several genes or protein would provide a precise evolutionary clock.

Is the neutrality theory of molecular evolution correct? Are the rates of nucleotide or amino acid replacement stochastically constant for any given protein? Early studies of protein evolution seemed to suggest that they were stochastically constant. More detailed and sophisticated studies have recently shown that molecular evolution is not stochastically constant. This, however, does not imply that protein evolution cannot be used as an evolutionary clock. Because protein evolution is not stochastically constant, we cannot use any single protein as an evolutionary clock. But as we shall see below, the study of *many* proteins provides a fairly accurate evolutionary clock. The *average* amount of molecular change over many proteins in many organisms appears to be sufficiently constant to be used as an approximate clock of evolutionary events.

The phylogeny provides an example of evidence originally thought to support the constancy of molecular evolutionary rates. The figure indicates the number of amino acid differences between chains of *a* hemoglobins. The most recent ancestor common to the carp, a fish, and the four mammals lived some 350 million years ago. Yet the number of amino acid differences between the carp and mice (68), carp and rabbits (72) and carp and horses (67). Each one of these comparisons reflects the evolution of α hemoglobin for 700 million years —350 million years between the last common ancestor and carp, and 350 million years between the last common ancestor and each mammal. At first sight, the degree of protein differentiation appears remarkably similar in all four comparisons; but this similarity is deceiving. The most recent ancestor common to the four mammals lived some 70 million years ago. For 630 of the 700 million years of the evolution of the α chain (350 million years in the carp lineage and 280 million years in the lineage leading to the mammals) any amino acid replacement would affect all comparisons between any mammal and the carp. Only during 70 million years (or 10 per cent of the total time involved) would heterogeneity of evolutionary rates affect the comparisons made above.

Information about the homogeneity or heterogeneity of evolutionary rates can be obtained by comparing the amino acid sequences of α hemoglobin of the four mammal species. The number of amino acid differences range from 17, between man and mice, to 28, between rabbits and mice. The number of amino acid differences is 65 per cent greater between mice and rabbits than between mice and man. Since one of the "legs" (the lineage going to mouse) is the same in these two comparisons, it follows that 11 more amino acid replacements have occurred in the lineage leading to rabbits than in the lineage going to man. The different rates of evolution in these two lineages might conceivably be accounted for if it were assumed that rates of amino acid replacement are constant *per generation* rather than per year. More generations have occurred in the rabbit lineage than in the human lineage. But even when rates of replacement are assumed to be constant per generation, the agreement is far from good. The number of differences between man and rabbits is 25, while between man and mice it is only 17. The lineage leading to man is the same in these comparisons. Therefore eight more replacements must have occurred in the rabbit lineage than in the mouse lineage, although approximately the same number of generations must have passed in both lineages, or if anything, fewer in the rabbit lineage.

Are the differences in evolutionary rates just discussed compatible with a *stochastically* constant rate? The neutrality theory predicts that the *probability* of replacement should be constant through time, but stochastic variation is to be expected. The question raised is far from definitely solved. Yet, as more and more homologous sequences are compared and as more sophisticated methods of analysis are used, evidence is gradually accumulating against the idea of constancy of evolutionary rates at the molecular level. As an example, we may consider the phylogeny of the globin molecules shown in Figure. *Goodman, Moore,* and *Matsuda* (1975) have calculated that for the first 380 million years of evolution of vertebrate globin (from the divergence between invertebrates and vertebrates to the divergence between birds and mammals), the hemoglobin genes evolved at an average rate of about 46 nucleotide replacements per 100 nucleotides per 100 million years (per cent MTR/100 MY). However during the next 300 million years (from the mammal—bird split to the present) the average rate of hemoglobin evolution appears to be 15 per cent NTR/100 MY, or only one third of the previous rate. Statistical tests indicate that the hypothesis of a constant rate of hemoglobin evolution has to be rejected.

Langley and *Fitch* (1974) have devised an ingenious method to test the constancy of evolutionary rates. They have applied the method to the amino acid sequences of four proteins in 18 species of vertebrates. The four proteins are hemoglobin α and β, cytochrome *c*, and fibrinopeptide A. The 18 species included one fish (carp), one frog (*Rana*), one bird (chicken), and 15 mammals. The method is based on the minimum number of nucleotide substitutions required to account for the observed differences in amino acid sequence. The data for all four proteins are examined together by adding up the number of nucleotide replacements in each of the four. This method maximized the amount of information that can be obtained from the data.

Table 9.8: Statistical test of the constancy of evolutionary rates of four proteins (hemoglobins alpha and beta, cytochrome c, and fibrinopeptide A) in 18 vertebrate species

	Chi-square	Degrees of freedom	Probability
Overall rates (comparisons among branches over all four proteins)	48.7	24	2×10^{-3}
Relative rates (comparisons among proteins within branches)	102.7	62	10^{-3}
Totals:	151.4	86	10^{-4}

Two kinds of comparisons are made. First, the total number of substitutions per unit time is examined for different times. The hypothesis tested is whether the *overall* rate of change is uniform over time. Because the departure from expectation is statistically significant, it cannot be attributed to stochastic variation alone. The conclusion follows that the proteins have not evolved at a constant rate. It might be possible to argue at this point that the rate of molecular evolution is constant *per generation* rather than per year. The heterogeneity in evolutionary rates might be due

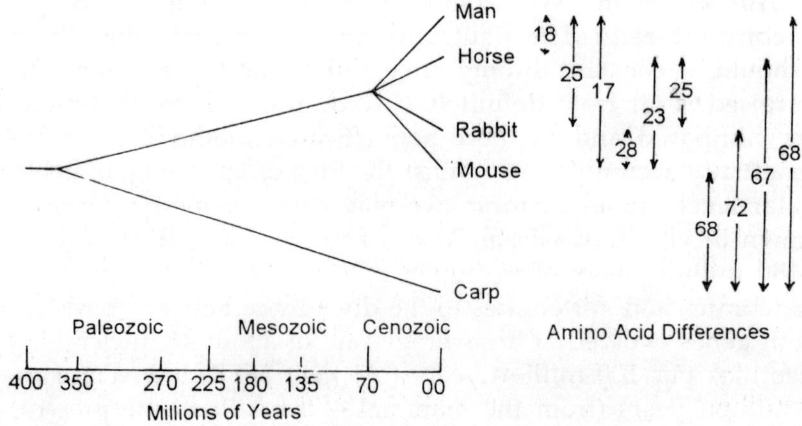

Fig. 10.19: Phylogeny of carp and four mammal species, indicating the number of amino acid differences between them in the alpha chain of hemoglobin.

to variations in generation length through evolutionary time. However, this has to be rejected once a second type of comparison is taken into consideration. The hypothesis tested now is whether the *relative* rates of evolution are constant through time. If the heterogeneity in overall rates of evolution were due to changes in generation length through time, the relative rates of change among the proteins should remain constant, since all the proteins would change their rates proportionately. As shown in Table 9.8 the relative rates of change are also not constant. More recently the analysis has been extended to seven proteins from 17 mammals. The overall as well as the relative rates of protein evolution deviate significantly from constancy.

Proteins do not evolve at a stochastically constant rate. Nevertheless the *average* rates of evolution over long periods and over many proteins may be used as an approximate evolutionary clock. The minimum number of nucleotide substitutions that have occurred in all four proteins named above (α and β hemoglobins, cytochrome c, and fibrinopeptide A) is plotted against paleontological time. The overall correlation is fairly good for all organisms except the primates, which have evolved at a substantially lower rate than the average of the other organisms.

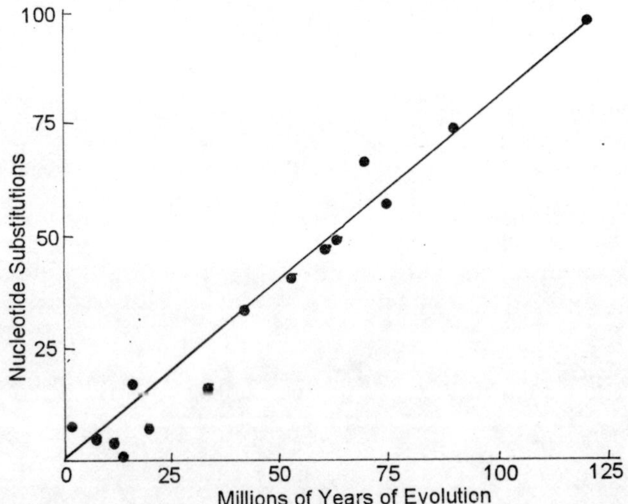

Fig. 10.20. Nucleotide substitutions *versus* paleontological time. The total nucleotide substitutions for seven proteins (cytochrome *c*, fibrinopeptides A and B, hemoglobins alpha and beta, myoglobin, and insulin C-peptide) have been calculated for comparisons between pairs of species whose ancestors diverged at the time indicated in the abscissa. The solid line has been drawn from the origin to the outermost point, and corresponds to a total rate of 0.41 nucleotide substitutions per million years (or 98.2 nucleotide substitutions per 2 × 120 million years of evolution) for the genes coding for all seven proteins. The fit between the observed number of nucleotide substitutions and the expected number (as determined by the solid line) is fairly good in general. However, in the primates (points below the diagonal at lower left) protein evolution seems to have occurred at a slower rate than in most other organisms.

In conclusion, informational macromolecules such as proteins and nucleic acids contain significant information about phylogeny. Rates of molecular evolution are not constant, yet protein changes may be used as an approximate evolutionary clock whenever they are averaged over many proteins and organisms. The application of protein-sequencing to the study of phylogeny is a fairly young scientific field. There is little reason to doubt that in the future the study of proteins and nucleic acids will make increasingly significant contributions to our knowledge of the evolutionary record.

Chapter—11
Evolution of Unicellular Organisms

The classical division of the world of life into two kingdoms, animals and plants, has such a low informative and instructional value as to be almost useless for our purpose. At the level of unicellular organisms the distinction between animals and plants is based upon a single characteristic, the presence or absence of photosynthesis, and so is highly artificial. Much more instructive are the various modern schemes that recognize several kingdoms, and use several different characteristics for distinguishing them. That of *Whittaker* (1969), which recognizes five kingdoms, is most satisfactory. These kingdoms represent three grades of advancement; prokaryote (Monera), unicellular eukaryote (Protista), and multicellular or coenocytic eukaryote. The first two grades are each represented by a single kingdom, but at the multicellular level three kingdoms are recognized: plants, animals, and fungi. The reasons for doing this are discussed later in this chapter.

First, however, the historical reason for revising the biologist's conception of kingdoms must be reviewed. It results directly from the greatly increased knowledge of microorganisms acquired by biologists during the past century. From about 1870 to 1940 the relationships between different groups of unicellular eukaryotes or protista were intensively investigated and the modern discipline of protozoology was founded. These investigations confused rather than clarified the problem of the relationships between different kinds of microorganisms. In spite of the presence of cholorophyll and the ability to perform photosynthesis, many newly discovered unicellular organisms were recognized by zoologists as animals and classified as Protozoa because of their motility, their ability to ingest food particles, and their resemblance to other colourless Protozoa. These same organisms, of which the best known example is *Euglena,* were classified by botanists as plants, because of their photosynthetic ability. Each systematist could present logical arguments to support his claim. In this way a sort of "no man's land" of organisms arose, containing species and genera that were classified in one way, with one set of terms, by zoologists, and in an entirely different way by botanists.

During the same period the position and relationships of the bacteria became equally confused as a result of two conflicting tendencies. Since they do not ingest food and are very different from Protozoa, bacteria have never been recognized by zoologists as animals. They are, however, sufficiently plant-like to have attracted the attention of botanists, who have placed them in the plant kingdom as a separate division or phylum in most textbooks and general treatises. At the same time, a study of bacteria has become an essential part of medical science, and a large school of medical bacteriologists have devised their own methods of investigating bacterial relationships. In general, they have had little communication with botanists, even with mycologists, interested in bacteria. The resulting cross currents and lack of communication have led to a confusing situation.

The problem was clarified and progress was made toward its solution by two important developments in the middle of this century: the rise of molecular biology and the invention of the electron microscope. Modern research has revealed a previously unsuspected cleft between two major groups of organisms, one that cuts right across the preexisting classification of animal and plant kingdoms. With respect to their biochemistry, genetics, and cellular and nuclear fine structures, organisms can be divided into two sharply distinct groups, *eukaryotes* and *prokaryotes*. The prokaryotes include the bacteria and blue-green algae; all other cellulose organisms are eukaryotes. Table 11.1 summarizes the principal difference between these two groups. On the basis of biochemical differences, McLaughlin and Dayhoff (1970) have concluded that the evolutionary distance between prokaryotes and eukaryotes is twice that between animals, plants, and fungi.

The distinctive characteristics of prokaryotes are the absence of important structural features and reproductive cycles, such as mitosis, meiosis, and syngamy, which are present in most or all eukaryotes. That this absence in prokaryotes is a primary condition rather than a secondary loss is evident from the Precambrian fossil record, which shows that prokaryotic cells appeared long before eukaryotes. Consequently, evolutionists are now faced with a major problem: How did the first major ascent in evolutionary grade, the origin of eukaryotes from prokaryotes, take place? This problem is discussed later in this chapter.

The separation of prokaryotes as a kingdom distinct from eukaryotes does not solve the problem of the distinction between plants and animals. *Whittaker's* solution is to recognize at the level of kingdoms a second advance in grade, that from unicellular to multicellular or coenocytic eukaryotes. This procedure localizes all the previous artificial separations between autotrophic and heterotrophic organisms either at the level of unicellular eukaryotes (Protista) or of prokaryotes (Monera), i.e., bacteria vs. blue-green algae. The difference at these levels between autotrophy and heterotrophy is no more significant than other cytological, physiological, and biochemical difference, and in most instances is useful chiefly for distinguishing between orders, families, or even genera within a family. The logic of recognizing Protista as a single and separate kingdom becomes clear when the distribution of autotrophy *vs.* heterotrophy among them is compared with that of other characteristics, such as the chemistry of the cell

Table 11.1: Principal differences between prokaryotes and eukaryotes.

	Prokaryotes	Eukaryotes
Nucleus		
Nuclear membrane	No	Yes
Mitotic spindle	No	Yes (no)
Meiotic reduction of chromosome number	No	Yes (no)
Recombination by union of entire gametes	No	Yes (no)
Chromosomes		
Number of different nonhomologous chromosomes in basic genome	1	2-600+
Histones and acidic proteins complexed with DNA	No	Yes (no)
Regular condensation and relaxation of chromosomes during mitotic cycle	No	Yes (no)
Cytoplasm		
Mitochondria	No	Yes
Chloroplasts	No	Yes or no
Metabolic enzymes in cellular membrane	Yes	No
Cilia and flagella with 9-plus-2 fibre arrangement	No	Yes or no
Endoplasmic reticulum	No	Yes
Vacuoles	No	Yes (no)
Golgi apparatus	No	Yes or no
Phagocytosis, pinocytosis	No	Yes or no
Amoeboid movement, cytoplasmic streaming	No	Yes or no
Molecular synthesis		
Nature of cytoplasmic ribosomes	70S	80S
Nature of organellar ribosomes	No	70S

wall, the position and fine structure of flagellae, and the organization of the mitotic apparatus. For instance, with respect to these other characteristics the facultatively autotrophic *Euglena* resembles various heterotrophic flagellates and is completely different from the autotrophic flagellate *Chlamydomonas*. Calling *Euglena* a plant and placing it with *Chlamydomonas* rather than with the flagellates it resembles is a highly artificial procedure. Another course adopted by some botanists is to classify *Chlamydomonas* with plants because it resembles them in having a cellulose wall, and exclude *Euglena* because the chemical structure of its cell wall is completely different.

This procedure creates other problems. Red and brown algae, which are plant-like in many respects, including photosynthetic ability, nevertheless have walls that do not contain cellulose or that differ considerably from those of typical green plants in other respects.

Unicellular eukaryotes, therefore, are here recognized as a single, probably monophyletic kingdom, the Protista, and kingdoms that are parallel branches or clades of major dimensions are recognized only at the grade of multicellular or coenocytic organisms. This procedure raises two additional problems: where to draw the boundaries between Protista and the kingdoms of multicellular organisms, and how many kingdoms to recognize at the multicellular or coenocytic grade. The first problem does not arise with respect to animals, since the gap between Protista and the animal kingdom, which according to the present treatment is synonymous with the Metazoa, is so great that the origin of Metazoa is still obscure. On the other hand, the different phyla of algae and fungi are connected to various groups of unicellular forms through a series of intermediates, which may be colonial cell aggregates, small coenocytes, or have low degrees of cellular differentiation. One might justly ask, therefore, why reject the artificial distinction between autotrophs and heterotrophs at the level of Protista, while erecting more or less artificial boundaries between unicellular and multicellular autotrophs?

Our first answer is that the world of life is not divided into a few sharply separated kingdoms, which can be easily delimited. If plants are to be separated from ciliate Protista, which are surely much more like animals than plants, artificial boundaries must be established somewhere. Second, the advantage of recognizing kingdoms at the somewhat indistinct boundary between the unicellular and the multicellular or coenocytic grade is admittedly heuristic. It calls attention to the great importance of this advance in grade for the evolution of life, and invites studies directed at learning how this advance came about.

Once the policy is adopted of recognizing several parallel kingdoms at the multicellular grade, the question arises: How many should be recognized? This question is not easy to answer. If one adopts the rule that each kingdom should be monophyletic, including only a single advance to the multicellular grade, two difficulties arise. First, the number of these advances is relatively large; an estimate of 16 is given later in this chapter. Second, their actual number is largely a matter of inference based upon comparisons between modern forms, since multicellular microorganisms lack a significant fossil record. Clearly the phylogenetic approach, because of its uncertainties, is a weak base upon which to found categories as broad as kingdoms.

The approach adopted here separates three kingdoms of multicellular or coenocytic organisms on the basis of fundamental differences in their mode of nutrition. Plants are autotrophic, except for a few saprophytes or parasites among seed plants, which are obviously of secondary origin. Fungi are heterotrophic and, like bacteria, absorb their food in a predigested form. Animals are heterotrophic ingesters. All three of these kingdoms are probably polyphyletic. Among animals, sponges have probably had a different origin from the remaining phyla. Among fungi, the water moulds, or oömycetes, probably arose from heterotrophic flagellates independently of terrestrial fungi. The plant kingdom is probably polyphyletic even if, following Margulis (1970), the brown and red algae are excluded from it and only green algae plus land plants are included. There are good reasons for believing that the coenocytic green algae were derived from Protista independently of the cellular forms. Even the filamentous green algae now appear to have had at least three separate origins from unicellular forms: the Conjugales, Ulotrichales, and *Klebsormidium* group, which are the only green algae that form typical cross walls from a cell plate. Once polyphylesis is admitted, the addition of the brown and red algae to the plant kingdom as separate and independent phyla becomes reasonable because of the numerous parallelisms between their morphological and ecological evolution and that of green algae. In our view, kingdoms of multicellular organisms are aggregates of phyla that, considered together, have a heuristic value because they represent similar ways of achieving advances in grade.

The principal heuristic value inherent in this system is as follows. First, the origin of multicellular organisms having differentiation or division of labour among the parts of their bodies involved two major ascents in grade. The first was the evolution of the eukaryotic cell. Without the division of labour among the organelles and membranes evolved in this kind of cell, cellular differentiation such as exists in animals and plants could not evolve. Some of the essential characteristics of the eukaryotic cell evolved only once. Many writers have pointed out the similarity between all eukaryotic cells with respect to: cilia and flagella, when present; ribosomes and other protein-synthesizing machinery of nuclear origin; and with few exceptions, the mitotic apparatus and molecular organization of the chromatin of their chromosomes.

The second advance in grade, the origin of multicellular or multinucleate differentiated organizms, took place many times in various evolutionary lines, and followed the first advance only after tens or hundreds of millions of years had elapsed and extensive adaptive radiation had taken place at the unicellular level. The initial evolution of the eukaryotic type of cellular organization was a necessary, but not sufficient, condition for the origin of multicellularity and differentiation.

A second kind informational content inherent in this system is the recognition of the three principal types of nutrition: autotrophy, heterotrophy with absorption of soluble food, and heterotrophy with ingestion of solid particles and internal digestion.

The earliest true cells were probably heterotrophic, and surrounded by firm cell walls. They excreted enzymes into the surrounding medium and absorbed soluble food that had been externally digested. This form of nutrition is predominant in prokaryotes, since only the relatively advanced and, for prokaryotes, structurally complex blue-green algae plus a few groups of photosynthetic and sulfur bacteria have evolved autotrophy. On the other hand, all three forms of nutrition exist in unicellular eukaryotes or Protista. Not infrequently two different methods, such as autotrophy or ingestion, as well as ingestion or absorption, can be carried out by the same cell. At the unicellular level differentiation with respect to these three basic methods of nutrition does not require extensive differentiation with respect to structure, physiology, or gene content.

The situation is completely different in multicellular organisms. Both autotrophy and heterotrophy with absorption require extensive contact between cell surfaces and the external medium, whether it be water, air, or the cellular environment of a parasite's host. In a multicellular organism this can be achieved only by extending its external surface area more or less proportionately with its size. Large autotrophs and absorbers are, by their nature, sedentary, so that compactness of bodily form is not an advantage to them. Furthermore, because of their sedentary nature they can attain wide distributions and can colonize new habitats only with the aid of propagules: cells or multicellular structures especially adapted for dispersal. This explains the origin of spores, motile gametes, and, in the more advanced forms, seeds. Many sedentary ingesters have evolved similar propagules. The situation in heterotrophic ingesters is completely different. Since they find their food by active motion, they must remain relatively compact and symmetrical, no matter what their size. As size increases, each organ increases the surface area of its internal membranes or cellular layers relative to its size. If they are predators, efficiency of nutrition usually depends upon activity and keenness of the senses, hence the successive evolution of more efficient sense organs and locomotor apparatus. If they are small and nonpredatory, their compactness makes them the logical prey of predators. Increasing efficiency of sense organs and locomotor apparatus is adaptive for most multicellular ingesters, i.e., animals, the chief exceptions being those that acquire, secondarily, armoured coverings or noxious, unpalatable characteristics. At the same time, the sense organs and locomotor apparatus that are vital for nutrition and escape can be modified with relative ease to serve as aids in sexual union. Hence, the distinction between adaptations for survival and for reproduction is far less marked in animals than in plants.

At the multicellular level the distinction between plants and animals is sharp and clear. Moreover, fungi are almost as distinct from multicellular plants as are animals. They are, like bacteria, heterotrophic absorbers, so that with respect to their ecology and many of their biochemical properties the smaller fungi are more like bacteria than higher plants. In addition, the chitinous cell walls of most fungi are completely different from those of plants, and certain genetic characteristics, such as the heterokaryotic condition in Ascomycetes and the dikaryotic condition in Basidiomcetes, are also distinctive. The recognition of fungi as a kingdom separate from animals and plants, and of mycology as a discipline separate from botany, has both a phylogenetic basis and a justification in terms of its informational content.

Some groups of organisms do not fit into a system of five kingdoms based upon three grades. The best known of these are the slime molds, both cellular (Acrasiales) and plasmodial (Myxomycetes), and the Volvocales. The slime molds, during their stages of growth and metabolism, retain the properties of heterotrophic, ingesting unicellular Protista. Nevertheless, they have evolved the ability to form propagules, differentiated cells that enable them to reproduce in a manner essentially like that of fungi, by means of spores. The most highly evolved Volvocales, particularly the genus *Volvox*, have many properties of colonial autotrophic Protista, but their mature colonies have a polarized differentiation, and carry out a process of "gastrulation" not unlike that of animals.

Two ways of handling these anomalous groups seem reasonable. One way would be to stretch somewhat the definition of Protista, and treat the anomalous groups as unusual members of that kingdom. The other would be to erect little "principalities" for them, and thus to exclude them from all of the five kingdoms. The latter procedure has greater heuristic value, since it calls attention to the fact that multicellularity and differentiation have evolved many times, in response to immediate adaptive advantages in particular environments, and that only some of these multicellular organisms have undergone extensive further evolution.

The earliest known cellular organisms, bacteria of the African strata known as Fig Tree Chert, are about 3.3 billion years old. The first appearance of unicellular eukaryotes was "earlier than 1,300 million years ago and possibly earlier than 1,700 million years ago". Consequently, for more than 1,500 million years the only living organisms on the earth were prokaryotes. Since the oldest fossils that could be regarded as multicellular or multinucleate organisms are not more than 700 million years old, the period when evolution was exclusively at the prokaryote level was twice as long as that encompassing the entire evolution of multicellular eukaryotes. Since the evolution of prokaryotes is largely biochemical rather than structural, it remains, to a large extent, a closed book. Nevertheless, on the basis or reasonably reliable information about the nature of the earth during the era of prokaryote evolution, some inferences about the nature of prokaryote evolution can be made.

The first important point is that when life first arose the oxygen content of the atmosphere was much lower than in more recent times, so that aerobic metabolism, if it existed at all, had to be at a much lower rate than in modern aerobes. The greater part of the oxygen in the modern atmosphere was and is produced by the activity of photosensitizing autotrophs. The origin of a truly oxidizing atmosphere took place, probably, not much more than a billion years ago. The greater part of prokaryote evolution, therefore, was based upon anaerobic or weakly aerobic metabolism.

The most significant single event of this evolution was the origin of photosynthesis. It took place relatively soon after the origin of cellular life. The fossil record of blue-green algae is instructive in this connection. Their modern forms have molecules of chlorophyll that are identical with those of higher plants, and even the carotenoid pigment associated with their chlorophyll resembles that found in eukaryotes. Although some of them can carry out anaerobic metabolism, most of them are aerobic, and perform photosynthesis with H_2O as the hydrogen donor. Hence, by the time blue-green algae appeared, photosynthesis similar to that of higher plants either existed or was represented by a similar process. The oldest fossils of blue-green algae are calcareous deposits known as stromatolites, similar to the mineral layers that surround colonies of marine forms in some parts of the modern world, particularly Australia. The oldest stromatolites are in the Buluwaya formation of South Africa, 2.7 billion years old. They are 500 million years younger than the oldest known bacteria, but 1.4 billion years older than the earliest eukaryotes.

The complex, sophisticated kind of photosynthesis carried out by blue-green algae must have ben preceded by a whole series of more simple, primitive methods of autotrophy. The only methods known to us are those of three groups of photosynthetic bacteria: the green bacteria (*Cholorobium, Pelodictyon*) purple sulfur bacteria (*Thiospirillum, Lamprocystis*) and purple nonsulfur bacteria (*Rhodospirillum, Rhodopseudomonas, Rhodomicrobium*). All these forms have chlorophyll molecules that resemble those of other photosynthesizers with respect to the tetrapyrrole porphyrin ring surrounding a central atom of magnesium, but differ with respect to certain side chains. None of the forms has a strictly aerobic kind of photosynthesis characteristic of most blue-green algae and all higher plants. The green bacteria and purple sulfur bacteria are strictly anaerobic, and require sulphide, thiosulfate, or perhaps hydrogen as an electron donor. The purple nonsulfur bacteria carry out photosynthesis anaerobically, but use organic compounds as the principal electron donors, and can become aerobic and heterotrophic in the presence of oxygen and low illumination. Slow, inefficient methods of photosynthesis like that found in these bacteria must have preceded aerobic photosynthesis of the modern type. Although fossil evidence may never be adequate to solve the problem, the assumption that forms similar to photosynthetic bacteria evolved prior to blue-green algae is quite reasonable.

The total range of adaptive radiations carried out by bacteria before the evolution of eukaryotes may never be known. The ecology of the extinct species cannot be inferred, and one cannot expect to find among contemporary bacteria representatives of all groups that must have existed. If the relative diversity of modern forms is any kind of a guide, one must conclude that this radiation was much more restricted in its earlier stages, when oxygen was limited and only anaerobic forms could exist, than at later periods, when the principal radiants must have been aerobic. The contemporary anaerobic free-living bacteria are, besides the photosynthetic bacteria, principally of three groups. The methane bacteria inhabit the bottoms of ponds or quiet streams, where they obtain energy by the anaerobic oxidation of molecular hydrogen. The *Clostridium* group is characterized by the ability to form highly resistant spores, and live in moist soil. Finally, there is a large series of relatively little-known anaerobic bacteria living in the mud at the bottom of the sea, chiefly at depths of 1,000 to 2,000 metres. In addition, some pathogenic bacteria, including members of the *Closrtridium* group, as well as some of the bacteria that inhabit the mouth and digestive tract of animals, are obligate anaerobes. These latter groups, obviously, are relatively recent origin. In addition, many groups of bacteria, such as the lactic-acid bacteria, are facultative anaerobes, which are able to utilize oxygen for respiration in an aerobic atmosphere but which, in its absence, can derive energy from anaerobic metabolism.

The extent to which aerobic metabolism enabled bacteria to undergo additional adaptive radiation even before the evolution of eukaryotes may be estimated by the wide range of ecological niches in the contemporary world inhabited by aerobic prokaryotes. The blue-green algae have already been mentioned; their range of habitats extends from hot springs, at temperatures of 73° to 75°C, to the icy wastes of Antarctica. They are also much more tolerant of polluted water than are eukaryotic photosynthesizers. Derivatives of the blue-green algae that have reverted to heterotrophy are the filamentous gliders, such as *Beggiatoa* and *Thiothrix,* which live in sulfur springs. Other strictly aerobic free-living bacteria are the nitrogen-fixing soil bacteria (*Azotobacter, Beijerinckia*), the colonial actinomycetes, the highly salt-tolerant *Holobacterium,* and the free-living members of the group of spirochaetes. These organisms are remarkable for the long flagella-like structures contained in their cell wall, which enable them to carry out their characteristically rapid spiral motility. Judging from the diversity of these modern forms, the advent of oxygen in the earth's atmosphere greatly expanded the possibilities for adaptive radiation and evolution at the prokaryote level.

The advent of an oxygen-rich atmosphere required many readjustments of cellular metabolism. One of them must have consisted of ways by which cells could overcome the toxic effect of oxygen on obligate anaerobes. Of the many evolutionary strategies for acquiring this adaptation, one was particularly striking: the origin of bioluminescence in bacteria, and presumably in their protistan descendants. McElory and Seliger (1962) have produced a good argument to support their hypothesis that the biochemical reactions responsible for bioluminescence or fluorescence evolved from the vestiges of mechanisms that in the earliest aerobic organisms reduced molecular oxygen directly and quickly, thus preventing it from exerting toxic effects, which oxygen has when it accumulates in the cells that are normally anaerobic. The cells that produce luminescence produce a compound, luciferin aldehdye, which reacts with the potentially toxic peroxide produced by free oxygen to form the corresponding organic acid and water. The organisms responsible for the phosphorescent glow so often seen in seawater, the dinoflagellate *Noctiluca,* may have acquired its property via an evolutionary descent from a luminous prokaryote.

Evolution in Viruses

For many years viruses were not regarded as organisms at all, so that their evolution need not have been considered in connection with that of cellular organisms. Now, however, molecular biologists generally accepted the hypothesis that viruses have been derived from detached portions of cellular organisms. They contain typical nucleic acids, either DNA or RNA or both, and with the aid of enzymes provided by the host synthesize proteins, just like cellular organisms.

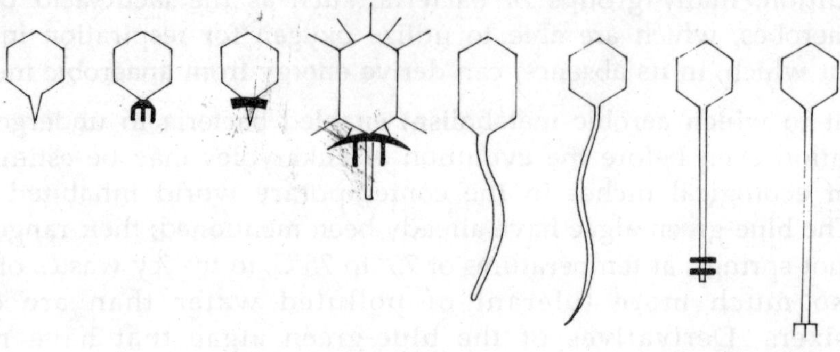

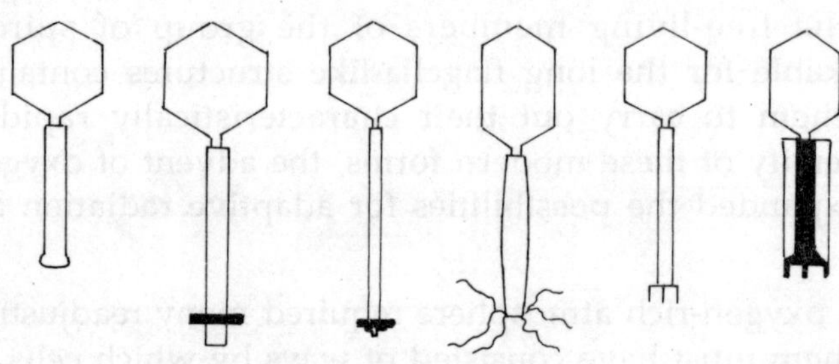

Fig. 11.1: Comparative morphology of bacteriophage viruses, showing various degrees of complexity. Among the simplest are the noncontractile viruses having short wedge-shaped tails, such as bacteriophage *T3* of *Escherichia coli* (top left). The most complex is the contractile pahge *T4* (lower right see Figure 11.2 for an electron micrograph). Among viruses having intermediate complexity, one of the best known is lambda (top row, third from right).

Three further points established by Joklik (1974) in his review of virus evolution are significant to evolution in general. First, groups of viruses differ so radically from each other with respect to both the cross affinities of their nucleic acids and the immunological relationships of their proteins that they almost certainly have been derived separately from cellular organisms. Second, the members of the same viral group, with similar nucleic acids and proteins, include parasites of very different organisms. Among the closely knit group of Reoviridae, for instance, some attack mammals, other birds, while still others have plants and/or insects as hosts. In most of these groups with heterogeneous host affinities, insects appear to be the common denominators, so that their role as vectors may have been highly significant in the origin of the group.

Third, some groups of viruses show clear evidence of evolutionary lineages. This is evident partly from their diversity form. Particularly important in this connection are the capsid-forming bacterio phages. The most highly evolved of these, such as the

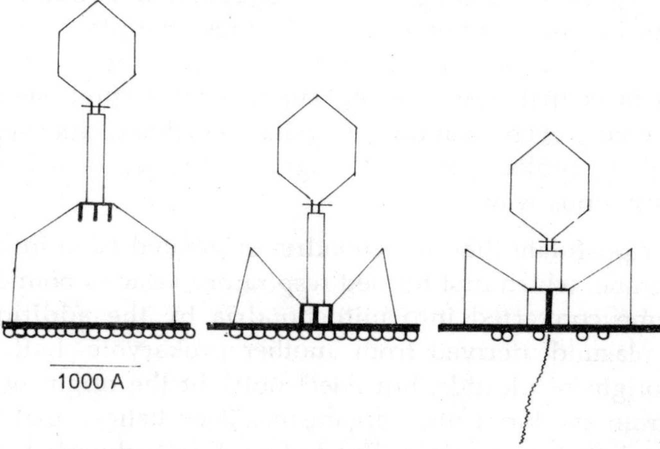

Fig. 11.2: Virions of bacteriophage *T4* adsorbed on a cell wall of the bacterium *Escherichia coli* and in the process of injecting their DNA into the cell.

well-known *T4* phage, have elaborate tails and tail fibers that act like a hypodermic syringe in injecting DNA into the host cell. In different phages of this group a whole series of intermediate kinds of tail mechanisms have been recognized. These viruses have clearly undergone extensive evolution of the conventional sort after originating as detached portions of cells.

The Origin of Eukaryotes

The first advance in grade after the origin of cellular life was that of the eukaryotic cell. Its origin can be regarded as the most significant single advance of organic evolution, since until cells had acquired the complexity and the degree of division of labour among membranes and organelles that are associated with the eukaryotic condition, multicellular organisms, having differentiated cells and tissues, could not evolve.

There is general agreement among biologists that the basic organization of eukaryotic cells evolved only once. This is because eukaryotic cells possess certain complicated organelles that are strikingly similar to each other in their supramolecular structure. The most conspicuous of these are the mitochondria, the plastids of autotrophic eukaryotes, and flagella or cilia having the 9+2 organization of fibrils. Among eukaryotes lacking these organelles, only the red algae can be postulated with reasonable probability as never having had such flagella. A second similarity is the mitotic apparatus in nearly all eukaryotes including red algae. Biochemical characteristics common to eukaryotes are the Embden-Meyerhof pathway of glucose metabolism, the universal presence of Krebs-cycle oxidations, and particularly the protein-synthesizing machinery. The ribosomes of eukaryotes are larger than those of prokaryotes, and their activity is insensitive to chloramphenicol, which inhibits protein synthesis in prokaryotes, while it is sensitive to cyclohexamide, to which prokaryotic cells are resistant.

Not unexpectedly, many biologists have speculated about the origin of the eukaryotic cell, a cardinal event of evolution. The most complete hypothesis is that of Margulis (1970), who has marshalled a great wealth of morphological, biochemical, and paleontological facts that, she believes, support the hypothesis that the cellular organelles of eukaryotes, even including flagella, centrioles, and spindle fibers, are derived from symbiotic prokaryotes that originally existed is a host cell and later became modified in various ways.

Raff and Mahler postulate that mitochondria originated from invaginations of the inner cellular membrane, which first formed respiratory vesicles bound by membranes. These vesicle became converted into mitochondria by the addition of a protein-synthesizing DNA plasmid, derived from another prokaryote. Raff and Malher say nothing about the origin of plastids, but reject outright the origin of flagella and the mitotic apparatus from symbiont microorganisms. They believe that the difference in modern eukaryotes between nuclear-directed and mitochondria-directed protein synthesis and other metabolic activities is due to divergent selective pressures that have acted since the acquisition of mitochondria.

Bogorad offers an alternative hypothesis for the origin of the eukaryote nucleus and the DNA contained in plastids and mitochondria. Emphasizing the well-known fact that some of the proteins of which these organelles consist are coded by nuclear genes, while others are coded by genes located in the organelles themselves, he has suggested the "cluster-clone" hypothesis. He believes that all of the DNA found in eukaryote cells originally existed in a nuclear area, perhaps surrounded by a common nuclear membrane. Later, clusters of genes derived from the proto-nucleus became separated and surrounded by their own membranes, thus forming the plastids and mitochondria.

At first sight, the hypothesis of symbiosis and the suggested alternatives appear to be diametrically opposed. Actually, they are not so contradictory when one examines the qualifications made in each theory to harmonize with the facts. An intermediate theory, containing elements of both extreme hypotheses, may be acceptable. The biochemical similarities between plastids, mitochondria, and prokaryotic cells cannot be denied. One striking similarity is between the enzyme superoxide dismutase of chicken liver mitochondria and the corresponding enzymes in *Escherichia coli:* both are very different from the dismutases of nuclear origin found in bovine erythrocytes. Another similarity, which concerns the origin of chloroplasts, is the extensive homology between sequences of oligonucleotides found in the chloroplast DNA of the red alga *Porphyridium* and corresponding sequences recognized in prokaryotes, such as *E. coli, Bacillus subtilis,* and the blue-green alga *Anacystis nidulans*. Nevertheless, the similar sequences could have been introduced into the evolving prokaryotic cell in the form of plasmids, as Raff and Mahler have suggested. Whichever theory is accepted, evolutionary modification of both nuclear and organellular DNA during early eukaryote evolution must be postulated.

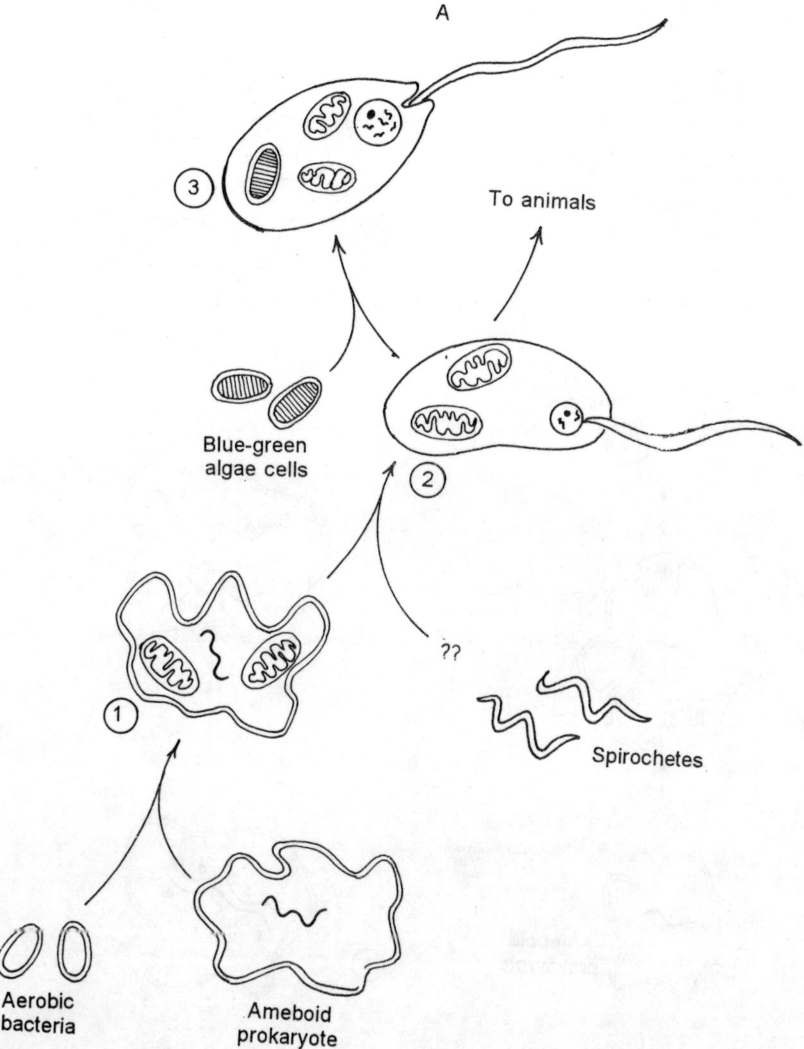

Fig. 11.3: The origin of the eukaryotic cell.

A. The *symbiosis theory* of the origin of the eukaryotic cell. *A-1*. A prokaryotic cell having a flexible cell wall, capable of ingesting bacterial cells. *A-2*. An intermediate cell, in which ingested bacteria are evolving into mitochondria. According to the theory, spirochetes and converted into the nuclear fibrils, centrioles, and flagella of a primitive unicellular eukaryote (shown immediately above the spirochetes). *A-3*. Blue-green algal cells became incorporated into the colorless unicellular eukaryote to form a simple green flagellate (top). Animals were derived from more highly evolved descendants of a colorless unicellular eukaryote, or flagellated protozoan.

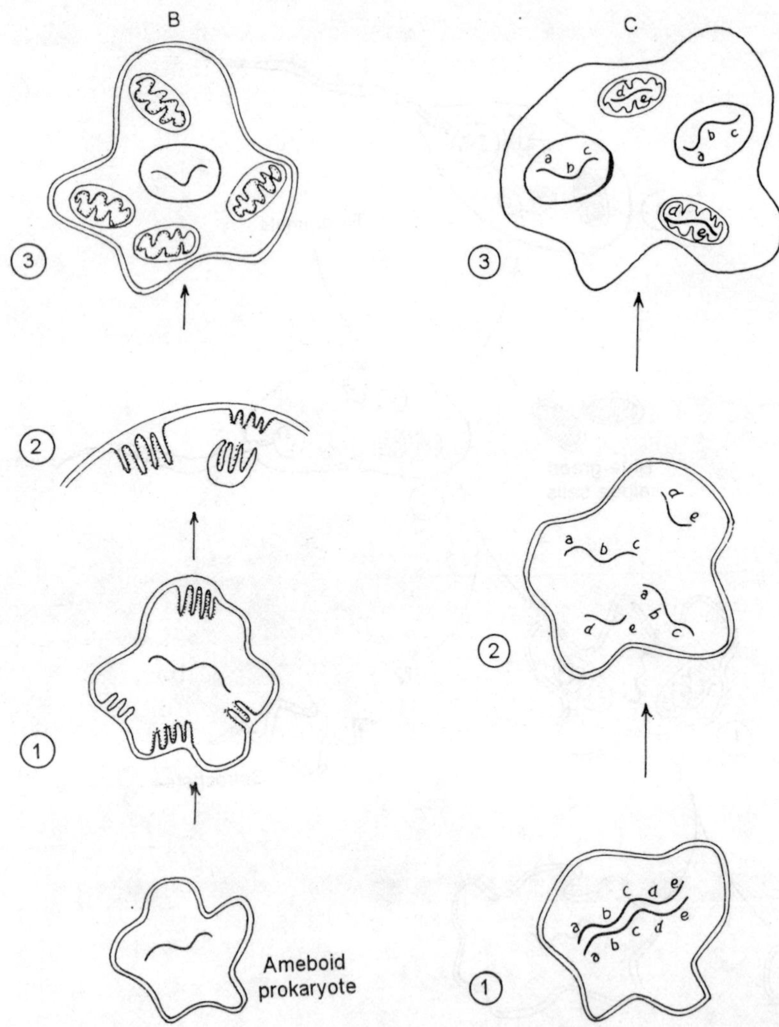

B. The origin of mitochondria according to the *invagination theory.* B-1. A prokaryotic cell similar to that shown in part *A.* Extensive invaginations are evolved, derived from the inner cell membrane. *B-2.* Enzymes associated with glycolysis and respiration become associated with these invaginations. *B-3.* The invaginated membranes lose contact with the cell membrane and become converted into mitochondria.

C. Bogorad's *cluster clone theory. C-1.* A prokaryote having a single duplicated strand of DNA on which are five genes (indicated by letters). *C-2.* A similar cell in which the DNA strand or chromosome has broken into two parts containing different genes. *C-3* Retention of some of these genes in a nucleus, which has acquired a nuclear membrane, and conversion of other DNA segments into mitochondria through the acquisition of separate membranes. Neither the method nor the source of these membranes is explained by Bogorad.

The strongest case for the symbiosis theory concerns the origin of plastids. The chlorophylls of eukaryotes are very similar to those of the prokaryotic blue-green algae; in addition, certain accessory pigments, particularly the phycobilins of red algae, are also similar. Furthermore, the cells of blue-green algae and the chloroplasts of red algae contain organelles known as phycobilisomes, which are so similar in their fine structure that they are almost certainly homologous. There are numerous other examples of unquestioned symbiosis between heterotrophic eukaryotes and blue-green algae. Two genera of unicellular flagellates, *Glaucocystis* and *Cyanophora,* which have typical eukaryotic nuclei, contain photosensitizing organelles (or perhaps symbionts) that in some respects are intermediate between blue-green algae cells and plastids. Furthermore, the chloroplast DNA of another autotrophic flagellate, *Euglena,* has been shown by hybridization techniques to be very similar to that of blue-green algae.

There is every reason to believe that some or all autotrophic eukaryotes have acquired their photosynthetic apparatus by evolution from blue-green algae. If one rejects the symbiosis hypothesis and assumes that plastids have originated by invagination, one must assume a direct connection via cell lineages from blue-green algae to autotrophic flagellates. This connection is very hard to accept, chiefly because of the great differences in the chemistry and the ultramicroscopic structure of cell walls between blue-green algae, the most primitive autotrophic eukaryotes, and typical green plants. The rigid walls of blue-green algae, like those of bacteria, have a thick network consisting of murein, which includes, covalently bonded together, N-acetylmuramic acid and a sugar, N-acetylglucoseamine. This complex is in no way related to the cellulose of higher plants, or to the mucilages based on galactose that form the cell walls of red algae. Furthermore, autotrophic eukaryotes described in the next section, such as dinoflagellates and *Euglena,* which have chromosomes and a mitotic apparatus intermediate between prokaryotes and eukaryotes, do not have rigid cell walls at all, but thin, somewhat gelatinous pellicles. One can logically postulate that organisms having thin flexible walls evolved from the lower grade of rigid-walled autotrophic prokaryotes and later gave rise to typical plants, algae as well as land plants.

Since in autotrophic organisms flexible cell walls are less adaptive than rigid ones, the loss of the rigid cell wall during the evolution of an autotrophic eukaryote would be very hard to explain. If, however, the first eukaryotic cell was heterotrophic and thin-walled, as postulated by the symbiosis hypothesis, the origin of the primitive autotrophic flagellates by ingestion of small blue-green algae and their conversion to plastids would be a plausible phenomenon, and one analogous to many well-documented evolutionary sequences. The acquisition of rigid cell walls having a different structure would certainly be adaptive in a number of ecological situations.

On the other hand, the internal membranes of eukaryotic cells, particularly the nuclear membrane and the endoplasmic reticulum, are best explained as invaginations.

Since the prokaryote chromosomes are attached to the nuclear membrane, at least when they divide, an invagination of this part of the cellular membrane would automatically place the chromosomes inside the nucleus. The connections and molecular similarity between flagella and centrioles could be explained by assuming that in the earliest eukaryotes the nuclear membrane was still attached to the outer cell membranes at the time when both the mitotic apparatus and flagella were evolving simultaneously. The DNA of the basal body of the flagellum could then be explained by the addition of a plasmid, as suggested by Raff and Mahler.

The most plausible explanation of the complex series of changes required for the origin of the eukaryotic cell is therefore that both symbiosis and invagination were involved. The most disputed problem, that of the origin of mitochondria, may become settled when more biochemical data become available. As *Taylor* (1974) has maintained, the origin of flagella and the mitotic apparatus in symbiosis is highly improbable. Indeed, at the present state of knowledge no plausible hypothesis exists about the origin of flagella and the mitotic apparatus.

Both the invagination and symbiosis hypotheses can provide equally well a clue to the ecological conditions that prevailed during the prokaryote-eukaryote transition, and the type of organism that carried it out. Either of these methods would occur most easily in a relatively large cell having a thin wall and flexible cell membrane, i.e., an amoeboid prokaryote. The only organisms of this kind now alive are the mycoplasmas. It is highly unlikely these organisms are the ancestors of eukaryotes, since they inhabit chiefly the blood plasma of vertebrates, but wall-less forms of conventional bacteria, known as the L-phase, have been induced by a number of cultural techniques. Some of them breed true, apparently as a result of appropriate mutations. When derived from bacteria having the less specialized gram-positive type of cell wall, these forms are adapted to media having an osmotic strength of NaCl as high as 0.25 to 1.1 M. One might imagine that free-living gram-positive marine bacteria, living in the ooze at the bottom of seashore pools, would find a medium of the right osmostic strength, and a habit rich in available organic matter. If a prokaryote should begin occupation of this habitat, the ability to ingest the organic particles, derived from decaying marine organisms, would have a high adaptive value, so that the selective pressure for losing the murein mesh and acquiring the amoeboid form of nutrition would be strong. At the same time, a strong selective pressure would exist for transferring the enzymes responsible for glycolysis and respiration, which in bacteria are embedded in the external cellular membrane, to an internal position as associates of separate organells. This could be done efficiently either by invagination or by ingestion of smaller bacteria. Subsequently, selection would favour increased activity of enzymes located in the internal organelles and inactivation of those remaining on the external cellular membrane, so that the latter would become devoted solely to such functions as selective permeability and extensibility in connection with the formation of pseudopodia and the ingestion of food.

Since the marine environment postulated would have a low oxygen content, particularly if the atmosphere above it contained less oxygen than the contemporary atmosphere, these first amoeboid prokaryotes would be anaerobes. However, once they had acquired the division of labour between external and organelle membranes postulated in the last paragraph, they would be ready for further evolution toward more efficient metabolism by cellular activity. For this new phase of evolution, aerobic metabolism, motility by means of flagella, and autotrophy, either separately or combined with each other, would have a high adaptive value. If the ameboid prokaryotes already had mitochondria or proto-mitochondria, the number of these organelles per cell, and hence the rate of aerobic metabolism, could be increased by their differential reproduction relative to cell division. Flagella could be acquired by ingesting various other prokaryotes and by digesting all of their parts except the DNA of their chromosomes, which could be broken up into plasmids that could either be rejected or incorporated into the nucleus depending upon the adaptive value of the genes contained in them. As already mentioned, autotrophy would be acquired by ingestion of small unicellular blue-green algae and, their modification into plastid symbionts.

The suggestion of Sonea (1972), that bacterial plasmids derived from ingested cells contributed greatly to building up the chromosomal complements of eukaryotic cells, is one of the most plausible explanations yet offered for this fundamental part of the evolution of the eukaryotic nucleus. The transitional forms may well have evolved endonuclease enzymes specially adapted to the conversion of bacterial DNA's into such plasmids. The existence in viruses of endonucleases having this kind of specificity has recently been demonstrated by several workers.

The partial endosymbiosis hypothesis, in the ecological version just proposed, is supported further by the reclassification of amoeboid Protozoa or Sarcodina proposed by Jahn, Bovee, and Griffith (1974). This greatly alters the phylogenetic position of the genus *Pelomyxa*, which Margulis singled out for particular attention because its cells contain large numbers of inclusions of various sorts. In earlier classifications of the Sarcodina this genus was regarded as relatively specialized, but according to Jahn, Bovee, and Griffith, it may be primitive.

The present version of the symbiosis hypothesis has a direct bearing upon transpacific evolution in general for the following reasons. First, it postulates that the event that triggered off this major advance in grade was one of many adaptive radiations carried out by anaerobic prokaryotes, and one that was distinctive because of the highly specialized and unusual environment the radiant entered. Second, it postulates that the one major way of food-getting that does not exist in free-living prokaryotes—ingestion of solid particles—was the primary method in eukaryotes. This is in accord with the fact that ingestion is more common among contemporary heterotrophic protists than is the bacterial method of external digestion and absorption. Finally, it postulates that the evolution of autotrophic organisms did not follow a direct line from blue-green algae to eukaryotic protists and algae, but involved the intermediate stage of heterotrophy followed by symbiosis.

These ideas are all completely compatible with the hypothesis of Margulis as well as with that of Raff and Mahler, at least concerning the mitochondria, the only organelles the latter authors considered. It is incompatible with the cluster-clone hypothesis of Bogorad, which postulates a direct continuity from blue-green algae to photosynthetic eukaryotes.

The Origin of Mitosis and Sex

After the eukaryotes originated they evolved for at least 600 million years, and probably longer at the unicellular level. This evolution included the invasion of a large number of habitats, although present data do not permit definite conclusions as to what this adaptive radiation was like. Nevertheless, the perfection of the mitotic cycle and the origin of meiosis and the sexual cycle most certainly took place during this period.

Analyses of cell division in some group of flagellates give clues to the origin of the mitotic cycle, including chromosome splitting, condensation of chromosomes during prophase, relaxation at the end of telophase, and the formation of the mitotic spindle. Of greatest interest are the dinoflagellates, which have a kind of nucleus designated as mesokaryotic. Their nuclear membrane contains pores, and they have nucleoli like eukaryotes. The dinoflagellates contain several chromosomes rather than a single "genophore" as in typical prokaryotes. Their chromosomal fibres, however, are densely packed together in a very different way from those of eukaryote chromosomes. Furthermore, the metabolic nuclei, between mitosis, contain chromatin that is nearly or quite as condensed as that of metaphase and anaphase chromosomes; the cycle of condensation and elongation of characteristic of other eukaryote chromosomes is absent. They contain acidic proteins, but little or no basic protein or histone. During cell division, typical mitotic spindles are not formed, and the chromosomes remain within the nuclear membrane. The separation of the daughter chromosomes apparently results from elongation of this membrane, and so is somewhat similar to the separation of chromsomes or genophores in a dividing bacterial cell. The cytoplsmic organelles of dinoflagellates are typical of eukaryotes. The situation in *Euglena* is somewhat similar. On the basis of these observations, as well as a less anomalous but still aberrant situation in a parasitic protozoan, *Syndinium*, Ris and Kubi (1974) distinguish three successive stages in the origin of typical eukaryotic mitosis. Whether or not this scheme has general validity will not be clear until more transitional examples have been studied by means of modern electron-microscopic techniques.

With respect to the origin of sexual reproduction, two challenging questions present themselves. First, in what kinds of organisms did sex first arise? And second, what was the adaptive advantage that caused sexual reproduction to become predominant in higher organisms?

Until very recently no clear answer could be given to the first question. A survey of modern unicellular organisms conducted that those having the least-specialized cell structure—Chrysomonadina, Cryptomonadina, Euglenoidea, Dinoflagellata, and Amoebina—are not known to possess sex. Recently, however, earlier, doubtful records of a sexual cycle have been confirmed for the dinoflagellate genus *Crypthecodinium* (Gyrodinium) by clearcut morphological, cytological, and genetic evidence. Since this organism is mesokaryotic, its sexuality suggests strongly that sexual cycles evolved contemporaneously with the perfection of the mitotic cycle. Moreover, the sequence of meiotic divisions in this genus is not fully developed. Most probably, flagellate and ameboid protists with a fully evolved mitotic cycle but not known to possess sex either have an undetected sexual cycle or are asexual derivatives of sexual ancestors. The presence of fully developed sexual cycles in more specialized protists, such as *Chlamydomonas* and its relatives *Actinosphaerium* and *Actinophrys*, polymastigoid flagellates, Foraminifera, Sporozoa, and Ciliates, is in harmony with this conclusion. Paleontological data suggest that both mitosis and meiosis were present in microorganisms, probably algae, preserved in the Bitter Springs Formation of Australia, about 900 million years old.

The adaptive value of the sexual cycle has recently been debated, and the conventional statement, that it favours new adaptive combinations of genes, has been challenged. Williams points out that a well-adapted diploid genotype, in order to produce offspring that are equally well adapted to its particular environment, must pay a "50 per cent cost of meiosis." This is because each gamete produced by meiosis contains only 50 per cent of the parental alleles, including, presumably, those that compose the organism's adaptive gene combination. They are normally replaced in the zygote by genes derived from another individual, which may or may not be equally well adapted to the environment of the parent being considered. Obviously, if an organism occupies a constant, homogeneous environment, natural selection will favour those offspring that are genetically most like their parents. Since such offspring are always produced by asexual reproduction, the initial steps of evolution toward sexuality will lead to offspring with lowered reproductive fitness relative to the asexual offspring, and so will be eliminated by selection. The argument that sexuality can evolve because of its long-term advantages, particularly in a changing environment, is regarded as inadmissible by Williams, since natural selection is not predictive but operates from generation to generation.

According to Williams, the solution to this dilemma lies in the hypothesis that the first sexually reproducing organisms possessed an alternation between sexuality and asexuality, each process serving a somewhat different function. Asexuality served to maintain the population in a particular habitat, while sexuality made possible the exploration of new and slightly different habitats, thus increasing the ecological and geographic amplitude of the species. Examples of organisms having this duality are hydroid corals and molluscs among animals, and perennial plants, particularly those that, like the strawberry, reproduce efficiently by such vegetative structures as stolons, rhizomes, and bublets. Efficient exploration of the environment calls for cross-fertilization at random and high fecundity. This produces a great excess of gametes and zygotes with slightly different genotypes. If these can move about actively or passively, each habitat similar to the parental one can be colonized by a large number of genetically different zygotes or young offspring. Natural selection can then weed out those that are poorly adapted and favour those that are particularly well suited to the new environment. If the latter become capable of asexual reproduction, they can "capture" the new habitat and continue the cycle.

The early phylogenetic history of multicellular organisms support this hypothesis. Most of the phyla and classes of Metazoa, except for terrestrial vertebrates, arthropods, and molluscs, include some organisms that have efficient asexual reproduction, or at least the capacity to produce a great excess of gametes during their lifetime. Sponges (Porifera), colonial coelenterates, pelecypod molluscs, most algae, bryophytes, most spore-bearing vascular plants, and the great majority of seed plants are all good examples. Unfortunately, the exceptions—animals having either shorter lives or lower fecundity or both, as well as annual flowering pants—include nearly all of the multicellular organisms that have been studied intensively by geneticists. Hence, according to Williams, our knowledge of population genetics is biased toward those organisms with relatively low fecundity. Recent information about the gene pools of the more fecund organisms, based upon allozyme studies, should help to remove this bias.

Williams' hypothesis could explain why sexual cycles are rare or absent in free-living protists, such as *Euglena* and the majority of flagellates that make up the nannoplankton of the oceans. Once an organism has become adapted to the planktonic mode of life, exploration of new habitats is no longer necessary, and natural selection favours maximum production of genotypes similar to the successful parent and obtained by asexual reproduction.

Combining Williams' hypothesis and the one previously suggested for the origin of eukaryotes themselves, the following hypothesis may explain the very first origins of sexuality in primitive eukaryotes. These ameboid cells were asexually able to reproduce genetically similar ameboids, and thus could saturate the habitat of a particular tide pool. But without sex their future success depended either upon considerable phenotypic flexibility or the ability of passively dispersed propagules to find a new habitat sufficiently similar to the old one to make new colonization possible. Colonization might be aided by the evolution of active movements propelled by flagella. Actively motile propagules could also acquire the ability to attract each other by chemical means and so produce zygotes. If the latter then should settle down in a new pool and give rise by meiosis to four ameboid progeny, the chances that one of them could successfully colonize this new habitat by asexually produced progeny would be greatly increased. The hypothesis suggests that the first sexually reproducing eukaryote protists may have resembled the modern cellular slime molds.

It must be emphasized that the "cost of meiosis" argument applies only to diploid organisms. Protists, such as *Chlamydomonas,* as well as many green algae, lower fungi, and probably the earliest sexual organisms, such as *Crypthecodinium,* are strictly haploid or monoploid during their asexual or vegetative reproduction; the only diploid cell is the zygote. Such organisms can evolve mechanisms for genetic exploration of new habitats without paying any cost at all. Consequently, the really critical question to ask is: What was the initial advantage of diploidy, and why is it almost the only condition present among all phyla of Metazoa?

The most plausible answer to this question is that the first diploid organisms possessed marked heterosis or hybrid vigour. The advantage of delaying meiosis until many cell generations after fertilization (so that the somatic cells became diploid) was a particular rather than a general one. It existed only in the progeny of two parents having good combining ability. Diploidy helped to stabilize this highly adaptive heterotic condition. If diploidy and heterosis then became regularly associated in the population, reversions to haploidy would have been less fit and would have been eliminated. Later, the diploid condition made possible the enrichment of the gene pool through mutations, particularly those that gave rise to recessive alleles that were eliminated with difficulty or were retained indefinitely because of their contribution of superior heterozygotes. A high correlation exists between the diploid state and the existence of complex developmental patterns. This is most easily recognized by comparing parallel lines of evolution, one of them diploid (or predominantly so) and the other haploid. Examples are comparisons of the larger diploid brown algae with the haploid green algae; dikaryotic basidiomycete fungi with the monokaryotic ascomycetes; and the diploid vascular archegoniates ("Pteridophytes") with the haploid gametophytes of nonvascular ones ("Bryophytes"). These comparisons justify the conclusion of Raper and Flexer that the superiority of diploidy lies in the fact that two similar genomes, intimately associated, can exert subtle control upon minute steps of complex sequences of development and differentiation.

Another question raised by Williams (1975) is, "Why have most higher animals, which have acquired reduced fecundity and no longer perpetuate themselves by means of genetic exploration of the environment, nevertheless retained sexual reproduction, and are able to withstand the cost of meiosis?" He answers this question in part by noting that once sexuality has evolved into a highly complex and harmoniously integrated sequence of events, the pathway to successful asexuality passes through several intermediate stages that will usually be less adaptive. This adaptive "valley" is difficult to cross. A second answer for land vertebrates as well as many insects is that maternal care can provide a buffer that greatly reduces the significance of small genetic differences between sibling offspring. In birds and mammals the entire "cost of meiosis" is probably paid by parents in the form of feeding and otherwise taking care of their offspring.

The Origins of Multicellular Organisms

Even a cursory review of the simpler multicellular organisms indicates that the grade of multicellularity was achieved many times independently, and probably for rather different reasons in different groups. Recent intensive studies—particularly of cell-wall structure and chemistry; cytoplasmic organelles, including cilia and flagella; the number of nuclei per cell or coenocyte; the chromosomal or sexual cycle; and the plastic pigments in autotrophic forms—have tended to increase our estimate of the number of separate origins. The following list is a minimal number (17) of simple multicellular kinds of organisms that we believe have arisen independently from unicellular Protista.

Autotrophic. Red algae; brown algae; filamentous golden algae; Vaucheriales; siphonalean green algae; *Acetabularia* and other Dasychadales; Volvocales; Conjugales; Ulotrichales and other uninucleate green algae; phragmoplast-forming green algae (*Klebsormidium* line).

Ideally, one would like to explain each of these separate origins as the outcome of a particular kind of adaptive radiation at the level of unicellular eukaryotes. Unfortunately, this is impossible, owing to the absence of fossils of the delicate organisms the earliest members of each line must have been, and the absence from the contemporary biota, probably through extinction, of most transitional forms. In the case of the earliest plant groups a reasonable hypothesis can be formed based upon two facts. First, the common denominator of differentiation with respect to the vegetative parts is polarity. All groups that evolved beyond diversification within a single order acquired filaments or other cellular aggregates in which one or more cells are adapted to anchoring the plant to a substrate and others are differentiated into the tip of a filament or the margin of a flat, sheet like structure. Second, the simpler aquatic algae having polarity reproduce asexually by means of motile cells that are themselves polarized and resemble unicellular protists in all cytological details. When these cells settle down to form a filament, their front end, bearing flagella, becomes attached to the substrate, and the first cross wall of the future filament is formed at right angles to the polarized axis of the motile cell. Hence, the polarization of the filament, which eventually in higher plants became transferred to the polarization root-shoot, did not arise *de novo*, but was derived directly from that of the motile unicellular ancestor.

On the basis of these facts a plausible hypothesis for the origin of the multicellular filament has been suggested. Unicellular autotrophs could colonize turbulent of flowing water only if they became attached to a firm substrate, to avoid being swept away. The flaggella of many unicellualr forms are viscid and adhesive, and so would be the natural organelles for attaching the cell to the substrate. Natural selection under these conditions would favour cells having the most viscid flagella, and mutations that replaced typical flagella with more elongate, branching structures similar to the rhizoids of algal cells would have a particularly high adaptive value. Once this happened, reproduction would depend upon the ability to produce, by mitosis, cells capable of differentiating into motile zoospores, which would have a very high selective value. On the other hand, the most successful attached filaments would be those in which a considerable amount of photosynthetic surface was produced before zoospores were initiated. This could be acquired by cellular elongation before the onset of the first cell division, or by mitotic divisions that produced cylindrical, photosynthesizing filament cells before producing cells capable of differentiating into zoospores. Ever since the nineteenth-century observations of Julius Sachs, botanists have known that when a plant cell becomes relatively long and narrow, the mitotic spindle becomes oriented parallel to its long axis and the first cross wall is formed at right angles to this axis. Hence, selection for increased photosynthetic surface would result directly in the evolution of a polarized filament. Further evolution forming branched filaments, flat plates, or three-dimensional structures would depend upon processes that as yet are not understood.

There is no reason to believe that the origin of multicellular organisms, any more than the origin of the eukaryotic cell, depended upon processes other than natural selection of genetic combinations fitting the earliest multicellular plants and animals to particular environmental niches. The most general cause favouring the later success of multicellular organisms was the adaptive value of increased size.

Chapter—12

Origin of Metazoa

Metazoa may be defined as animals that share a common multicellular ancestry with coelenterates and flatworms. Although there is a good fossil record of the major groups that have well-mineralized skeletons, the origins and earliest evolution of the metazoan phyla cannot be documented from fossil evidence. Nevertheless, from the fossils that we do have, from comparative studies among living organisms, and from interpretations of the functional significance of primitive characters, it is possible to produce models of metazoan origins and diversifications.

As noted already that there are several types of fossil remains, and each has a different probability of preservation. We shall place them in three classes here: trace fossils (trails, burrows, tracks), soft-bodied fossils (chiefly impressions or carbon films); and rigidly skeletonized fossils, which are usually impregnated with minerals (chiefly carbonates or phosphates). Although soft-bodied animals frequently posses hydrostatic skeletons, fossil skeletons are of mineralized or protein-tanned structures. Both soft-bodied and rigidly skeletonized fossil are called *body fossils*. So far as the record is now understood, trace fossils appear earliest, then soft-bodied forms, and finally mineralized skeletons.

Traces left by the activities of organisms are difficult to discriminate from sedimentary structures left by many inorganic processes. Although there are occasional claims of very old animal fossils, these can usually be clearly ascribed to inorganic causes; a few remain doubtful. The earaliest traces that appear likely to be valid organic traces were recently discovered in Zambia. They are probably near 1,000 million years old. They are horizontal burrows, up to about 10 mm in diameter, that have been back-filled with sediment, presumably by detritus-feeders that ingested the sediment itself. The next earliest known burrows may be less than 700 million years old. The long barren interval between these fossils is puzzling. Sediment deposits from this interval, confidently interpreted as shallow marine deposits, have been searched for fossils without success. The burrows that appear sometime after 7 million years ago (their ages are not closely determined as yet) include horizontal and vertical burrows and sea-bottom trails. The abundance and diversity of traces increase progressively in younger rocks until, during the Cambria Period (beginning 570 million years ago), they reach a richness in shallow water that they maintain for hundreds of millions of years.

Among the earliest authenticated records of soft-bodied invertebrate fossils are those of the Ediacaran fauna, named for an unusually large assemblage from the Ediacara formation, South Australia. This fauna seems to have lived between at least 680 and 580 million years ago and is found on several continents that are at present widely dispersed: Europe, Africa, North America, and Australia. The fauna is principally composed of medusae, pennatuloids, and other forms that appear also to have been coelenterates, and a small variety of worms of uncertain affinities, but perhaps including annelid-like forms.

The next younger assemblage of soft-bodied fossils of importance is from the Burgess Shale, British Columbia, Canada. It is of Middle Cambrian age and postdates the appearance of mineralized skeletons by a few tens of millions of years. For over 65 years this fauna has been known to exist and is now undergoing a much needed reappraisal. The new results are most impressive. About a quarter of the fauna represents groups that have already appeared by mid-Cambrian time, many with mineralized skeletons. These include coelenterates, molluscs, trilobite arthropods, brachiopods, echinoderms, and hemichordates. Another quarter is represented by non-trilo-bite arthropods, including sorts that are otherwise unknown; sponges constitute another quarter. Of the remainder, many specimens represent the earliest undisputed records of living phyla: Annelia (which may be represented in the Ediacara fauna), Priapuloida, Nemertea, and Chordata. Finally, a few specimens belong to groups that are distinct from living phyla; they represent extinct phyla that do not seem ancestal to any living animals. There are at leas three of these; only one has as yet been adequately described and is interpreted as a lophophorate coelomate.

The contrast between this soft-bodied assemblage, teeming with invertebrate phyla, and the depauperate late Precambrian fauna thus far known, is striking. There are only a few other occurrences that provide us with important glimpses of some of the soft-bodied associations of later times. An important one is in the lower carboniferous of Illinois, which has yielded a form that is not readily assignable to any living phylum.

The earliest mineralized skeletons appear slightly earlier than the generally accepted base of the Cambrian, below the Tommotian Stage of the Russian platform. The fossils are minute and of uncertain affinities. Additional skeletons appear in the Tommotian, including small coiled shells that are probably molluscs, and small conical elements that may be associated with tentacle of primitive suspension feeders. In the next stage many more phyla appear, so that within the first two Cambrian stages, seven skeletonized phyla are known. The last skeletonized phylum to appear, the Ectoporocta, does so in the earliest Ordovician. Figure 12.1 summarizes our knowledge of the appearance in the fossil record of those phyla that have been formally described.

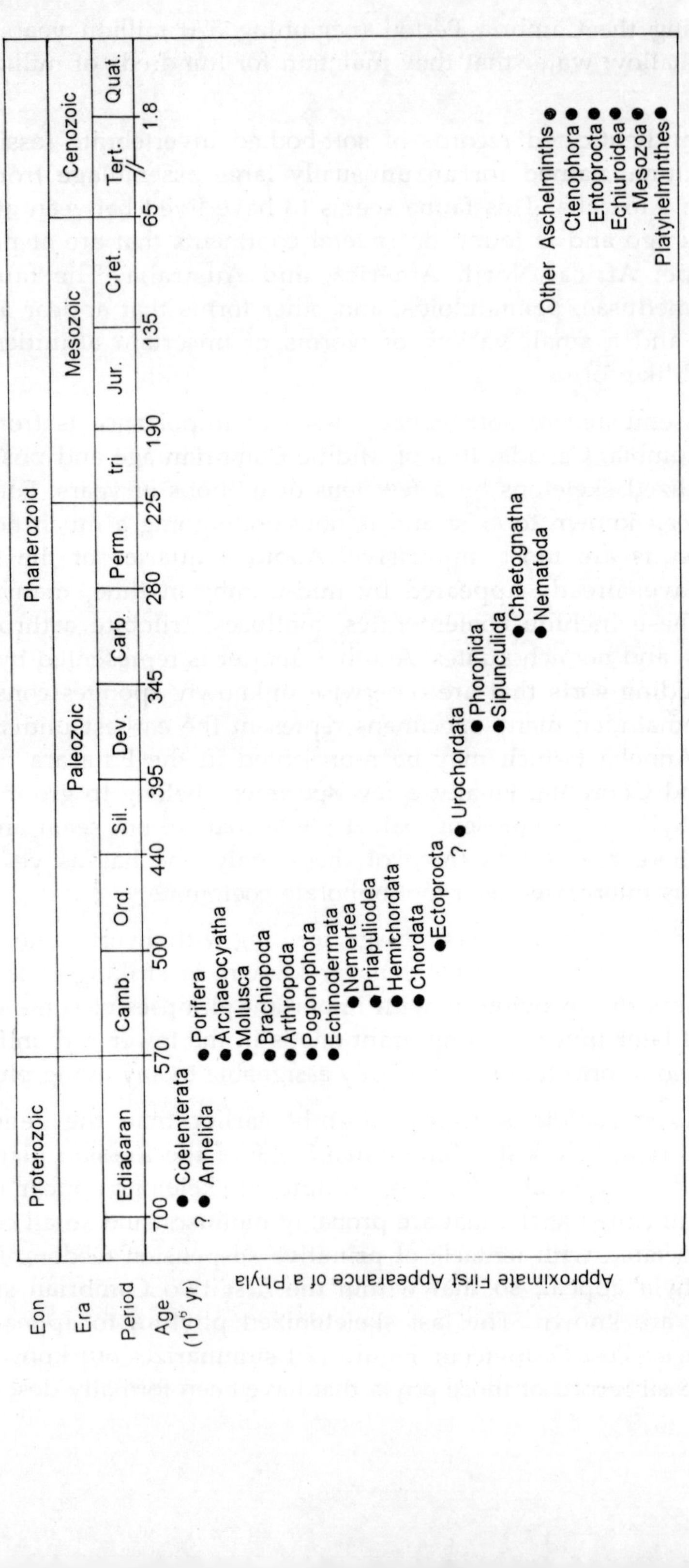

Fig. 12.1: Approximate times of the first appearances of animal phyla that are well-described. Undescribed phyla are also known from the middle Cambrian period (at least three) and the Pennsylvanian period in the Carboniferous (one).

During the Cambrian and Ordovician, radiations among the durably skeletonized phyla led to the establishment of numerous classes and orders, so that the number of taxa in those categories was considerably greater than at present for these same phyla. There is no reason to believe that the phyla for which records are poor—the soft-bodied groups—were not represented also by large numbers of classes and orders at that time.

To summarize, the earliest animal fossil (traces) may date from one billion years. Body fossils appear later and include coelenterates, possibly annelids, and a number of enigmatic forms. Then shortly after 570 million years ago, durably skeletonized forms that include several phyla of coelomate metazoans appeared in quantity. The next younger soft-bodied assemblage known is also rich in advanced metazoan types. By the early Ordovician, all living phyla that have durable skeletons are present; a large number of phyla that have no hard parts at all, including some extinct ones, also have been discovered. It appears that among marine invertebrates diversity of anatomical architecture as represented by phyla, and major modifications of these basic body plans as represented by classes and orders, reached higher levels in the early Paleozoic than ever again.

Evolution of the Tissue Grade

The earliest metazoans are unknown; the earliest yet described had already attained a tissue grade of organization. The adaptive and phylogenetic pathways that led to this grade have been inferred from functional and morphological studies of living organisms, which can be interpreted in a number of ways. As a result, there are a number of different historical models to explain both the evolution of metazoan grades and the radiations of body plans within grades.

It is widely accepted that metazoans evolved from protozoans, though there is a dispute as to the ancestral type. The first significant advance towards metazoans must have been the development of a multicellular organism. As we have seen, the step to multicellularity occurred several times. The ancestral type usually suggested for the metazoans is a flagellate protozoan, which developed a colonial habit not unlike living *Volvox*, leading eventually to obligate multicellularity. Hadzi (1963) suggests that the protozoan ancestors were ciliates and that multicellular metazoans arose from the division of a multinucleate syncytium. Among the difficulties faced by this hypothesis is the highly specialized nuclear behaviour of living ciliates, which is unlike that of metazoan cells.

The advantages of multicellularity per se flow chiefly from the sizes, shapes, and repetitions of cellular machinery that are achieved through a modular construction. Among them are increased homeostasis, longevity, and reproductive potential. Like the primitive eukaryotes, the metazoans must have evolved in the sea, where there would be numerous ecological opportunities for these primitive animals. For example, important adaptive zones in the sea contain detritus- or suspension-feeders, which utilize particles in the water column by sweeping them from the water or by receiving a rain of descending particles on the sea floor. In shallow water particles are likely to be somewhat patchily distributed, so that a large surface area for particle capture is advantageous. One way the primitive organisms can increase their areas is for their replicating cells to remain in contact. Uneven distribution of food items can then be corrected by transferring ingested materials or their degraded products across cell walls; in this way each cell has access to a more stable food supply, enhancing homeostasis.

The adaptation of the colony could be enhanced if growth patterns and morphogenetic movements were specified within the genome of the colony to control its size and shape. To accomplish this, the regulatory repertory would have to enlarge from control of individual cell functions to the integration of many genetically identical cells. Evolution of multicellular control could proceed through the intercellular exchange of substances, such as hormones or larger molecules originally employed in internal cell regulation but which become inductors. Once some genetic multicellular integration was achieved, variations in the shapes of colonies could have arisen through recombination and mutation, and natural selection would establish an array of shapes, each adapted to a distinctive range of habitats. For example, a dome-like or hummocky shape can be advantageous in benthic colonies because it increases the surface area of the colony. Furthermore, water currents flowing over irregular surfaces become turbulent, so that a hummocky colony living in fairly quiet water is washed by a disproportionately larger volume of water than a flat sheet of cells; of course, increased current flow increases the probability of food. Cylindrical or cup shapes also permit colonies to obtain food particles from a larger segment of the water column. Strands of cells extending above the colony proper serve a similar purpose.

Establishment of even simple shapes, while making for more efficient exploitation of resources by a colony, often results in an imbalance in food-uptake at different locations within a colony. In domical shapes, for example, the cells towards the center tend to receive more food than peripheral ones. Natural selection would promote differentiation to increase the feeding efficiency of the central cells, those chiefly responsible for nourishing the colony. Reproduction might well be suppressed in such cells but selected in well-nourished cells at neighboring localities, while peripheral cells would be differentiated for strength in support of the colony. Patterns of differential selective pressures on cells would vary among colonies of different shapes. Suspension-feeding colonies would be modified in shape to promote efficient current patterns for feeding. Other forms, especially those feeding on benthic detritus, would be required to creep along in order to exploit new areas of the substrate. A rim or some pattern of ventral cells could be specialized for locomotion, while other cells specialized in ingestion of food. In some lineages feeding cells came to be located in partial invaginations in order to increase cell number and localize digestive capability. Floating or swimming modes of life are often postulated for primitive multicellular colonies, with food items trapped by secretions or swept into pockets or locations specialized for accepting them.

During the evolution of intercellular communication, regulation, and differentiation, a point is reached where the cells no longer form a colony, but an integrated individual organism. We can imagine an array of simple organisms. The early forms at thus grade are all extinct. Although reconstructions are speculative, they indicate the selective advantages of small steps along an adaptive pathway from unicellular to truly multicellular organisms. Perhaps the most interesting hypotheses concerning the earliest metazoans are those by Haeckel (1874) and by Metschnikoff (1833). Haeckel postulated that metazoans arose through a hollow spherical layer of cells, the *blastaea*. The posterior cells of this sphere are inferred to have become specialized for feeding and to have invaginated, thus producing a two-layered *gastraea*. The interior cells became endoderm and the hollow of the invagination became a primitive gut. The evolution of a jellyfish-like coelenterate from such a primitive form can easily be imagined. Metschnikoff, however, believed that the primitive metazoan was solid, and that digesting (which is intracellular in lower metazoans) occurred in cells that filled the interior by ingression from the external cell layer. The planula larvae of most coelenterates is in fact a living analog of such a form. Invaginating to produce an endoderm and a primitive gut space is generally rare in metazoan ontogeny. To the extent that such ontogenies indicate phylogenetic pathways, the evidence favour Metschnikoff. However, if the very early ontogenetic stages evolved simply as effective developmental steps rather than reflecting phylogenetic development, then we are left without direct evidence of the precise phylogenetic and adaptive pathways during earliest metazoan evolution.

There are three living multicellular animal groups with especially primitive features; the Porifera (sponges), the Coelenterata (jellyfish, corals), and the Ctenophora (comb jellies). It is worthwhile to speculate briefly upon plausible pathways leading from simple multicellular forms to these groups.

The multicellular sponges display cellular differentiation, although they lack nervous systems or well-defined organs. Some are solitary, while others are considered "colonial" because they have numerous openings or osculae and can be considered as collections of individuals. However, there seems to be no differentiation of function whatsoever among individuals within a "colony." The body walls are composed of outer and inner cell layers separated by a gelatinous substance containing amoebocytes. The cell layers are not homologous with the ectoderm and endoderm of other animals; sponge development is quite unique. Most sponges secrete mineralized skeletal spicules, which are first found as fossils in the Lower Cambrian.

Since cells of the inner layer closely resemble certain flagellate protozoans, it is commonly suggested that sponges have evolved from such ancestors. Since their unique development suggests that they evolved independently of other animal-groups, they are often classified as Parazoa, separate from the Metazoa. It is quite possible that they evolved well after the Metazoa, which appear in the fossil record at least 100 million years earlier. Sponges illustrate that animals are perfectly viable at the simpler multicellular grades of organization.

Coelenterates possess two cell layers, ectoderm and endoderm—the *diploblastic* condition. Space between the layers is filled with a fibrous, gelatinous substance, the *mesoglea,* the jelly of jellyfish. Cell differentiation from the cell layers to form tissues is well advanced; gonads are present, and a relatively simple nerve net is developed. A number of evolutionary routes could lead to be coelenterate condition. A conical, detritus-feeding, multicellular animal with either internal or invaginated digestive tissues, which we have previously considered, has essentially reached a diploblastic state. However, such an animal would be a detritus feeder or a grazer and somewhat vagile, and would tend to be polarized anteroposteriorly and develop bilateral symmetry. Nevertheless, a coelenterate ancestor of this general sort has been postulated.

A diploblastic state could also develop via the evolution of a cup-shaped suspension feeder, the cells located centrally within the cup becoming specialized for ingestion and then becoming internalized to promote digestive efficiency. A sessile benthic habitat commonly correlates with a radial symmetry. There is evidence from comparative coelenterate embryology and anatomy, however, that the more primitive form is a medusa, a pelagic animal. Hand (1959, 1963) has argued that the coelenterates arose from a pelagic planula-like organism and that their radial symmetry is adaptive to pelagic life. The mesoglea in medusae is highly functional, serving to antagonize muscles and to provide support, shape, and floatation for the body. Rees (1966) has suggested a simple coelenterate ancestor that has four hollow tentacles and a tetraradiate stomach. He postulates a bentho-pelagic habitat for this form. One can derive such an animal from any of the more generalized multicellular organisms discussed above. Hadzi (1963, and references therein) has sought to derive coelenterates from bilateral flatworms, but this notion has not stood up well to critical examination.

Ctenophora or comb jellies are another diploblastic group that are sometimes treated as a subphylum of Coelenterata or as a separate phylum. However they may be classified, their origin appears to be tied up with that of the coelenterates, with which they share numerous common features. Clearly, the precise adaptive pathway and the precise succession of direct ancestral forms of the coelenterates is not known. However, our problem lies not in imagining a reasonable adaptive route, but in determining which among a wide variety of plausible possibilities is the historical pathway. In the absence of a fossil record we may never be able to decide this question definitively, although information on molecular evolution and increased understanding of the early environmental framework may limit our speculations. At any rate, the evolutionary processes reviewed in previous chapters are clearly adequate to develop the diploblastic condition.

One trend among the coelenterates is particularly notable as an illustration of the evolutionary tendency toward complexity. This is the development and elaboration of colony formation among individual multicellular organisms. It has occurred a number of times in Coelenterata as well as in other metazoan phyla. The general advantages of coloniality for coelenterates are similar to those that we outlined for primitive

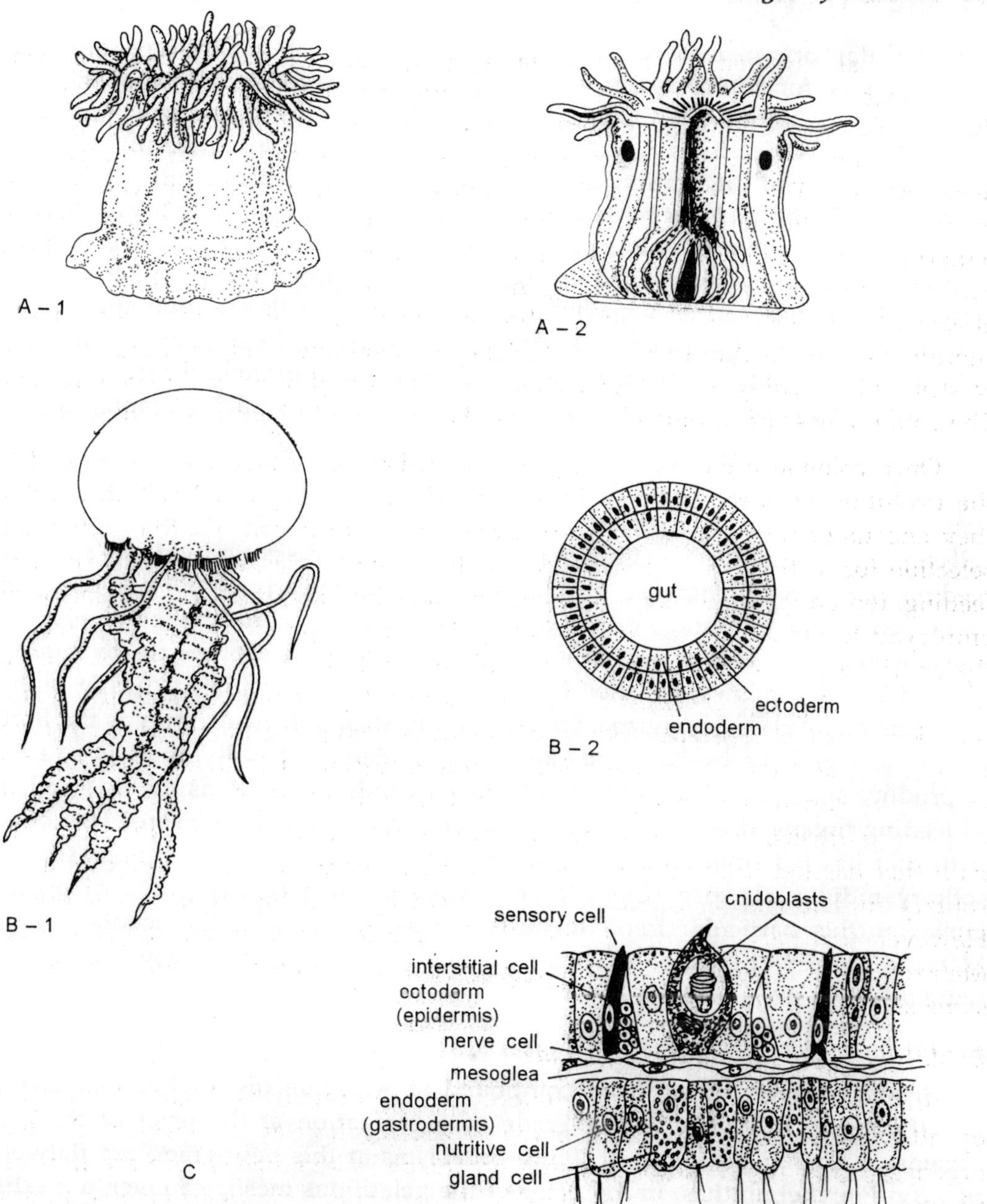

Fig. 12.2: Generalized architecture of some coelenterates. A. (1) A sea anemone (class Anthozoa); (2) vertical cross-section of a sea anemone showing tentacles surrounding mouth, gastric cavity, and an internal mesentary with associated muscles and gonads. B. (1) A jellyfish (class Scyphozoa); (2) horizontal cross-section of a simple jellyfish showing two-layered body wall. C. Close-up of a segment of a coelenterate body wall with detail of layers and cell types.

multicellular organisms. Noncolonial aggregations of individuals are commonly advantageous, for clumping can lead to useful surface topographies of the aggregates that are advantageous to the individuals, especially in suspension feeding and in reproduction. One disadvantage of aggregation is that it frequently results in overcrowding. The integration of aggregates into colonies can mitigate this problem by controlling the population density; the most advantageous spacing distances and patterns between individuals can be evolved, once these features are under genetic control. Furthermore, coordination among colonial individuals, which are physically attached by tissues, can be achieved through metabolites or common nerve paths. Such coordination can lead to improved efficiency in obtaining food, in eliminating metabolic wastes, or in warning of predatory attack or of the onset of other threatening conditions. The commonness of coloniality can easily be explained by such advantages.

Once colonial organization appears, a further advantage may be gained through the evolution of specialization. Since individuals are so placed within a colony that they encounter food, competitors, or predators more frequently than other cells, and selection for distinctive functions follows. In extreme cases, individuals specializing in feeding, reproduction, defense, or other functions lose nearly all but one major function, employed to the advantage of the colony. This development implies an elaboration of the machinery of genetic regulation. Colonial individuals originate by budding and have identical genomes, yet they become extremely differentiated during ontogeny. In the coelenterate class Scyphozoa, which includes such pelagic animals as the Portuguese man-of-war and the by-the-wind-sailor, differentiation of individuals is so extreme as to produce superorganisms of individuals. One individual forms the float, others serve as feeding organs, digestive organs, or reproductive organs. The same sort of adaptive path that has led from single cells to multicellularity has led a diploblastic line from solitary individuals to a colony that is virtually at a higher grade of organization. However, this particular trend towards complexity ends in the scyphozoans; higher Metazoa arose along other adaptive pathways, although in response to some of the same selective pressures.

Evolution of the Triploblastic Grade

Instead of proceeding from compound superorganisms, higher Metazoa evolved by attaining a truly new major grade of organization at the level of the individual organism. Among the simplest living organisms at this new grade are flatworms, the phylum Platyhelminthes. In the place of the gelatinous mesoglea layer, a meshwork of cells is developed to form a tissue, the *mesenchyme*. This is considered to represent a third tissue layer, or mesoderm, developed between ectoderm and endoderm in all higher Metazoa; the triple layered structure is *triploblastic*.

Flatworms are small elongate animals that employ one of three chief methods of locomotion. First, smaller species move by the activity of cilia packed on the flattened ventral surface. At larger body size the increase in mass is disproportionate to the increase in area and therefore the cilia become too few to provide adequate propulsion. Instead, locomotion is enhanced by the body wall musculature, which can provide great force. The second method is for these muscles to "inch" the worms along. A third method is for these muscles to cause waves (pedal waves) of alternate swellings and hollows in the body. The swellings adhere to the substrate, and the waves are passed backwards by coordinated muscular contractions. New waves form as the anterior portion of the body extends forward beyond the previous leading wave. The elongate, bilaterally symmetrical flatworm is well adapted to directional creeping on reasonably solid substrates. The relatively solid mesenchyme provides the internal rigidity required for locomotion by pedal waves. Flatworms cannot burrow well, although they can penetrate loosely packed sediments, such as sands, making their way within particle interstices.

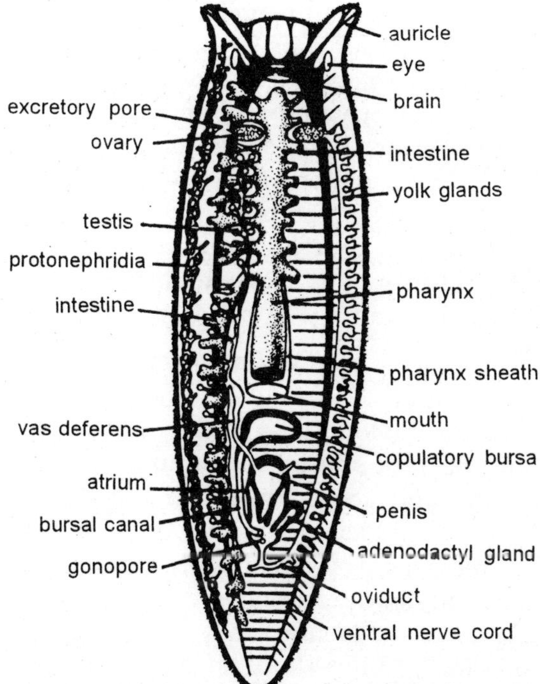

Fig. 12.3: Architecture of a typical flatworm. Note the serial repetition of intestinal diverticulae and consequent placement of series of gonads between them; nervous and excretory systems are also seriated.

The only simple living organism that much resembles the imagined flatworm ancestor is the coelenterate planula larva. It has been suggested that flatworms developed from early multicellular organisms resembling planulae. These ancestors may be imagined as metazoan larvae that underwent neoteny, thereby dropping the adult stages out of their life cycles (diploblastic forms, perhaps) and remaining on the substrate as worms.

A benthic diploblastic adult has also been postulated as the flatworm ancestor; some living coelenterates creep, and some ctenophorans, e.g., *Coeloplana,* are highly convergent with flatworms in general body form. However, the ctenophoran and platyhelminth ground plans are very distinct, and this pathway is unlikely. Hadzi. (1963) and some previous workers have proposed a phylogeny in which flatworms developed from ciliate synctia, which in turn gave rise to the coelenterates. Some difficulties with this hypothesis were mentioned earlier. At any rate, the rise of the triploblastic grade does not present any conceptual difficulties, although again we, cannot trace a precise phylogenetic route.

The development of the flatworm ground plan involved still other elaborations. Although the flattened form does not require specialized breathing apparatus (since oxygen exchange can occur over the general body surface), the elongate shape and triploblastic construction give rise to problems of nourishment and waste elimination. As body size increases, the interior cells become increasingly remote from food sources in the absence of a circulatory system. The endoderm-lined digestive tract opens ventromedially and runs both posteriorly and anteriorly to service the length of the organism. In some turbellarian flatworms the endodermal lining of the gut is greatly increased in area by a series of laterally paired diverticulae, which extend the digestive spaces into the lateral body regions. Excretion is provided by protonephridia, and reproduction is provided by special tissues that are concentrated into gonads. These organs are placed in the spaces between gut diverticulae, one on each side in series. This arrangement is of obvious advantage in serving an elongate organism.

The *pseudometamerous* body plan of flatworms, with a bilaterally symmetrical serial arrangement of internal organs, is found in other phyla, such as Nemertea. These are larger triploblastic worms that possess a few circulatory vessels beneath the body-wall musculature but still lack special respiratory organs. Many are far less flattened than the platyhelminths and some lineages have thick cores of mesenchyme. The thickened mesenchyme is associated with a type of locomotion that foreshadows developments in later metazoans: peristalic locomotion. Annular waves of contractions, passing entirely around the body, run consecutively backward along the body length. The dense mesenchymal tissue is not easily compressible, and when squeezed in by say a constricting crevice, it bulges out to each side, so that it forms a hydrostatic skeleton to antagonize the body-wall muscles. The worm flows forward as the waves pass backwards;this process is aided by ciliary activity on the body surface. Nemertines are thus able to burrow, for the peristaltic system permits the pushing aside of enclosing sediment. Nemertine burrowing is probably restricted by the relatively limited hydraulic efficiency of tissues, and by the lack of respiratory and advanced circulatory systems.

Evolution of the Coelom

A number of other triploblastic lower metazoan phyla are known, each presenting a fascinating puzzle in adaptation. For our purposes, however, it is best to examine the next evolutionary advance in grade, which freed the invertebrates from many ecological constraints and permitted their diversification into all the extant higher metazoans. This was accomplished by the development of the coelom, an internal, fluid-filled space surrounded by mesoderm and lined with a special tissue, the peritoneum. Unlike the gut, which represents a space that is continuous with the external environment, the coelom is entirely internal. It does communicate with the exterior through special ducts that transmit wastes (renal ducts) or reproductive products (gonoducts).

The origin of the coelomic cavity is uncertain; common theories are that it arose first as gonoducts, which were later expanded, or that it arose as out-pockets of the gut that migrated into the mesoderm. Clark (1964) has argued that the primitive function of the coelom is as a hydrostatic skeleton, and that the coelom arose during the evolution of peristaltic locomotion. Since there are a number of different invertebrate phyla, the coelomates may be polyphyletic.

Four or five basically different coelomate body plans can be recognized among living organisms, but we do not know which, if any, is the most primitive. Although the early coelomates have left fossil traces, these are not distinctive enough to permit the erection of phylogenetic trees. The simplest coelomate architecture consists of a hollow-bodied worm. Among living phyla, Sipunculida may have had a longitudinally unregionated coelom from the beginning. Other worms with annelidan affinities have little segmentation, but this is probably a secondary condition. Unsegmented worms are commonly detritus feeders, living in shallow burrows or nestling in borings or crevices and gathering food by ingestion of sediments or by trapping detritus on mucous nets. These worms are relatively sedentary. Introverts are commonly developed for burrowing and feeding.

A second coelomic architecture is found in forms that have two longitudinal coelomic compartments separated by a septum. This situation occurs among the lophophorate phyla Phoronida, Brachiopoda, and Ectoprocta. The phoronids are vermiform and may be closest to the common ancestral type of lophophorate. These worms are sessiel suspension feeders. Most form a burrow, from which they extend their lophophores above the substrate to feed. The lophophore consists of a set of tentacles surrounding the mouth. Within each tentacle is a coelomic lumen; these spaces are continuous within the tentacular crown, but are separated from the trunk coelom by a septum. The tentacles are manipulated by intrinsic muscles that are antagonized by the tentacular coelom. The cylindrical trunk with its coelomic cavity can antagonize body-wall muscles in burrowing. The hydrostatic systems of the trunk and lophophore function somewhat independently.

The deuterostome invertebrate phyla, the group from which the chordates have arisen, possess a coelomic plan similar to that of the lophophorates. These phyla include Echinodermata, Hemichordata, Urochordata, and some minor groups. The coelom in this alliance is primitively divided into three longitudinal compartments. Presumably this plan arose because three separate hydrostatic functions evolved. One compartment was probably used in burrowing, and another to manipulate tentacles or some similar structure for feeding. It has been suggested that the third compartment may have originally been associated with locomotion, perhaps to move up and down a burrow. Whatever their primitive function, the coelomic architectures of the deuterostomes are much modified during ontogeny.

The division of the coelom into two or three regions is termed *oligomerous,* a name that applies to both the lophophorate and deuterostome phyla. Rudiments of a third coelomic space may be present in some members of lophophorate phyla. Therefore, these phyla are sometimes considered to be descended from a stock with three coelomic regions and to share a common coelomate ancestry with the deuterostomes.

A much higher degree of coelomic partitioning, a *metamerous* condition, is found among Annelida. The body is divided into numerous longitudinal segments, each of which may contain a complement of organs, usually including a pair of nephridia, a pair of coelomoducts, a pair of nerve ganglia, a pair of lateral blood vessels, and frequently a pair of parapodia. In forms that are most vagile the segments are nearly all separated by transverse septa that prevent the transfer of coelomic fluid from one segment to the next, and insure that each segment deforms somewhat independently. Segments are tied together in groups of three or more by bundles of muscle fibres in the body wall. The internal fluid pressures resulting from muscular contraction during peristalsis are thus localized within a few segments, and succeeding peristaltic waves attain a large measure of mechanical independence, which makes for a relatively high locomotory efficiency.

In annelid lineages that have become largely sedentary, such as the deposit feeding burrowers (*Arenicola*) and tube-dwellers (*serpulids*), many of the septa are reduced or obsolete. These animals pursue a way of life reminiscent of the phoronids; their coelomic architecture is simplified although the regularized repetition of organ systems and musculature is retained, indicating that complete segmentation was the primitive condition. Annelid ground plans seem to have evolved primitively for efficiency in more or less continuous burrowing, a habit that is useful to infaunal detritus feeders and predators. Arthropods, which have descended from proto-annelidan stocks, have developed jointed exoskeletons and do not require a hydrostatic skeleton. Accordingly, there is no septation between their body segments.

A final type of coelomic architecture is displayed by the molluscs. Their coelom is neither regionated nor segmented; yet the primitive molluscs (Monoplacophora) display extensive serial repetition of organ systems, including gills (five pair), nephrida (six pair), pedal, musculature (eight pair each of two types), gonads (two pair), and a number of other organs. This sort of pseudometamerism recalls the flatworm condition. The monoplacophorans have rather extensive coelomic spaces. There are no data on monoplacophoran locomotion; presumably they move by pedal waves, like large flatworms and gastropods. If so, their coelom is not involved. A thorough functional account of the monoplacophorans is greatly needed.

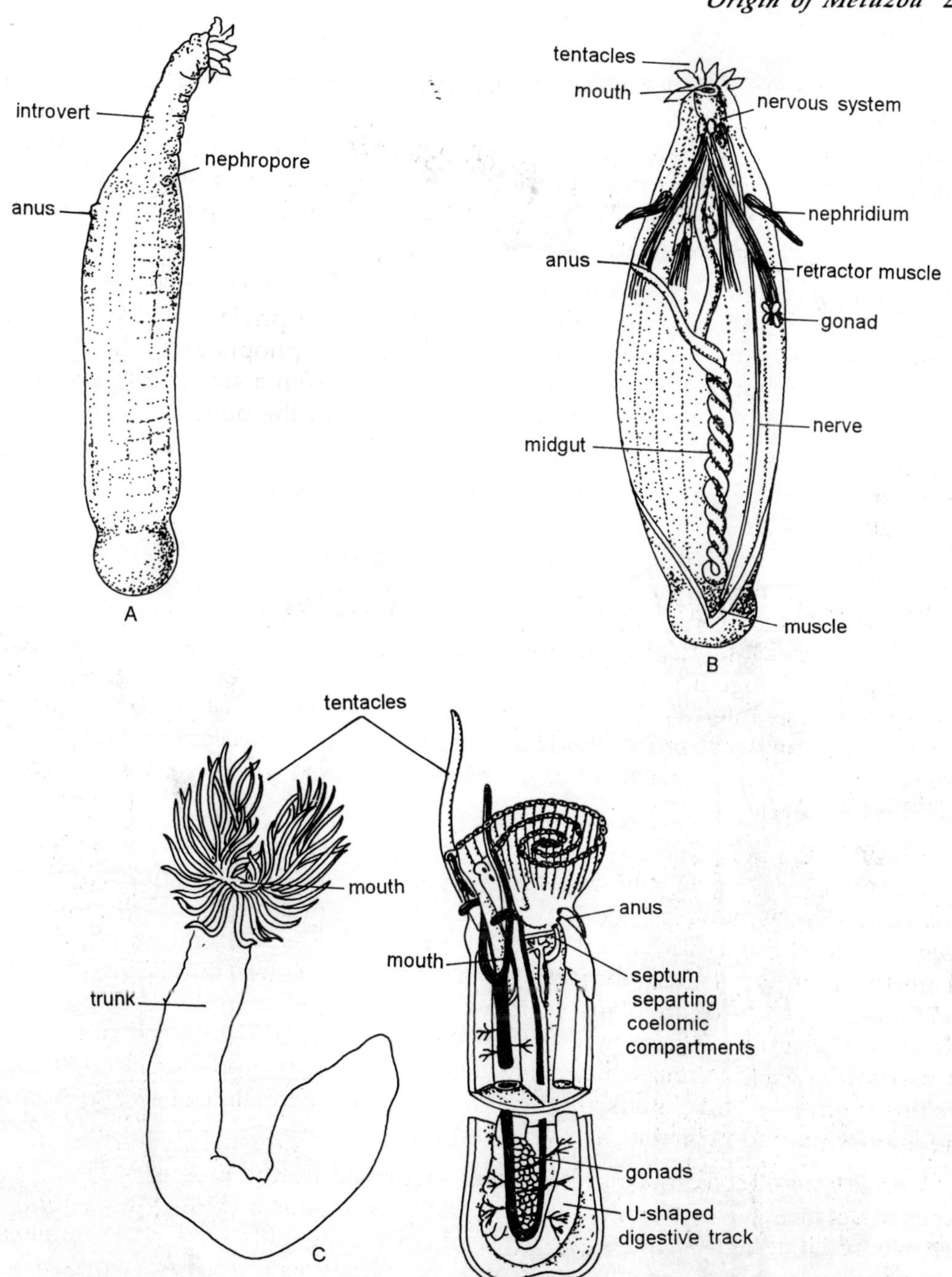

Fig. 12.4: Some coelomate architectures. A and B. Sipunçulids (*Sipunculus*), which entirely lack coelomic regionation (amerous architecture). C and D. Pheronids (*Phoronis*), which have a coelom divided into tentacular and trunk portions by a septum (oligomerous architecture). E and F. Annelids (*Arenicola* and *Nereis*, respectively), which have a highly

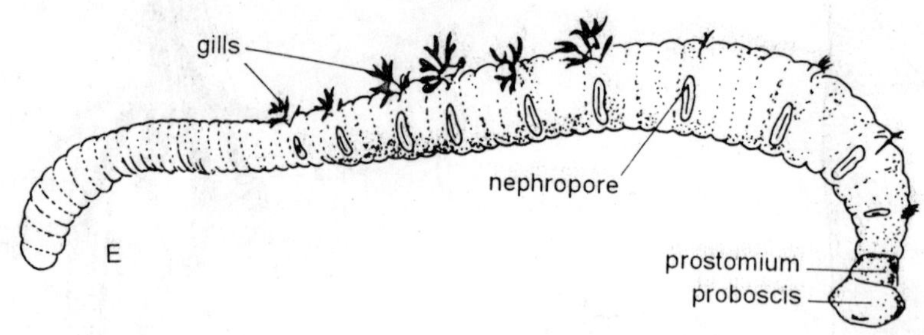

segmented coelom and regularized repetition of organs (metamerous architecture). G. A generalized arthropod, showing metamerous architecture; intersegmental septae are absent. H. A primitive mollusc (*Neopilina*, a monoplacophoran) showing serial repetition

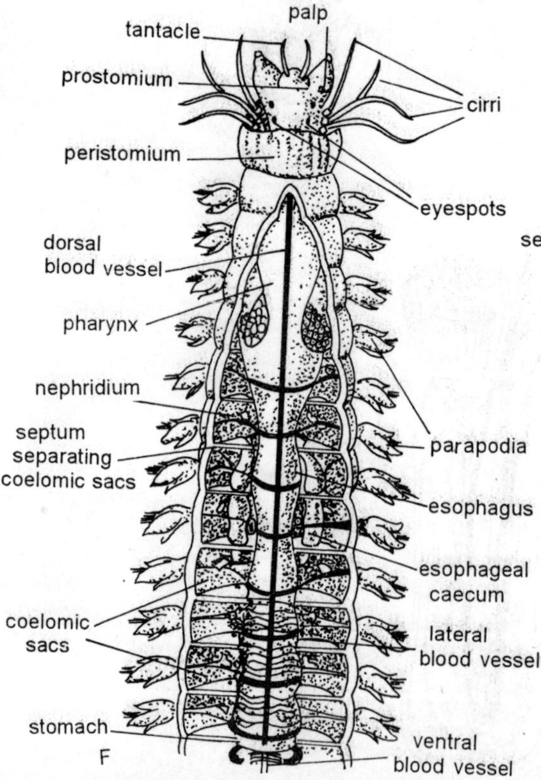

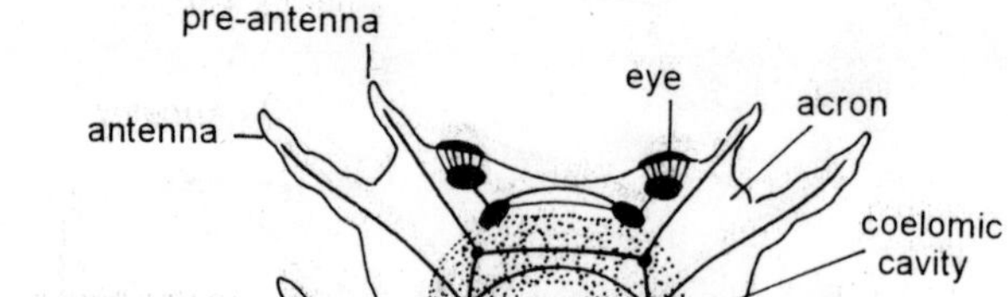

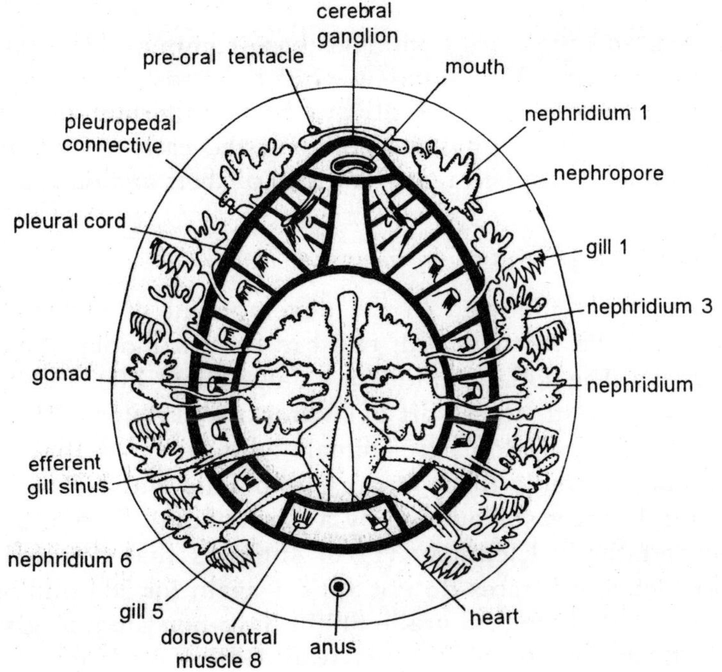

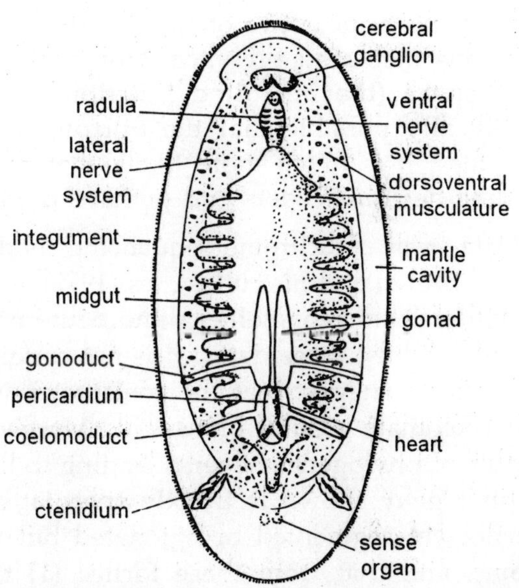

Fig. 12.4 (continued)
of some organs; there is however no longitudinal coelomic regionation or segmentation. I. A possible ancestral mollusc, with an architecture reminiscent of a flatworm.

Each of the distinctive coelomic plans can be explained as an adaptation to a distinctive mode of life. Each group has played an important role in the evolution of the biosphere, except perhaps the sipunculids. Lophophorates dominated many marine communities during the Paleozoic, and molluscs dominate these communities today. Arthropods are the most diverse terrestrial phylum today, while deuterostomes, which gave rise to the chordates and hence to man, have always been important marine forms. Even though we cannot yet certify a phylogeny among the early coelomate types, we can erect an adaptive model to explain many features of their anatomy and of the early fossil record.

Evolution of Early Metazoan Radiations

From the adaptive pathways likely followed in the development of metazoan ground plans, and from the data of the fossil record, one can erect a model of the early diversification of the metazoans. The horizontal burrows dated about 1,000 million years indicate the presence of animals with some sort of hydrostatic skeletons. These could be diploblastic but are more likely triploblastic animals with skeletons that are either mesenchymal (tissue), pseudocoelomic (body spaces with fluid but lacking a peritoneal lining), or coelomic (fluid spaces surrounded by a peritoneum). If so, much metazoan evolution had preceded them. Clemmay (1976) suggests that the early burrowers may have become extinct, since traces do not appear again for 300 million years or so. Of course, metazoans at the flatworm grade might have persisted, to give rise eventually to higher grades again. This problem must await further evidence.

From the appearance of the traces near 700 million years ago there is a record of continuous metazoan presence with increasing complexity of type and expansion of the ecospace (the inhabited functional regions of the bisophere). Although it is mechanically possible that the burrows of Ediacaran time were all made by non-coelomates, it is much more likely that primitive coelomate worms are represented, including detritus feeders and perhaps predators and suspension feeders.

Data from cytochrome sequencing tend to support this judgement. According to work by Brown and his colleagues (1972), annelids and deuterostomes separated about 750 million years ago, molluscs and deuterostomes about 720 million years ago. Further work may adjust these dates. It is not certain that the difference between them is real. The dates do suggest that the coelomates were present and that at least three of the major coelomate ground plans had diverged during the late Precambrian. A tentative schedule of phylogenetic events leading to living coelomate phyla is depicted in figure. The phylogeny shown is frankly speculative. We reason that the earliest coelomates were not yet segmented or egionated but were seriated, resembling flatworms with coeloms, and that from these forms: (1) the *metamerous annelids* diverged through coelomic segmentation, which promoted improved detritus feeding habits; (2) *oligomerous lineages* diverged as suspension feeders (lophophorates) living within the sediments of the seabed, and perhaps as sessile particle trappers (deuterostomes); (3) the *pseudometamerous molluscan* lineages diverged by development of a dorsal visceral mass, freeing the foot for creeping; and (4) *amerous sipunculids* diverged by suppression of seriation, developing an introvert for feeding. Whatever the actual pathways of descent may be, it is clear that the development of the coelom paved the way for new grades of organization. One immediate effect was to permit efficient borrowing into the sediments of the sea floor—an *infaunal* mode of life—opening up a region of biospace that today is densely populated by a large variety of animals. The diversification of coelomic architecture in the late pre-Cambrian is implied by the diversity of coelomic plans in Cambrian faunas.

With the widespread occurrence of durable skeletons in the Cambrian, the fossil record becomes much better. In all phyla the early skeletonized forms (for which modes of life have been worked out) were essentially *epifaunal,* that is, they lived upon rather than within the seafloor. Although they must have descended from ancestors with one or another of the coelomic architectures described, they were no longer vermiform, but had developed novel body plans.

Peristaltic locomotion is very inefficient on a surface, when only a short arc of a worm's cross section would actually contact the substrate. The energy required to maintain body shape and generate annular peristaltic waves would not be repaid by locomotory progress. One solution, followed by arthropods, was to develop a chitinous jointed exoskeleton with intrinsic musculature to operate appendages for locomotion. Arthropodization may have been the first adaptive pathway leading to a novel epifaunal coelomate. Another solution, favoured by unsegmented lineages, was to employ pedal waves, thereby consolidating body-wall muscles and moving the viscera into a dorsal position to free the locomotory foot. Body shape and muscular purchase is afforded by a consolidated body-wall musculature and a rigid exoskeletal cap. A seriated vermiform ancestor would result in a primitive monoplacophoran, creeping on or at shallow depths within the substrate.

Infaunal worms that were nearly or quite sessile seem also to have given rise to epifaunal descendants. Lophophorate oligomerous worms similar to the phoronids probably developed into solitary epibenthic forms, such as the brachiopods, along several different pathways, and also into the minute colonial epibenthic phylum Ectoprocta. Although the brachiopod skeleton seems to have developed in part for protection, it developed chiefly to enhance suspension feeding, first by framing and supporting the lophophore and later by bringing the feeding currents completely inside the mantel cavity, providing for increased homeostasis.

It is not necessary here to pursue hypothesis as to the rise of each epifaunal lineage. The central point is that the appearance of skeletons nearly 580 million years ago signals the rise of relatively diverse epifaunal communities, and suggests that a number of living phyla had their origin at about that time. As Cloud (1949) has pointed out, the ground plans of many of the durably skeletonized phyla do not make sense in the absence of the skeleton. The anatomical modifications associated with the shifting ecospaces of these lineages were coadapted with skeletonization. The brachiopods, for example, have a soft-part anatomy that is precisely adapted to reside within a bivalved exoskeleton, with pedicle, visceral, and lophophoral spaces and a complicated

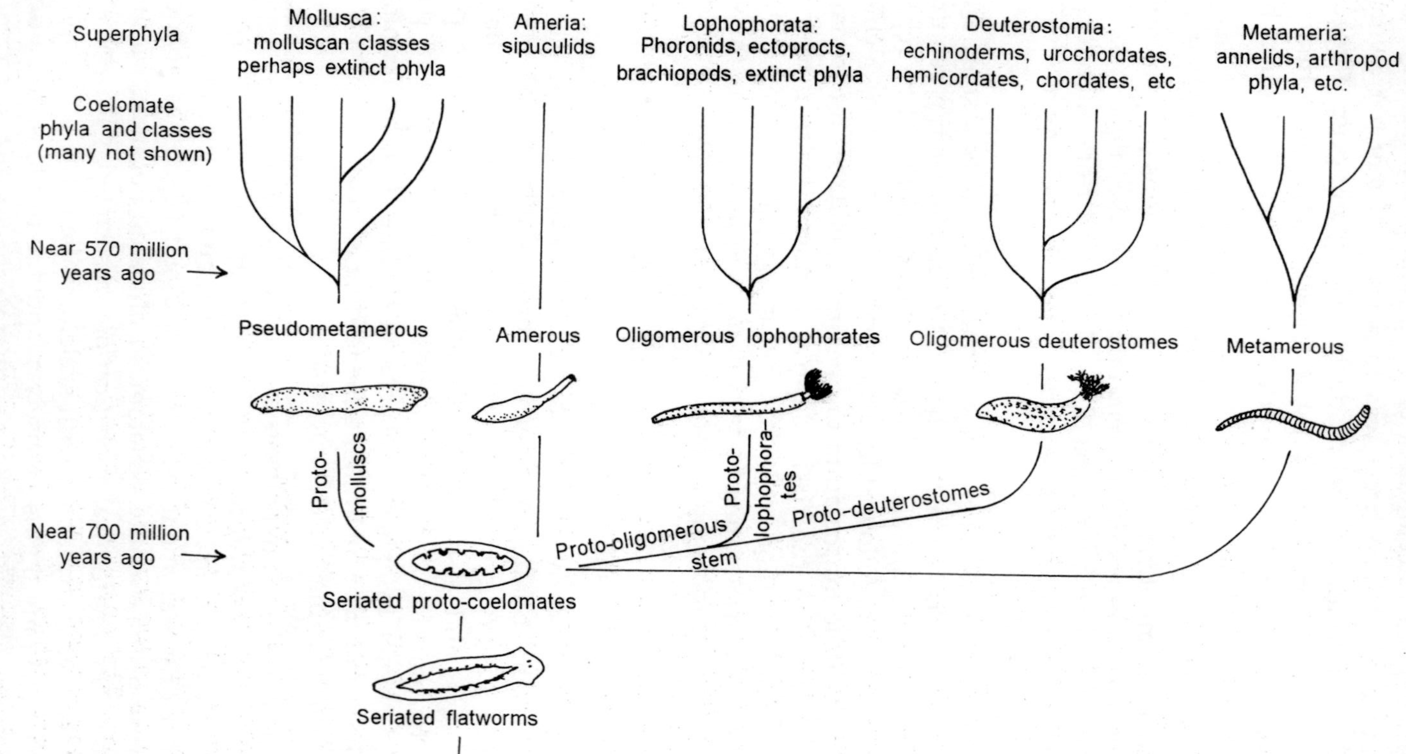

Fig. 12.5: Radiation of the major coelomate stocks from a flatworm ancestor. Primitive coelomates were presumably seriated like flatworms but lacking in segmentation, suggesting that some sort of proto-molluscan form may have been the earliest coelomate. From some such stock evolved several distinctive ground plans, including the metamerous one of annelids, the bi- or tri-regionated oligomerous ones of lophophorates and deuterostomes, and the unsegmented, unseriated one of sipunculids. The radiation of phyla and classes within the major lineages is highly diagrammatic.

musculature associated with valve functions. Without a skeleton, a brachiopod would have to be redesigned. The simplest such design would closely resemble a short species of phoronid, which may closely resemble the brachiopod ancestor. We can be reasonably certain that the appearance of the first primitive brachiopod skeleton in the fossil record is very close to the time of evolution of the association of features that constitute the brachiopod ground plan.

A similar line of reasoning holds for many other phyla and classes, although not for groups like corals and bryozoans, in which mineralized skeletons form no part of their anatomical ground plans. These two colonial groups do require rigid skeletons in order to occupy all of the habitats and to perform the array of ecological functions they have achieved, but their ground plans could have been evolved long before their skeletons appear in the record. Indeed, their skeletons first appear well after the bulk of invertebrate groups; perhaps they originated near the early Cambrian but did not become rigidly skeletonized until later. The arthropods present another special case. A rigid skeleton forms an integral part of their ground plan; but it is commonly formed only of organic material that does not preserve well. Aside from a few questionable impressions from Ediacaran time, we first see arthropods when a mineralized carapace was evolved by trilobites, very likely for shallow burrowing or grubbing to form superficial depressions or pits; possibly this was a feeding behaviour.

It seems, then, that quite a number of invertebrate phyla originated near the beginning of the Cambrian. It is not known just what events localized the important radiations at this time. A significant environmental change must have occurred. One possibility is that oxygen levels, which had been rising as a result of growing photosynthetic activities, reached a critical level that permitted the appearance of metazoans in the late Precambrian, and then a higher critical level that permitted the appearance of mineralized skeletons near the lower Cambrian boundary. Another suggestion is that plate-tectonic activity resulted in a geographic configuration that permitted a reduction in seasonality, with a consequent increase in the stability of trophic resources. This might have permitted the rise of diverse epifaunal communities. In any case, as the epifaunal invasion proceeded there were radiations of each major coelomate type into a number of lineages, many of which have skeletons. The novel body plans of many phyla with mineralized skeletons, and probably of soft-bodied phyla, must have evolved relatively rapidly, and this must have been due in large part to changes in the regulatory portion of the genome. Three aspects of regulatory change make this plausible. First, old regulatory functions may continue to operate as new ones evolve, thus maintaining adaptation of the lineage during important evolutionary trends. Second, repatterning of the regulatory apparatus may achieve new architectural relations by employing components that are already well developed, thereby altering the relationships among structural genes, batteries of genes, and sets of gene batteries through a hierarchical apparatus. Third, the morphological changes that arise from spatial or temporal additions or repatterning of gene batteries commonly involve the ground plans of the organisms, and the resulting morphological distances between ancestor and descendant are therefore considered to be large. This is especially true when such changes are compared to the morphological results of structural gene evolution, which tends to develop only general morphological variations. The results of regulatory evolution appear to be greater and to occur more rapidly because they can achieve novel biological architectures and may affect numerous gene simultaneously.

Evolution of the Chordates

For our purposes it is best to restrict the phylum Chordata to the subphyla Cephalochordata and Vertebrata. The most primitive living chordates are probably members of the subphylum Cephalochordata, typified by amphioxus. Amphioxus has a solid dorsal supportive structure, the notochord, which is regarded as a precursor to the vertebral column. The notochord is neither cartilaginous nor bony, but consists of a sheathed series of vacuolated cells that clearly function as a supporting structure. It is present in vertebrate (including mammalian) embryos, and perists into the adult form of some fishes, such as lampreys and even sturgeons. Despite its notochord, amphioxus is a rather sessile filter feeder. It burrows in sand and ingests water through the mouth, expelling it through gill slits—the feeding method inferred for primitive fishes.

A similar feeding system is found in two invertebrate deuterostome phyla: Hemichordata (enteropneusts, pterobranchs), which lack a notochord, and Urochordata (tunicates), which have tadpole-like larvae that possess notochords in their tails. The gill slits of these invertebrates are very similar to those of the primitive vertebtates, and most workers have concluded that they are all inherited from a common ancestor.

The earliest known vertebrate fossils are scraps of bone of early Ordovician age that in all probability represent a group of jawless fishes, the ostracoderms. These primitive fishes have a well-developed dermal armour of bone. They do not have paired fins like more advanced fishes, and have very small mouths. Presumably they were filter feeders like amphioxus, ingesting water through the mouth and expelling it through gill slits. The ostracoderms became quite diverse and evidently gave rise to jawed fishes with paired fins, which appear in Devonian rocks. Thereafter the ostracoderms tended to become more restricted, and most groups died out. Modern lampreys may represent an ostracoderm lineage that has retained many primitive features—they are jawless and lack paired fins, for example—while losing their bony skeletons. It is likely that in lampreys and some other fishes cartilage is an embryonic adaptation retained in the adult stage and is therefore a neotenous feature in such forms.

A plausible deuterostome phylogeny begins with infaunal oligomerous worms that feed by means of a tentacular system resembling a lophophore. From this stock epifaunal forms evolved, probably in the late Precambrian or early Cambrian. These included Echinodermata, a phylum in which the tentacular system was elaborated into an extensive coelomic water-vascular system. This group developed a unique skeletal ground plan based upon numerous small plates or ossicles, and diversified into a large number of classes and orders during the early Paleozoic. Other early epifaunal deuterostome lineages remained without mineralized skeletons, and from these sprang the hemichordates. They developed gill slits, presumably to enhance the feeding system. The slits provided an outlet for feeding currents, which could be directed into the mouth by cilia associated with the tentacular system. Perhaps the tunicates evolved from a hemichordate stock; they have become completely sessile as adults and seem little more than filtering baskets.

The chordates seem to have arisen from an early deuterostome invertebrate. Since the tunicate larvae possess notochords and gill slits and somewhat resemble our expectations for a primitive chordate, it has been suggested that such a larva might have evolved into a chordate by developing the ability to reproduce in the larval stage—a classic case of neoteny, if true. It is also possible that adutimembers of an early deuterostome worm lineage became swimmers, and that the notochord was elaborated to enhance this function. The urochordates could be an offshoot of this hypothetical stock, and amphioxus a more direct descendant. Still another phylogenetic hypothesis, argued most forcefully by Jeffries (1967, 1968), derives the vertebrates from carpoids, bilaterally symmetrical organisms with strong calcareous platy skeletons that are usually assigned to Echinodermata. Carpoids appear in the earliest Middle Cambrian and disappear in the Middle Devonian. Jeffries views them as a transitional group between echinoderms and chordates, with their more advanced members resembling huge armoured tunicate larvae. However, the earliest known chordate is soft-bodied and is also of Middle Cambrian age.

Vertebrate bone is certainly a unique skeletal material, but many of the skeletonized invertebrate phyla also have unique shell structures. For epifaunal invertebates and for the chordates, the appearance of mineralized skeletons seems to have heralded a new level of success, or at least diversification. Skeketonizatons was usually associated with the establishment of a new anatomical architecture that was coadapted with the skeleton to function in a novel environment or novel mode. Many coelomates became skeletonized once the epifaunal ground plans evolved, and commonly the skeleton proved a key to much further elaboration and diversification. The most general reason for the success of mineralized skeletons seems to be that they employ much less energy and are simpler to maintain than skeletal systems based on fluids.

Thus, for functions that require support and rigidity, mineralized skeletons are superior. Functions that require a variable skeletal system, such as burrowing, are performed efficiently by hydraulic systems; fluid skeletons are often employed for such purposes. Mineralized skeletons can be put to many uses other than support, such as protection and thermoregulation, and so are often subjected to a variety of positive selective pressures. It is therefore not surprising that once a skeleton has appeared within a lineage it should be extensively developed in a short time, and that skeletonized lineages should be potentially versatile and frequently capitalize upon environmental opportunities to become highly diverse. Skeletonization is frequently a significant component of the adaptation that permits invasion and exploitation of a new region of biospace.

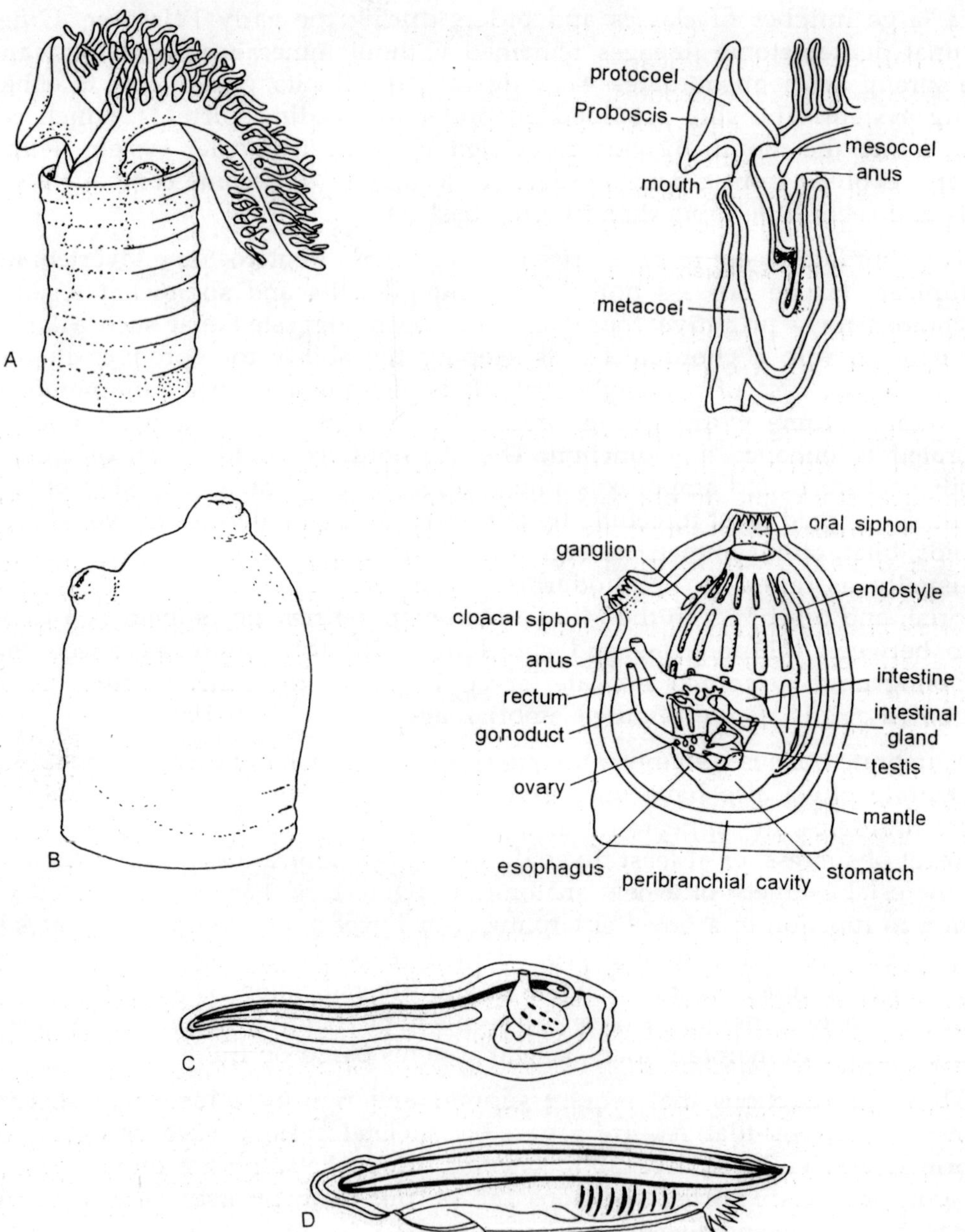

Fig. 12.6 Some deuterostome architectures. A. Hemichordate: a pterobranch and a cross-section through a pterobranch. The three coelomic compartments typical of deuterostomes (protocoel, mesocoel, and metacoel) are indicated. B, Urochordate: a tunicate and an anatomical scheme of a tunicate. C. Planktonic larva of a tunicate, resembling a tadpole, with notochord. D. Cephalochordate: an amphioxus.

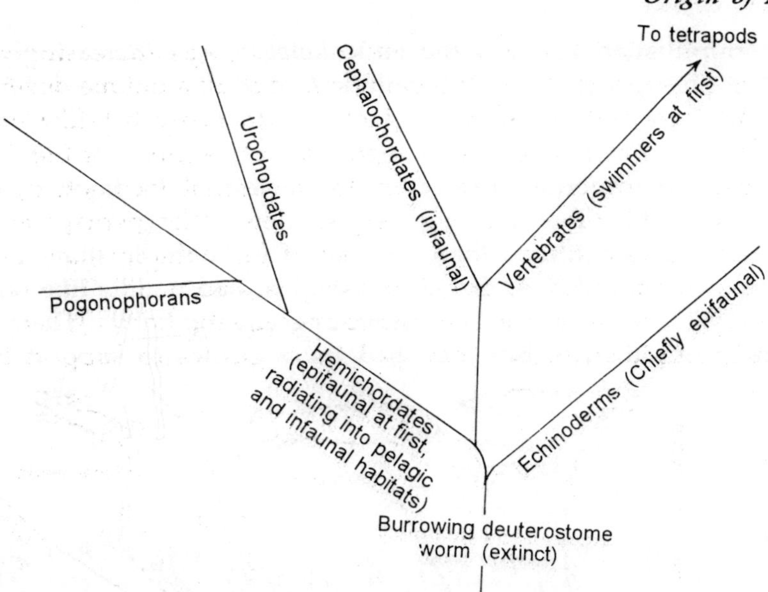

Fig. 12.7: A possible phylogenetic tree of Deuterostomia. Assuming primitive burrowing worms were the ancestors, radiation into the epifauna in late Precambrian or early Cambrian times produced one epibenthic group with well-mineralized skeletons (the echinoderms), another epibenthic group with a nonmineralized organic skeleton (the hemichordates), and perhaps a swimming group (proto-cephalochordates). The latter gave rise to cephalochordates and fishes, leading of course to the land vertebrates and man.

The Terrestrial Invasions

The development of a skeletonized swimming form, the fish, was an extremely successful event that led to several waves of subsequent radiation.A fish-like mode of life is obviously an apt way to cope with the problems of life in the sea and in terrestrial aquatic environments. It is not certain whether the first fishes arose in fresh or in salt water. Nevertheless, selection created a variety of fish forms in both fresh-water and marine environments, certainly in response to opportunities provided by the mode of life and probably to challenges afforded by environmental changes as well. As selection explored the adaptive potentials of fishes, two morphological innovations proved to have special consequences. These are the development of jaws (from gill arches) and paired fins. Jaws expanded the food sources of fishes, which must have soon become important predators. Paired fins improved swimming balance and orientation, which would promote efficient predation. Jawed fishes quickly diversified and eventually replaced the jawless ostracoderms, both in marine and fresh waters, so that by Middle Devonian time early ray-finned and lobe-finned fishes were well established. The ray-finned forms eventually gave rise to a vast majority of modern fishes. Lobe-fins were represented by lungfishes, coelacanths, and a group called rhipidistians. The rhipidistians became extinct near the end of the Permian, but they have an important place in evolutionary history because the amphibians evolved from early members of this group.

In some rhipidistian lineages the endoskeleton was increasingly ossified, legs developed from fin supports, and air breathing (not at all a unique development among fish) appeared. These features were coadapted with many fish-like features to form components of what must have been a highly adaptive mode of life. It is not certain that legs evolved in response to selection for terrestrial locomotion; they may well have originally served for locomotion in very shallow water, perhaps for quick plunges at the small fish that constituted the chief diet of early rhipidistians and amphibians. The earliest known amphibians, the ichthyostegids, had paddle-like tails and sensory systems associated with a lateral line, indicating aquatic habits. Their axial skeletons were essentially rhipidistian, but they had bony girdles to support both hindlimbs and forelimbs.

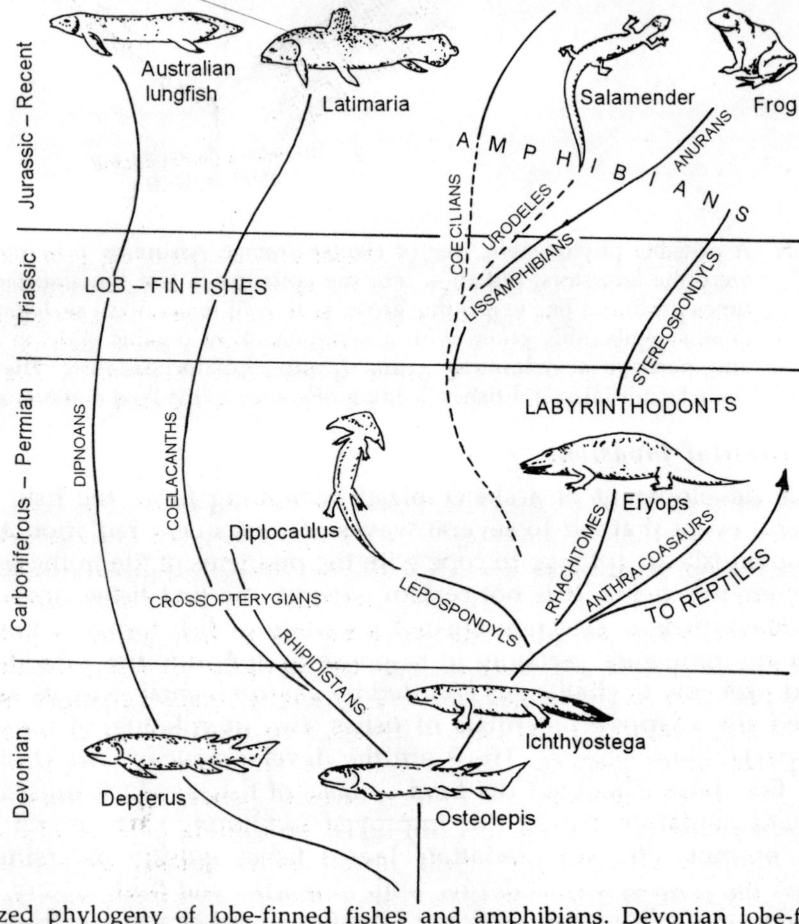

Fig. 12.8: Generalized phylogeny of lobe-finned fishes and amphibians. Devonian lobe-fins are classed in three groups: (1) dipnoans, represented today by lungfishes; (2) coelacanths, represented today by the "living fossil" *Latimeria* from deep water in the western Indian Ocean: and (3) rhipidistians, extinct as fishes although all land vertebrates are among their descendants. The tetrapods developed from primitive rhipidistians through such early amphibians as the ichthyostegids, radiating into disparate groups including large extinct amphibians, familiar living amphibians, and primitive reptiles.

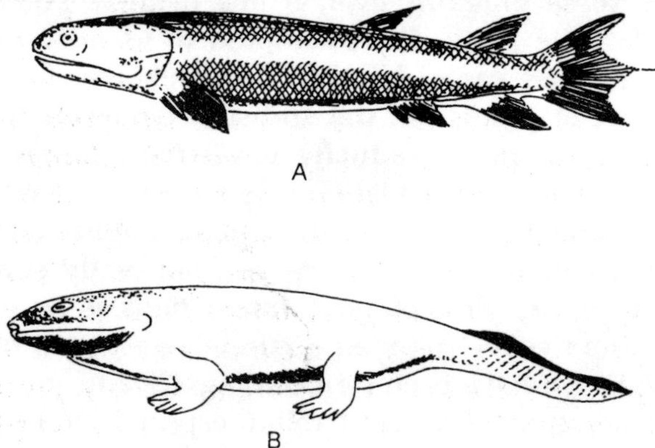

Fig. 12.9: The fish-amphibian transition. A. A progressive lobe-finned rhipidistian fish (*Eusthenopteron*) of the Devonian; animals of this sort were ancestral to the amphibians. B. Early ichthyo-stegid amphibian of the Devonian, probably highly aquatic descended from a *Eusthenopteron*-like stock.

Thus the ichthyostegids departed from a typical fish-like structure. The fishes have evolved numerous bizarre specializations. We would probably regard the ichthyostegids as merely another interesting evolutionary excursion of the fishes were it not for the fact that they gave rise to the tetrapods, which conquered terrestrial environments. Although many amphibian lineages remained aquatic (some were legless, perhaps neotenic), some groups became adapted to prolonged adult life on land.

As the fossil record of terrestrial habitats is generally poor, the course of the biotic invasions of the land is known only in broad outline. Plants must have been established in fresh water from at least the early Paleozoic, and invertebrates no doubt followed closely. Fish may have originated in fresh water, and if not, were represented there at least by Silurian time. During the middle Paleozoic or earlier, life moved beyond the edges of fresh-water lakes, ponds, and swamps to become established on land. The earliest known vascular (land) plants are Silurian, and probably lived in damp regions near fresh-water bodies. Arthropods must have evolved terrestrial habits soon after, perhaps living at first in subterranean cracks and feeding on the plant detritus then enriching the swampside soils; soon they were feeding on the plants themselves. By Devonian time the green belts were spreading widely into lowland terrains and trees appeared; also at this time flying insects developed. Some rhipidistians followed the burgeoning land biota into terrestrial environments, developing into amphibian tetrapods during the Devonian. Probably their terrestrial foods were chiefly arthropods.

The pressures that cause lineages to evolve in novel habitats are numerous. Populations suffer much internal competition between individuals for food and space, and commonly there is much competition from populations of other species as well. Furthermore, populations are usually heavily preyed upon, and may be subjected to periodic or regular inclement physical conditions. Given the opportunity, selection will act so as to minimize these pressure, even if this requires pioneering in an unusual environment. Scientists have frequently attempted to identify particular pressures as the most important in given cases. For early amphibians, for example, one theory suggests that desiccation of ponds led the ancestral tetrapods to migrate overland to larger pools, and thus to acquire gradually terrestrial adaptations, including limbs. Another theory suggests that overpopulation may have caused rhipidistian—amphibian stocks to evolve some independence from the aquatic habitat and to exploit terrestrial food sources for parts of their life cycles. We are not really sure what the pressures actually were; the complexity of ecological interactions is so great that even if the situation had been studied at the time, an accurate assessment of the contributions of all relevant factors might not have been forthcoming. Clearly, preadaptations commonly permit the successful invasion of a new habitat, especially whether creation of novel modes of life are involved. Partly for this reason, it is tempting to favour the idea that limbs were originally an aquatic adaptation.

The primitive amphibians became increasingly adapted to terrestrial life; improvements were achieved by different lineage at different times. The body could have been raised from the ground as a locomotory improvement and skin layers thickened to impede desiccation. Stahl (1974) gives an account of the functional modifications that occurred in early amphibians. The most diverse amphibian group to appear during the Carboniferous was the labyrinthodonts, which include the primitive ichthyostegids as a stem group and which diversified into a number of distinctive phyletic lineages. As has so commonly happened, it was one of the earlier lineages (the anthracosaurs) that eventually gave rise to the next major advance in grade.

Reptiles, Dinosaurs, and Mammals

Reptiles are known from the late Carboniferous, but may have descended from an amphibian stock that diverged from other anthracoasurs in the early Carboniferous. They possess eggs that can develop out of water, and skins that prevent undesirable water loss to the environment. In principle, these features free the reptile from necessary dependence upon life in water at any stage of the lifecycle. The detailed adaptive pathways that led to the reptiles are not yet understood. Important steps long that pathway seem to have included the development and elaboration of a membrane around the embryo, presumably for respiration at first, the subsequent acquisition of a horny protective covering around the egg to prevent desiccation, and finally the mineralization of the egg shell. Physiological and structural changes required of the embryo itself include the evolution of physiological systems to produce safe metabolites that are relatively nontoxic and can be stored in an enclosed egg. The reptilian egg may have evolved to prevent desiccation in impermanent ponds or drying swamp shallows, or to protect eggs from the numerous aquatic predators, which was accomplished by depositing them on land.

Many of the early reptiles, the captorhinomorphs, lived chiefly or entirely in water, but some seem to have inhabited forests fringing large swamps. They soon diversified, and from their basal stock arose distinctive lineages, including the pelycoasurs. These forms spread into moist terrestrial lowlands, which they eventually shared with advanced captorhinomorphs and other reptiles. Clearly, their integuments were proof against excessive desiccation. Numerous modifications of their skeletal morphology occurred as they radiated into a variety of habitats, developing modes of life that included feeders on plants, on fish, and on each other. Probably by early Permian time a group of active predators had split off from the pelycoasurs, in turn diversifying into herbivores as well as carnivores of several types and coming to dominate a wide range of middle and upper Permian terrestrial habitats. These were the therapsids. It is from predatory therapsids that mammals evolved. Developments in axial skeletons, limbs, and skulls in several therapsid groups in the Triasic produced features that were to be emphasized and continued in mammals. It is possible to reconstruct, from clues presented by therapsid skeletons, features of soft-part anatomy that are now

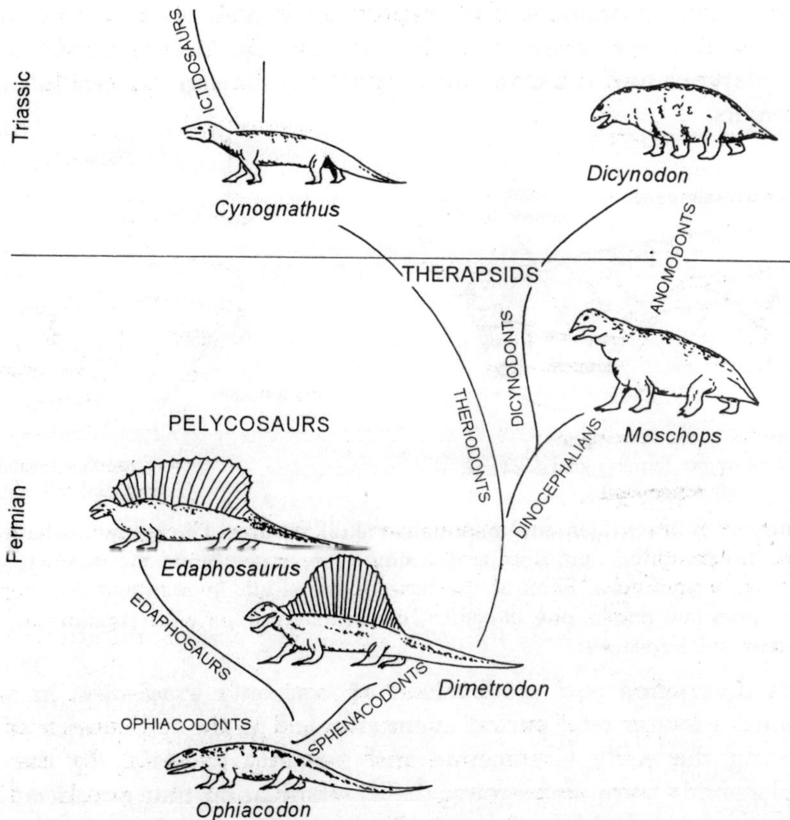

Fig. 12.10: Generalized phylogeny of pelycoasurs and the therapsid reptiles, from which mammals evolved. Primitive captorhinomorph reptiles (such as *Ophiacodon*) gave rise to pely-coasurs, of which one branch (characterized by *Dimetrodon*, a predator) produced the therapsids. The therapsids diversified, with one branch (characterized by *Cynoganathus*, a predator)' giving rise to mammals, possibly through a little-known group called ictidosaurs.

possessed by mammals, thought it is not really certain which of these were acquired by therapsids and which by their early mammalian descendants. Such features include the development of a diaphragm for breathing, hair, improved agility, and the facial skin and muscles that permitted suckling of young; even warm-bloodedness may have appeared in therapsids.

Because so many of these distinctively mammalian characteristics are difficult or impossible o infer from skeletal remains, and since they were doubtless acquired at different times, paleontologists have had to adopt arbitrary skeletal definitions of the Mammalia. Suggested requirements for mammalian status include jaws that contain a joint between dentary and squamosal bones, and teeth (especially molars) that are more complex than those of reptiles and that are not continually replaced during life. This definition is especially practical, since teeth and jaws are common fossil items. Based on such fragmentary fossils, we can construct a phylogenetic model for late therapsids and early mammals that depicts mammals as radiating from a single founding lineage. It is not certain that this is in fact the historical case. The egg-laying monotremes (platypus and echidna) may represent a lineage descended independently of other mammals.

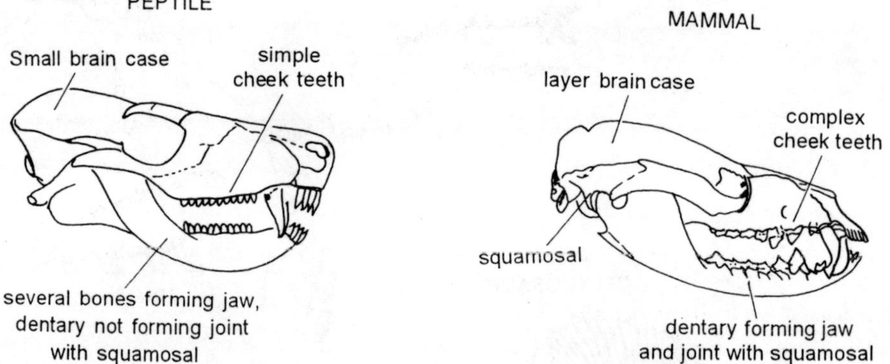

Fig. 12.11: Comparison of reptilian and mammalian skull features. The mammals have a larger brain case, differentiated dentition, and a single lower jaw bone, the dentary, which forms a joint with squamosal bone at the base of the skull. In mammal-like reptiles there are additional jaw bones, one of which, the articular, joins with squamosal. Two joints are present in ictidosaurs.

Mammals diversified and modernized at first only gradually; in some lineages the trend toward a longer fetal period eventually led to the appearance of the placenta, probably during the early Cretaceous and possibly in Asia. By the close of the Cretaceous, placentals were undergoing the diversification that produced the primates; this diversification extended into the early Tertiary and gave rise to modern mammalian groups. Throughout their long Mesozoic history, which embraces fully two-thirds of their existence, mammals were completely overshadowed in abundance, diversity, and size by reptiles, particularly by the dinosaurs. Dinosaurs probably arose from a lineage that diverged from the captorhinomorphs, only a few million years before the mammals appeared. The results of their spectacular radiations are well known, particularly the much depicted large brontosaurs, stegosaurs, ceratopsians, and the great predators. Birds also developed from a dinosaur stock. As mammals have now become far more successful than reptiles, the questions arise why reptiles were so successful in the Mesozoic, and why they have declined to the present status. There are now plausible, though unproven, answers.

These answers have emerged from studies, by Ostrum and others, that have reevaluated such features as posture and thermoregulation in reptiles and mammals. Bakker (1971) has proposed a model in which dinosaurs and early mammals each inherited a sprawling posture from their (separate) reptilian ancestors. By late Triassic time dinosaur lineages had achieved erect posture and probably became quite active, though mammals were still sprawling or semi-erect. Several of the larger dinosaur groups, including the famous predator *Tyrannosaurus*, are envisioned as having fast locomotory gaits. This indicates high body temperatures and suggests homoiothermy. If dinosaurs did attain this physiological grade, it would be easier to explain their numerous adaptive radiations, for we know that the level of activity provided by homoiothermy opens wide arrays of life modes similar to those later developed among larger mammals. It would also account for the apparent ease with which dinosaurs prevented mammals from encroaching upon their biospace; mammal homoiothermy would not have conferred any special competitive advantage. Indeed, visions of charging tyrannosaurs have led some placeontologists to be grateful that dinosaurs have become extinct.

Although more dinosaurs were rather small, down to about the size of a rooster, they were larger than contemporary mammals; even their hatchlings were larger than most of their adult mammalian contemporaries. Large animals have surface—volume ratios favourable to heat retention; and since dinosaurs lacked obvious insulating features, such as feathers or fur, they might have been unable to maintain body heat with small sizes. Early mammals, though small, were furry and could thus achieve homoiothermy despite their small sizes. According to this model, then, dinosaurs and mammals each achieved homoiothermy and high activity levels; the former radiated into biospace appropriate for large animals, the latter for small ones. It was not until dinosaurs declined that mammals radiated into the vacated habitats and took up many of the abandoned modes of life. Although the notion that dinosaurs were warm-blooded has been attacked, it has also been stoutly defended, and remains an attractive hypothesis.

What extinguished the dinosaurs? There is no lack of hypothesis, but none has been adequately tested. Since dinosaurs descended from reptiles, they traditionally have been considered as ecotootherms. Cold-blooded terrestrial animals are restricted in habits and habitats by thermoregulatory problems. Climatic changes to cool whether, to hot weather, or to highly seasonal climates, could in theory easily exceed dinosaur ectothermal potentials, and have often been suggested as a major factor in their extinction. Even if the dinosaurs were homoiotherms, their hatchlings would have been too small to be able to cope with cool temperatures. Climatic change thus remains a likely cause of dinosaur extinction, partly because it would have a wide variety of secondary effects that could increase dinosaur mortality and contribute to breakdowns in their population structure.

Primates

Although the placental mammals originated while dinosaurs still existed, the Mesozoic lineages were chiefly small, generalized insectivore-like forms. However, they began to differentiate somewhat near the end of the Cretaceous and underwent an explosive diversification that established most modern orders during the Paleocene, filling many places in terrestrial ecosystems that had been vacated by extinct dinosaurs. As the post-Pangaean continents became increasingly isolated owing to plate-tectonic processes, distinctive mammalian assemblages developed on different continents; although there were some subsequent interchanges.

The earliest primates appeared among the late Cretaceous orders of placental mammals, differing probably only a little from the insectivores from which they arose. Simons (1972) gives a careful resume of fossil primates. Cretaceous primates are represented only by teeth, so that little is known of their habitats, and Paleocene deposits have yielded few postcranial remains. In the Eocene rather modern, lemur-like prosimians appeared. These Eocene primates had shorter faces and larger eyes than their ancestors, and probably had arboreal habitats. As rodents were differentiating at this time with much apparent success on the ground, the adoption of arboreal habits by primates may have been due in part to their exclusion from forest floors.

Monkeys arose in the alter Eocene or early Oligocene. The living Old World (*catarrhine*) and New World (*platyrrhine*) monkeys seem to have been separated for a long time, and it is sometimes suggested that they arose independently from non-monkey ancestors, thus representing a case of parallel evolution. This is possible, but their common ancestor would have to have been well removed from the prosimian stocks that have living descendants; all monkeys and apes have proven to be rather similar when compared by techniques of molecular biochemistry, and rather different from living prosimians.

Apes are catarrhines; they are not known for certain until the Oligocene. The earliest known ape (*Aegyptopithecus*) had a small brain, a tail, and hands and feet that indicate an arboreal, quadrupedal life. By Miocene time several species assigned to the genus *Dryopithecus* had appeared. Modern chimpanzees, orangutans, and gorillas probably descended from this assemblage. Man seems to have descended from the catarrhines also.

There is nothing very unusual in the evolutionary processes that gave rise to the hominid lineage, although the evolution of the human brain has certainly had spectacular consequences—but so did the evolution of the reptilian egg, the development of the chordate skeleton, the epifaunal radiation of the invertebrates, and the invention of the coelom, the choose only a few of many landmark events in animal evolution. In each case the innovations evolved through selection and solved an immediate adaptive problem that opened the way into vast regions of biospace.

Evolutionary advances have been steadily accompanied by extinctions, which are clearly an integral part of the evolutionary patterns. They seem an inevitable consequence of adaptation to the ambient environment in a changing biosphere. Peak extincting levels probably correspond with reductions in the capacity of the environment to accommodate species. Today, diversity appears to be near a record high. Evolutionists therefore have a rich biosphere to study, furnished with a wide variety of organisms, communities, and provinces. One wonders what the future of this biosphere may be, and whether the potentials are as great as previously.

Chemical Similarities and Differences

All living bodies are astonishingly similar in chemical composition. Not only do they contain atoms of the same elements, but often in similar proportions. The same classes of compounds, particularly nucleic acids and proteins, are found everywhere. Proteins are composed of the same 20 amino acids, and the amino acids, with rare

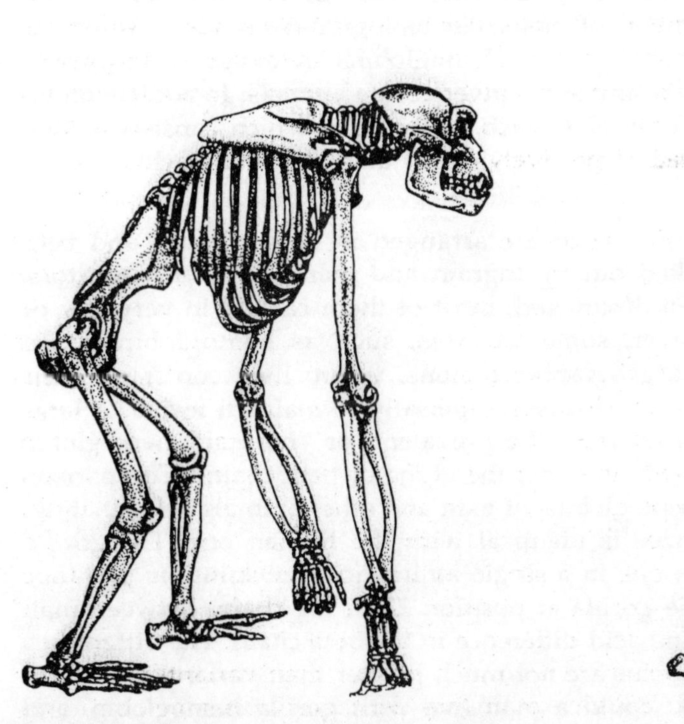

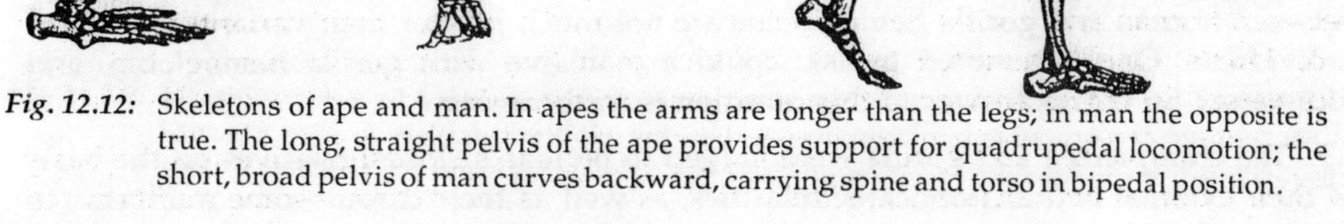

Fig. 12.12: Skeletons of ape and man. In apes the arms are longer than the legs; in man the opposite is true. The long, straight pelvis of the ape provides support for quadrupedal locomotion; the short, broad pelvis of man curves backward, carrying spine and torso in bipedal position.

Table 12.1: Means and approximate ranges of cranial capacities in higher primates.

	Mean (cm³)	*Range (cm³)*
Hylobates lar (gibbon)	103	82-125
Pan (chimpanzee)	383	282-500
Gorilla (gorilla)	505	340-752
Pongo (orangutan)	405	275-540
Australopithecus africanus	588	435-815
Homo erectus	950	755-1225
Homo sapiens	1330	1000-2000

exceptions, are represented only by left optical isomers. Energy carriers (such as adenosine triphosphate, ATP) and enzymes with identical functions (such as cytochrome *c*) are present in most diverse organisms. To an evolutionist, these chemical similarities are meaningful; they affirm that man is kin to all that lives.

More novel and more interesting are the differences between related compounds in man and other animals. Though some enzymes may play similar physiological functions in man and other organisms, they are often more or less distinct in their chemical composition. The achievements of molecular biology have made possible the detection and quantification of such distinctions. Hemoglobins are oxygen-transporting pigments in the blood of all vertebrate and some invertebrate animals. In adult humans the most abundant form of hemoglobin is A, each molecule of which consists of two alpha and two beta chains, composed respectively of 141 and 146 amino acids and an iron-containing heme group.

The sequences in which the amino acids are arranged in human alpha and beta hemoglobin chains have been worked out by Ingram and many other investigators. Many mutational variants have been discovered, most of them carried in very few or even in single individuals. However, some variants, such as hemoglobin S, are widespread in populations of large geographic regions, where they confer on their carriers a relative immunity to endemic diseases, especially to malarial fevers. A large majority of mutant hemoglobins differ from the prevalent, or "normal," hemoglobin A by substitution of single amino acids in either the alpha or beta chains. Comparison of the amino acid sequences in the hemoglobins of man and other animals is fascinating. The hemoglobin A of the chimpanzee is identical with the human one. The gorilla alpha chain differs from the human one in a single amino acid substitution: glutamic acid in man and aspartic acid in the gorilla at position 23 in the chain. Between man and the gorilla there is also one amino acid difference in the beta chain. The differences between human and gorilla hemoglobins are not much greater than variants in human individuals. One is tempted to ask: could a man live with gorilla hemoglobin, and vice versa? So far no answer to this question is forthcoming.

The chimpanzee and gorilla were judged to be man's closest relatives on the basis of their external and anatomical similarities, as well as their chromosome numbers (46 in man, 48 in the chimpanzee and gorilla). The amino acid sequences in their hemoglobins confirm this judgement. Zuckerkandl, Buettner-Janusch, and their co-workers found that among primates morphologically less similar to man than the chimpanzee and gorilla, the hemoglobins are also less similar. The alpha chain of the baboon differs from human alpha in several amino acids, and the beta chains show even more differences. A comparison of man and *Lemur fulvus* shows six substitution in the alpha chain and 23 substitutions in the beta chain. Studies by many workers disclose the following amino acid substitutions in the hemoglobin chains of various animals compared to man (these figures show-called mutational distances, i.e., the minimum numbers of nucleotide substitutions in the DNA of the genes needed to yield the amino acid changes observed):

Alpha Chains		*Beta Chains*	
Gorilla	1	Gorilla	1
Macaque	5	Macaque	10
Mouse	19	Mouse	31
Sheep	26	Sheep	33
Pig	20	Pig	28
Horse	22	Horse	30
Rabbit	28	Rabbit	16
Chicken	45	Kangaroo	54
Carp	93		
Lamprey	113		

The similarities as well as the differences in the above list are remarkable. As the animals examined become more and more obviously different from man, their hemoglobins become increasingly different. A single mutation could conceivably transform the alpha or beta chains of the gorilla into the respective chains of man; the coincidence of the several mutations needed to transform a human alpha chain into that of a mouse, or even of a macaque, has a probability indistinguishable from zero. It took the accumulation of many separate mutations over millions of years to build up the differences observed. And yet, even the hemoglobins of organisms as different as man and chickens or man and fish, have preserved similarities in amino acid arrangements that are far too great to be explained by chance. The similarities are marks of common ancestry.

Several similarities in amino acid sequence have also been observed in comparisons of other proteins of man and various organisms. The work of Margoliash, Fitch, Jukes, and their colleagues on cytochrome *c*, myoglobins, fibrinopeptides, and other proteins discloses increasing divergence of the amino acid sequences in more and more different organisms.

Man's Early Ancestors and Relatives

When Darwin put forward his conclusion that the human species evolved from ape-like ancestors, the evidence of fossils was almost wholly missing. Of course, in 1856 the skull of the Neanderthal Man had been unearthed in Germany, but the interpretation of this find was at once mired in controversy. Some authorities deemed the Neanderthal to be a pathological specimen of modern man, while others saw it as a "missing link" between *Homo sapiens* and his ape-like ancestors. The brain size of the Neanderthal was equal to or even slightly above the average for modern man, but the shape of the brain capsule, the presence of heavy brow ridges, and a strong and chinless mandible made him distinctive. Numerous Neanderthal skulls and skeletons have since answered to everyone's satisfaction. Some fragmentary remains of australopithecines have been found in strata of great antiquity—two to five million years old. Both *Australopithecus robustus* and *A. africanus* lineages seem to be present. On the other hand, some australopithecine remains are much younger, less than one million years. Most unexpected, however, is Leakey's discovery at Lake Rudolf of a more advanced hominid, allegedly a species of the genus *Homo,* dated about 2.6 to 2.9 million years. If this claim is validated, it would seem to follow that *Homo* and *Australopithecus* lived simultaneously and in about the same geographic region. Our ideas of the evolutionary sequence of hominids would then have to be revised.

The transition from *Homo erectus* to *Homo sapiens* also needs clarification. Several fragmentary remains which have been dubbed "pre-sapines," have been dated at the last interglacial period. The Neanderthals who inhabited Europe during the last period of the Ice Age seem to resemble modern *H. sapiens* rather less than the older pre-sapiens did, despite the fact their brain was as large as that of modern man. Some 35-40 thousand years ago the Neanderthals were abruptly replaced in Europe by a full-fledged *H. sapiens:* Cro-Magnon people, whose skeletal parts are indistinguishable from those of modern man.

Homo sapiens (or pre-sapiens) developed from *H. erectus* perhaps during the penultimate glaciation or even earlier, probably intropical and subtropical climates. It subsequently spread to Europe and during the last stage of the Ice Age formed the race *neanderthalensis,* adapted to the severe climatic conditions of that time. This race was finally replaced by a less rugged but culturally more advanced *H. sapiens* people. Thereafter, the success of human populations depended on the possession of superior technologies rather than greater bodily strength. That *neanderthalensis* was a subspecies of *H. sapiens* rather than a separate species has been substantiated by the discovery of intermediates in caves of Mount Carmel in Palestine. Some researchers have interpreted these intermediates as "hybrids," although this interpreation is not a necessary one if by "hybrids" one means the immediate progeny of parents who belong to different subspecies. A more plausible view is that Palestine was then a territory with a population intermediate between *H. sapiens sapiens* and *H. s. neanderthalensis,* just as at present we find territories inhabited by populations transitional between white and

Negroid races. The Neanderthals no longer exist at present. It is possible that they were mostly killed off by the invading *H. sapiens* —the earliest suspected case of genocide. It is just as plausible that they interbred with the invaders, and we may carry some of their genes.

Chapter—13
The Search for Life on Mars

Is there life on Mars? The question is an interesting and legitimate scientific one, quite unrelated to the fact that generations of science-fiction writers have populated Mars with creatures of their imagination. Of all the extraterrestrial bodies in the solar system Mars is the one most like the earth, and it is by far the most plausible habitat for extraterrestrial life in the solar system. For that reason major objective of the Viking mission to Mars was to search for evidences of life.

The two Viking spacecraft were launched from Cape Canaveral in the summer of 1975. Each spacecraft consisted of an orbiter and an attached lander. When the spacecraft arrived at Mars in July and August of 1976, each was put in a predetermined orbit around the planet, and the search for a landing place began. Cameras aboard the orbiters were the principal source of information on which the choice of the landing sites was based; important data also came from infrared sensors on the orbiters and from radar observatories on the earth. The sole consideration in the final selection of the sites was the safety of the spacecraft. It would be a mistake to suppose, however, that the sites were therefore without biological interest. Biological criteria dominated the initial decisions as to the latitude at which each spacecraft would land. Once the latitudes had been chosen there was relatively little difference between sites at different longitudes.

On command from the earth each lander separated from its orbiter. With the help of its retroengines and parachute it dropped to the surface of Mars. Both orbiters continued to circle the planet, operating their own scientific instruments and relaying to the earth data transmitted from the landers. Both landings were in the northern hemisphere of Mars, and the Martian season was summer. (Mars has seasons like those on the earth, but each season lasts approximately twice as long. The Martian year is 687 Martian days; each Martian day, named a sol by the Viking team to distinguish it from a terrestrial day, is 24 hours 39 minutes long). On July 20, 1976, the *Viking I* lander came to rest in the Chryse Planitia region of Mars, some 23 degrees north of the equator, Six weeks later the *Viking* 2 lander settled down in the Utopia Planitia region, some 48 degrees north of the equator. In longitude the two landers are separated by almost exactly 180 degrees, thus placing them on opposite sides of the planet. Since the instrumentation of the two landers is identical, the difference in their landing sites is the only distinction between them.

The first biologically significant task carried out by each lander was the analysis of the Martian atmosphere. Life is based on the chemistry of light elements, notably carbon, hydrogen, oxygen and nitrogen. To be suitable as an abode of life a planet must have those elements in its atmosphere. Spectroscopic observations from the earth and from spacecraft that had flown past Mars in previous years had already shown that carbon dioxide was the principal component of the Martian atmosphere. Small quantities of carbon monoxide, oxygen and water vapour had also been detected. Nitrogen had not been detected in any form, however, and atmospheric theory suggested that Mars had lost most of its nitrogen in the past.

Each Viking lander analyzed the atmosphere by means of two mass spectrometers. One spectrometer, operating during the descent to the surface, sampled and analyzed the atmospheric gases every five seconds. The second spectrometer operated on the ground. The results showed that the atmosphere near the ground was approximately 95 per cent carbon dioxide. 2.5 per cent nitrogen and 1.5 per cent argon, and that it also held traces of oxygen, carbon monoxide, neon, krypton and xenon. At both landing sites the atmospheric pressure was 7.5 millibars. (The atmospheric pressure at sea level on the earth is 1.013 millibars).

Since the Viking spacecraft revealed that nitrogen is indeed present in the Martian atmosphere, we can say that the elements necessary for life are available on Mars. Missing from the list of gases, however is one critically important compound: water vapour. Although earlier measurements had shown that traces of water vapour are present in the Martian atmosphere, the quantity varies with season and place. The Viking orbiters carried out a survey of water vapour over the entire planet with infrared spectrometers. The results showed that the highest concentration of atmospheric water vapour was at the edge of the north polar cap (the summertime hemisphere), and that the concentration fell off toward the south (the opposite of what is found on the earth). In the polar region the amount of water vapour in the atmosphere would form a film only a tenth of a millimeter thick if all of it were to be condensed on the planet's surface. At the landing sites the concentration of water vapour ranged between 10 and 30 per cent of the concentration at the pole.

These numbers put into quantitative terms a long-known fact about Mars: It is a very dry place. Mars has ice at its poles, but nowhere on its surface are there oceans or lakes or any other bodies of liquid water. The absence of liquid water is related to the dryness of the atmosphere through a fundamental law of physical chemistry: the phase rule. The phase rule states that for liquid water to exist on the surface of a planet the pressure of the water vapour in the atmosphere must at some times and in some places be at least 6.1 millibars. The Viking measurements imply that the vapour pressure of water at the surface of Mars in the northern hemisphere is at most .05

millibar, even if all the water vapour is concentrated in the lower atmosphere. At that low pressure liquid water cannot remain in the liquid phase: depending on the temperature, it must either freeze or evaporate. By the same token raindrops cannot form in the Martian atmosphere and ice cannot melt on the Martian surface.

The extreme dryness presents a difficult problem for any Martian biology. Liquid water is essential for life on the earth. All terrestrial species have high and apparently irreducible requirements for water; none could live on Mars. If there is life on Mars, it must operate on a different principle as far as water is concerned. If Mars had a more favourable environment in the past, however, and if the planet did not dry up too fast, species may have had time to evolve and adapt to present conditions. Pictures made by the *Mariner 9* spacecraft, which went into orbit around Mars in 1971, suggested that Mars may indeed have had running water on its surface in the past. The pictures from the Viking orbiters have confirmed that impression. The evidence consists of channels in the Martian desert that resemble dry riverbeds. There seems to be little doubt that the channels were carved by rapidly flowing liquid, and there is widespread agreement that the most probable liquid is water.

If liquid water once existed on Mars, could life have arisen on the planet? If the life evolved to meet changing conditions, could it exist there still? There is no way to settle these questions by deductive reasoning or even by experimentation in laboratories on the earth. They can be answered only by the direct exploration of Mars, and that is what the Viking spacecraft did.

Five different types of instrument on each Viking lander were involved in the search for evidences of life: two cameras for photographing the landscape, a combined gas chromatograph and mass spectrometer for analyzing the surface for organic material and three instruments designed to detect the metabolic activities of any microorganisms that might be present in the soil engineers and managers whose joint efforts made all the Viking projects possible. They work in universities, industrial laboratories and the National Aeronautics and Space Administration and its field centers. Their names are recorded in the growing technical literature dealing with this historic mission.

Each of the Viking landers carried two cameras of the facsimile type, which built up a picture of the scene by scanning it in a series of narrow strips. Such cameras make pictures slowly, but they are rugged and versatile. Their resolution was moderately high: a few millimeters at a distance of 1.5 meters. They produced pictures in black and white, in colour and in stereo. The two cameras on each lander could between them survey the entire horizon around the spacecraft.

As life-seeking tools cameras have inherent advantages and disadvantages. Their chief advantage lies in the fact that a picture contains a large amount of information. In principle it would be possible to prove unequivocally the existence of life on Mars with a single photograph. For example, if a line of trees were visible on the horizon or if foot-prints appeared on the ground in front of the spacecraft one morning, there would be no room for doubt that there is life on Mars. Another advantage lies in the fact that pictorial evidence is independent of all assumptions about the chemistry and physiology of Martian organisms. The organisms need not respond in certain ways to certain substances or treatments in order to be recognized. The cameras could identify, say, a mushroom made of titanium as a form of life if one were to sprout up from under a rock in the course of the mission. Of course, reliance on pictorial evidence rests on its own set of assumptions about the morphology of living things. The most obvious disadvantage of the camera as a life-seeking instrument is the fact that an entire world of life can exist below the camera's limit of resolution.

Of all the results of the Viking mission the wonderful photographs of the Martian desert at the two landing sites are the most impressive. The photographs have been eagerly scanned by alert and hopeful eyes, but no investigator has yet seen anything suggesting a living form.

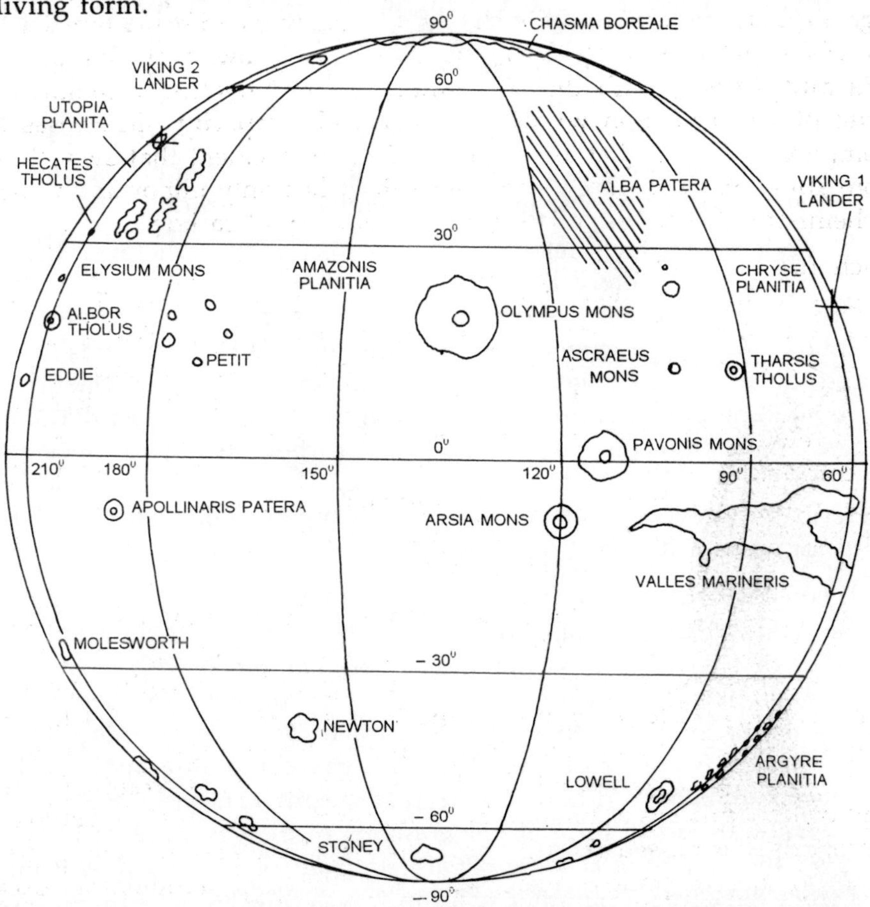

Fig. 13.1: Locations of the Two Viking Landers are indicated on this map of Mars, which shows some of the major geological features of the planet. The two spacecraft are on opposite sides of the planet, some 4,600 miles apart. Both are in the northern hemisphere at sites selected partly for their possible biological interest. *Viking* 1 lander is in Chryse Planitia region at latitude of 23 degrees; *Viking* 2 lander is in Utopia Planitia region at a latitude of 48 degrees.

The next step was to analyze the soil for any organic constituents. Among the elements carbon is unique in the number, variety and complexity of the compounds it can form. The special properties of carbon that enable it to form large and complex molecules arise from the basic structure of the carbon atom. The structure enables the carbon atom to form four strong bonds with other atoms, including other carbon atoms. The molecules thus formed are very stable at ordinary temperatures, so stable, in fact that there seems to be no limit to the size they can attain. The connection between life and organic chemistry (that is, the chemistry of carbon) rests on the fact that the attributes by which we identify living things—their capacity to replicate themselves, to repair themselves, to evolve and to adapt—originate in properties that are unique to large organic molecules. It is the highly complex information-rich proteins and nucleic acids that endow all the living things we know, even "simple" ones such as bacteria and viruses, with their essential nature. No other element, including that favourite of science-fiction writers silicon has the capacity carbon has to form large and complex structures that are so stable. It is no accident that even though silicon is far more abundant than carbon on the earth, it has only minor and nonessential roles in biochemistry. Biochemistry is largely a chemistry of carbon.

Such fundamental facts lead to the conclusion that wherever life arises in the universe it will most likely be based on carbon chemistry. That view has been strengthened by the discovery of organic compounds of biological interest in meteorites and in clouds of dust in interstellar space. Although these compounds are nonbiological in origin, they are closely related to the amino acids and the nucleotides that are the respective building blocks of proteins and of nucleic acids. The fact that they are formed in settings remote from the earth implies that carbon chemistry gives rise to familiar organic compounds throughout the universe. This fact in turn suggests that life elsewhere in the universe will be based on an organic chemistry similar to our own, although not necessarily identical with it.

Such considerations led to the decision to include an organic-analysis experiment *aboard* the Viking landers. The instrument used in the experiment was the mass spectrometer that had analyzed the atmosphere combined with a gas chromatograph and a pyrolysis furnace. A sample of the Martian soil was first heated in the furnace through series of steps up to a temperature of 500 degrees Celsius. Any volatile materials released were passed through the gas chromatograph. Since each of the different compounds has a different molecular weight, composition and polarity, among other properties, it passed through the columns of the gas chromatograph at a unique rate, and so the compounds were separated from one another. As each compound emerged from the chromatographic column it was directed into the mass spectrometer for identification. Since essentially all organic matter is cracked, or decomposed, into smaller fragments at 500 degrees C., the method is capable of detecting organic compounds that have a wide range of molecular weights.

Two soil samples were analyzed at each landing site. The only organic compounds detected were traces of cleaning solvents known to have been present in the apparatus.

The fact that the solvents were detected shows the instruments were functioning properly. The heated samples gave off carbon dioxide and a small amount of water vapour; nothing else was found.

This result is surprising and weighs heavily against the existence of biological processes on Mars. The combined gas chromatograph and mass spectrometer aboard each Viking lander is a sensitive instrument, capable of detecting organic compounds at a concentration of a few parts per billion, a level that is between 100 and 1,000 times below their concentration in desert soils on the earth. Even if there is no life on Mars, it has been supposed the fall of meteorites onto the Martian surface would have brought enough organic matter to the planet to have been detected. Because Mars is near the asteroid belt, from which meteorites originate, it is believed to receive a much larger number of meteorite impacts than either the earth or the moon. Indeed, a question that was frequently discussed before the Viking spacecraft were launched was whether or not it would be possible to distinguish biological organic matter on Mars from the meteoritic organic matter that was expected to be present. The absence of organic matter at the parts-per-billion level, however, suggests that on Mars organic compounds are actively destroyed, probably by the strong ultraviolet radiation from the sun.

The other experiments aboard the Viking landers searched not just for organic matter in the soil but for living organisms. On the earth microorganisms such as bacteria, yeasts and molds are the hardiest of all species. There are few places on the earth where microbial forms do not live; they are the last survivors in environments of extreme temperature and aridity. The reasons for their hardiness are interesting but need not detain us here. Suffice it to say that if there is life on Mars, the chance of detecting it would be maximized by searching for microorganisms in the Martian soil.

Each Viking lander carried three instruments designed to detect the metabolic activities of soil microorganism. First, the gas-exchange experiment was designed to detect changes in the composition of the atmosphere caused by microbial metabolism. Second, the labeled-release experiment was designed to detect decomposition of organic compounds by soil microbes when they were fed with a nutrient. Third, the pyrolytic-release experiment was designed to detect the synthesis of organic matter in Martian soil from gases in the atmosphere by either photosynthetic or non-photosynthetic processes. All three experiments analyzed portions of each sample of Martian soil.

All the experiments detected chemical changes of one kind or another in the soil. All the experiments are now completed, and some of the changes they observed suggest biological processes. There has been much discussion both within the team of Viking investigators and outside it as to the best way to interpret the findings. Are the changes due to biological responses or are they just chemical reactions we would like to believe are biological? Indeed, since life is a form of chemistry, how can the two be told apart?

One way to decide whether or not a process is biological is to test its sensitivity to heat. Living structures are highly organized and fragile, and they are destroyed by temperatures that leave many chemical reactions unaffected. A process that is

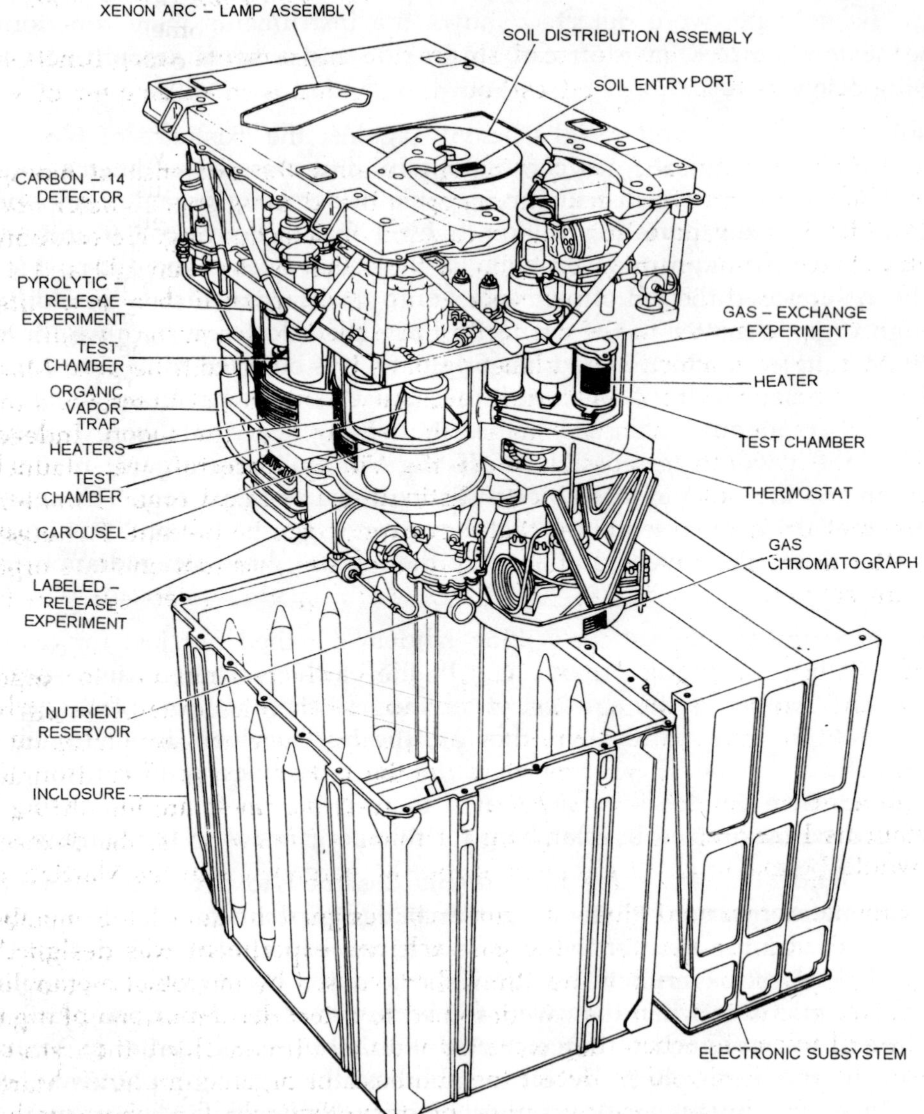

Fig. 13.2: Biological Laboratory aboard both Viking spacecraft occupies a volume of only one cubic foot. The three biological experiments were the gas-exchange experiment (*right*), the labeled-release experiment (*bottom left*) and the pyrolytic-release experiment (*top left*). Each experiment, shown cut away, had several test chambers on a carousel so that the experiment could test several samples of Martian soil. The soil was dumped into an entry port at the top of the laboratory, where it fell into a hopper. For each sample of soil one test chamber of each experiment was rotated under the hopper in order to receive a portion of the sample. All together approximately half a dozen samples of soil were tested at each landing site. The experiments were completed in April. Results are given on following pages.

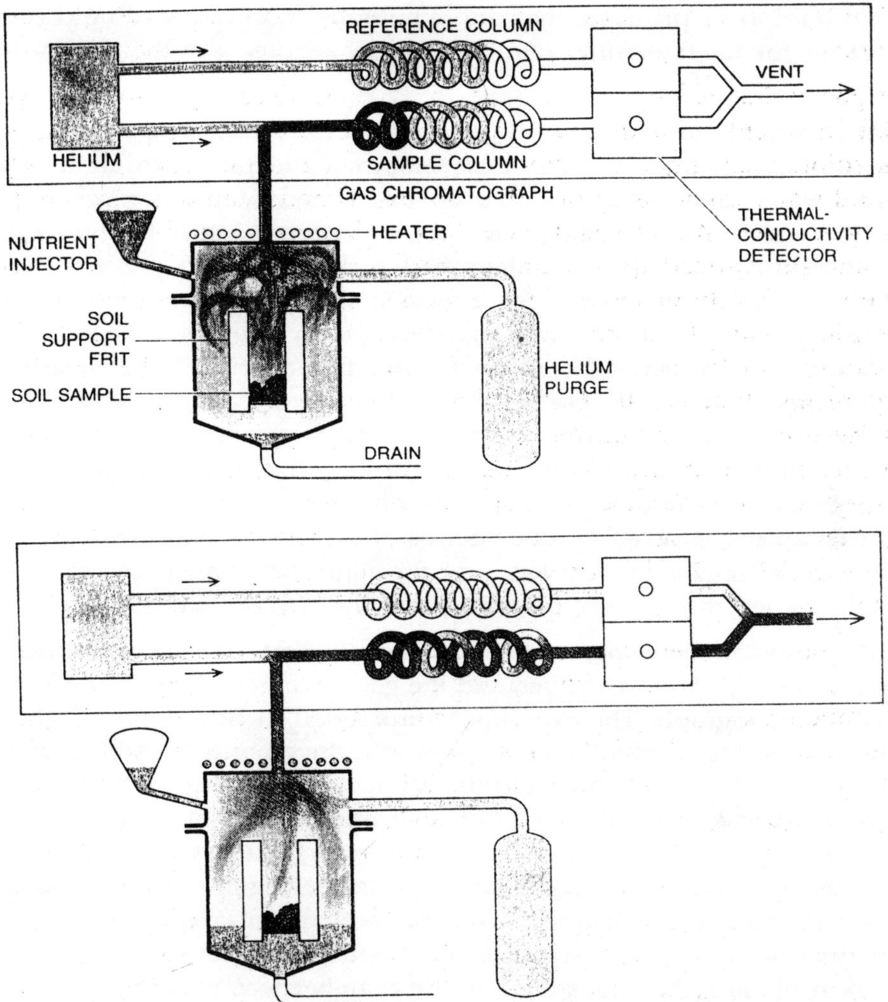

Fig. 12.3: Gases-exchange Experiment tested the Martian soil to see if there were any microorganisms in it that took in atmospheric gase and nutrients and gave off gaseous by-products. The experiment proceeded in two stages. In the first stage a small volume of a complex nutrient solution (*dark colour*) was injected into the test chamber in such a way that it humidified the chamber without wetting the soil (*top*). The gases evolved (*light gray*) were flushed into a gas chromatograph with a stream of helium (*light colour*), where they were analyzed for organic compounds and compared with results of reference analysis run as a standard. In second stage of experiment a large volume of nutrient was poured into chamber to wet the soil (*bottom*).

insensitive to heat is thus likely to be a nonliving chemical reaction, but a process that is sensitive to it could be either living or nonliving.

The decision as to whether a heat-sensitive process is biological or not must be based on additional evidence. In the end, however, the judgement is based on Occam's razor: the traditional principle that the hypothesis most likely to be correct is the one that accounts for the maximum number of observations with the minimum number of assumptions.

The gas-exchange experiment and the labeled-release experiment were frankly terrestrial in orientation. In both experiments a nutrient medium composed of an aqueous solution of organic compounds was mixed with a sample of Martian soil. Since liquid water cannot exist on Mars, the experiments could not be conducted under Martian conditions; the test chambers had to be heated to prevent the water from freezing and pressurized to prevent it from boiling. Both experiments were based on the universal property of terrestrial organisms to evolve gas as they metabolize food. If a sample of soil from the earth is moistened with a nutrient solution, the microorganisms in the soil take up the nutrients and convert them partly into more microorganisms (that is, the population of microorganisms grows) and partly into various by-products, including gases. Among the gases given off in microbial metabolism are carbon dioxide, methane, nitrogen, hydrogen and hydrogen sulfide. On the earth gases evolved by one species of organisms are eventually consumed by other species of organisms. In that way the light elements at the earth's surface are continually cycled through the biosphere and the atmosphere.

In the gas-exchange experiment a complex nutrient solution was added to a sample of Martian soil in a closed chamber, and the gases were analyzed periodically by means of a gas chromatograph. The experiment proceeded in two stages. In the first stage a small volume of the nutrient solution was introduced into the soil chamber in such a way that it humidified the chamber without actually wetting the soil, and the resulting gases were analyzed several times. In the second stage a large volume of the nutrient was poured into the chamber, saturating the soil. With the soil now in direct contact with the medium the main part of the experiment began. The soil was incubated for nearly seven months, so that whatever microorganisms might be in the sample had enough time to signal their presence by producing or consuming gases. During the period of incubation the atmosphere in the chamber was periodically analyzed.

The findings of the first stage of the experiment were both surprising and simple. Immediately after the soil sample was humidified carbon dioxide and oxygen were rapidly released. The release of the gases ceased soon after it had begun but not before the pressure in the chamber had risen measurably. At the Chryse site in a period of little more than one sol the quantity of carbon dioxide in the incubation chamber of the *Viking I* lander increased by a factor of five and the quantity of oxygen increased by a factor of 200. At the Utopia site the increases were less, but they were still considerable.

The rapidity and the brevity of the response recorded by both landers clearly suggested that the process observed was a chemical reaction not a biological one. The appearance of the carbon dioxide is readily explained. Carbon dioxide gas would be expected to be adsorbed on the surface of the dry Martian soil; if the soil was exposed

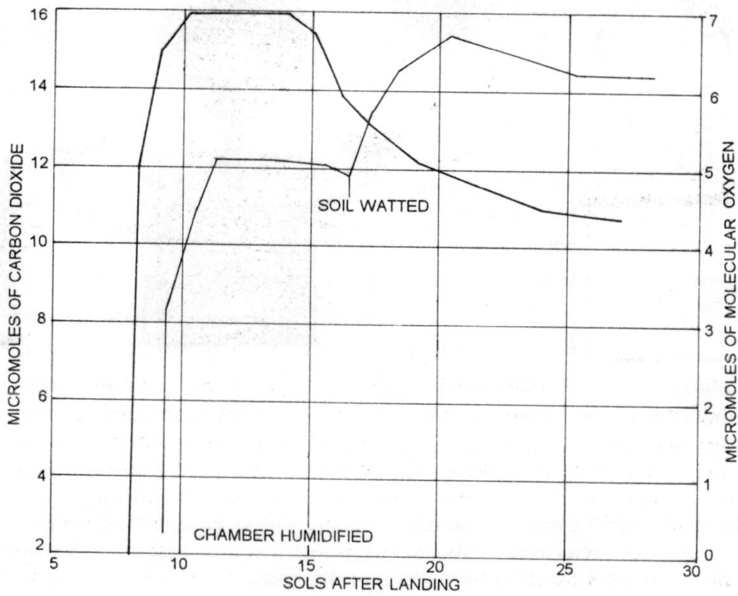

Fig. 13.4: Results of the Gas-exchange Experiment, according to data of Vance I. Oyama of the Ames Research Center of the National Aeronautics and Space Administration, showed that in the first humid stage of the experiment a large amount of carbon dioxide (*black*) and molecular oxygen (*colour*) surged into the test chamber. In the second wet stage the amount of carbon dioxide continued to rise at a decreasing rate and then declined. The amount of oxygen, however, quickly fell. It is believed the gases were released by physical and chemical processes, not by biological ones. One micromole is a millionth of a mole, where one mole is the amount of a substance that has a weight in grams equal to its molecular weight. Oxygen curve is displaced to the left by one sol so that the curves do not overlap. A sol is one Martian day.

to a very humid atmosphere, the gas would be displaced by water vapour. The appearance of the oxygen is more complex. The production of so much oxygen seems to require an oxygen-generating chemical reaction, not just a physical liberation of pre-existing gas. It is likely that the oxygen was released when the water vapour decomposed an oxygen-rich compound such as a peroxide. Peroxides are known to decompose if they are exposed to water in the presence of iron compounds, and according to the X-ray fluorescence spectrometer aboard each Viking lander the Martian soil is 13 per cent iron.

At both landing sites the second phase of the gas-exchange experiment was anticlimactic. When the soil sample was saturated with the nutrient medium and incubated carbon dioxide continued to be released. The production of the carbon dioxide gradually tapered off, however and the oxygen gradually disappeared. The slow increase in the amount of carbon dioxide was probably a continuation of the reaction in the humid stage of the experiment. The disappearance of the oxygen also

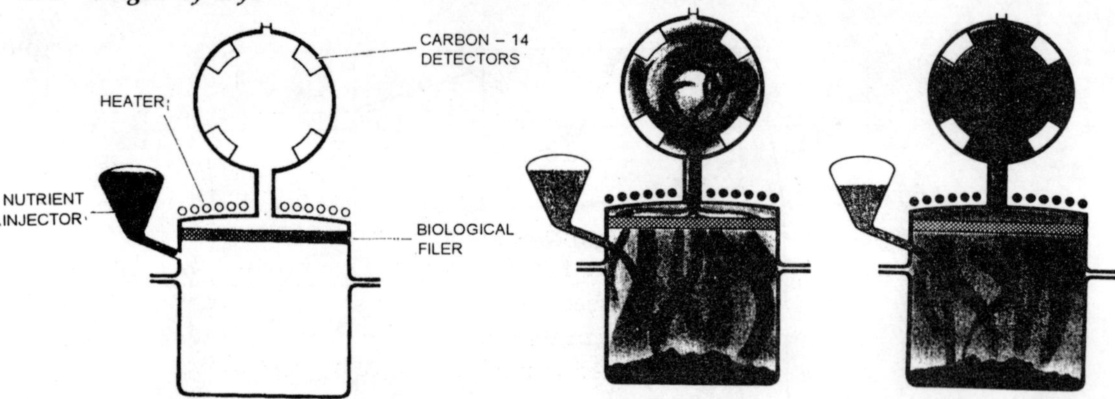

Fig. 13.5: Labeled-release Experiment on the Viking landers tested the Martian soil for microorganisms that could metabolize simple organic, or carbon, compounds. The nutrient medium was composed of several organic compounds that are widely abundant in the universe. The compounds were labeled with radioactive carbon. If microorganisms exist in the Martian soil, they might consume the labeled nutrient and give off radioactive gases (particularly carbon dioxide), which would be detected by the carbon-14 counters. Before the soil was tested the background level of radiation was measured (*left*). The soil was dumped into the test chamber, injected with a small amount of medium (*middle*) and incubated for up to 11 sols. The amount of nutrient in this first injection was planned to wet only part of the soil but to humidify the entire chamber. A subsequent injection of the nutrient (*right*), controlled by signals from the earth, thus brought the medium into contact with soil that had already been wetted and with other soil that had been humidified but not wetted. If the labeled-release experiment worked on Mars as planned, its results would serve as a check on the results of the gas-exchange experiment.

can be easily explained: one of the ingredients of the nutrient medium was ascorbic acid, which combines readily with oxygen. And so after seven months it became clear that everything of interest had happened in the humid stage of the experiment, before the soil came in contact with nutrient! What the gas-exchange experiment detected was not metabolism but the chemical interaction of the Martian surface material with water vapour at a pressure that has not been reached on Mars for many millions of years.

The labeled-release experiment differed from the gas-exchange experiment in several ways. The nutrient medium employed was a simpler one containing only a few cosmically abundant organic compounds such as formic acid (HCOOH) and the amino acid glycine (NH_2CH_2COOH). All the compounds were labeled with atoms of the radioactive isotope carbon 14. The labeled-release instrument was designed to detect radioactive gases, principally carbon dioxide, released when the nutrient medium was added to a sample of soil. The number of radioactive disintegrations in gases can be counted quite efficiently, so that the labeled-release experiment is faster and more sensitive than the gas-exchange experiment in detecting microbial activity in terrestrial soil. The labeled-release experiment's sequence of operations did not include a humid stage as such, but it attempted to accomplish the same end by injecting a volume of nutrient medium that was insufficient to wet the entire soil sample but sufficient to humidify the chamber. If the experiment worked on Mars as planned, subsequent injections of the medium, which were controlled by commands sent from the earth, brought the medium into contact with some soil that had previously been wetted and with other soil that had been humidified but not wetted.

As in the gas-exchange experiment, immediately after the nutrient medium was added to the soil in the labeled-release experiment, gas surged into the chamber. The release of gas tapered off soon after the first sol. The gas, undoubtedly carbon dioxide, was radioactive, showing that it had been formed from the radioactive compounds of the medium and not from compounds in the Martian soil. Non-radioactive gases, which also must have formed when the aqueous medium came in contact with the soil were not detectable in the experiment.

The production of radioactive carbon dioxide in the labeled-release experiment is understandable in the light of the evidence from the gas-exchange experiment suggesting that the surface material of Mars contains peroxides. Formic acid, one of the compounds of the labeled-release nutrient medium, is oxidized with particular ease: if a molecule of formic acid (HCOOH) reacts with one of hydrogen peroxide

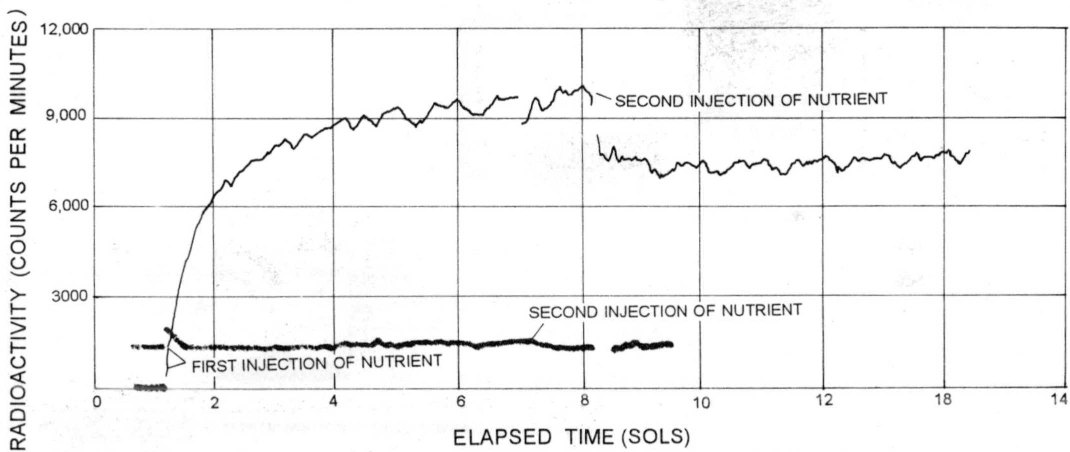

Fig. 13.6: Result of the Labeled-release Experiment for the first sample of soil analyze at the Chryse site (*curve in colour*) were indeed consistent with the results of the gas-exchange experiment, according to data from Gilbert V. Levin and Patricia A. Straat of Biospherics Inc. Immediately after the first injection of the nutrient, radioactive gases surged into the chamber. The radioactivity was measured at 16-minute intervals throughout the experiment except for the first two hours after the first injection, when measurements were made every four minutes. After the second injection the amount of gas in the chamber dropped, then remained at a nearly constant level until the end of the experiment. In order to test the sensitivity of reaction to heat a second portion of the soil sample was sterilized at a temperature of 160 degrees Celsius for three hours and the experiment was repeated (*curve in black*). The reaction was abolished. Although such behaviour is consistent with a biological process, it is more likely that experiment again detected only a chemical reaction.

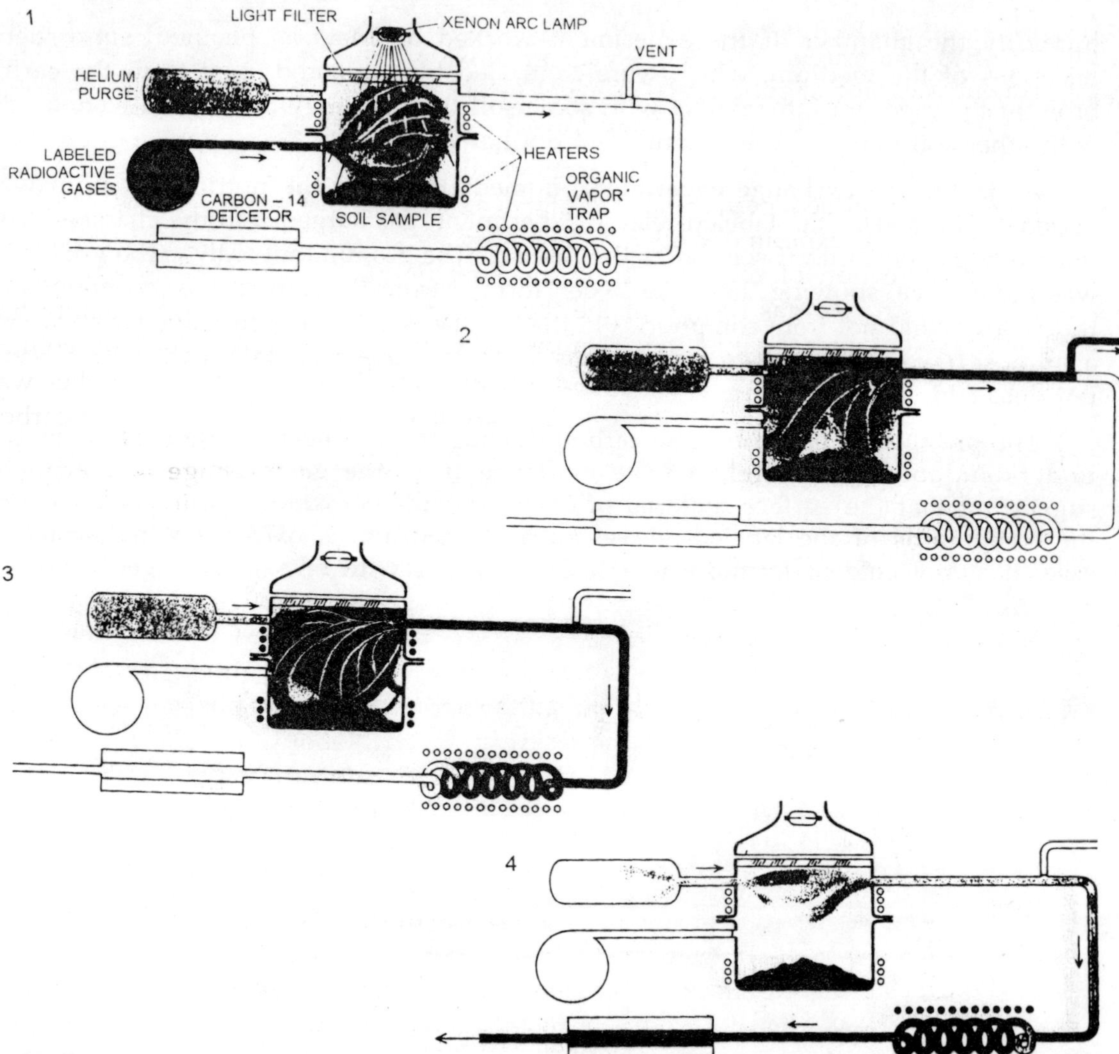

Fig. 13.7: Pyrolytic-release Experiment tested the Martian soil (*dark gray*) to see if there were microorganisms in it that would create organic compounds out of atmospheric gases by either a photosynthetic process or a nonphotosynthetic process. A sample of soil was sealed into a chamber along with some Martian atmosphere and a small amount of radioactive carbon dioxide and carbon monoxide (*light gray*). A xenon arc lamp irradiated the soil with simulated Martian sunlight (1). After five days lamp was turned off and the atmosphere was removed from the chamber (2). Soil was heated to a temperature high enough to pyrolyze (decompose) into small volatile fragments any radioactive organic compounds produced. Fragments (*light gray*) were swept out of the chamber (3). By a stream of helium (*light colour*) into a column designed to trap organic molecules but pass carbon dioxide and carbon monoxide. In column trapped radioactive organic molecules were released by raising column's temperature; the molecules were oxidized to form carbon dioxide, which was carried into a radiation counter (4).

(H_2O_2), it will form a molecule of carbon dioxide (CO_2) and two molecules of water ($2H_2O$). The amount of radioactive carbon dioxide given off in the labeled release experiment was only slightly less than what would have been expected if all the formic acid in the medium had been oxidized in this way.

If the source of the oxygen released in the humid stage of the gas-exchange experiment was indeed peroxides in the soil decomposed by water vapour, then in the labeled-release experiment all the peroxides should also have been decomposed by the first injection of nutrient. Thus the next injection should have evolved no additional radioactive gas inspite of the fact that part of the sample presumably had not yet been wetted by the medium. That proved to be the case. When a second volume of medium was injected into the chamber, the amount of the gas in the chamber was not increased; indeed, it decreased. The decrease is explained by the fact that carbon dioxide is quite soluble in water; when fresh nutrient medium was added to the chamber, it absorbed some of the carbon dioxide in the head space above the sample.

This result was obtained with all the samples tested by the labeled-release experiment at both Viking sites. In that respect the results of the labeled-release experiment did not parallel the results of the gas-exchange experiment. At both sites both experiments tested soil gathered from the ground's exposed surface; at the Utopia site the experiments also tested soil gathered from under a rock. Although the labeled-release experiment found essentially no difference in the amount of gas released by any of the samples, the gas-exchange experiment recorded about three-fourths as much carbon dioxide from the surface samples at the Utopia site as it had from the surface samples at the Chryse site, and it recorded even less carbon dioxide from the sample from under the rock.

The gas-exchange experiment also recorded less oxygen from the samples from the Utopia region, but the interference of the ascorbic acid in the complex nutrient medium of that experiment makes it difficult to quantify the difference. In every case, however the gas-exchange experiment detected considerably more gas than the labeled-release experiment did with portions of the same sample. Those results, however, do not contradict the thesis that the production of oxygen detected by the gas-exchange experiment and the production of radioactive carbon dioxide detected by the labeled-release experiment are simply different measurements of the same surface chemistry. The gas-exchange experiment measures the total amount of oxidant in the surface; the labeled-release experiment measures only a fraction of it.

The labeled-release experiment also tested the stability of the reaction to heat. When the soil was reheated to 160 degrees C. for three hours before incubation, the reaction was abolished. When it was heated to 46 degrees for the same length of time, the magnitude of the reaction was reduced by about half. These results have been regarded by some as evidence in favour of the hypothesis that the reaction is biological. The results are of course consistent with such a hypothesis, but they are also consistent with a chemical oxidation in which the oxidizing agent is destroyed or evaporated at relatively low temperatures. A variety of both inorganic peroxides and organic peroxides could probably have produced the same results.

The third microbiological experiment, the pyrolytic-release experiment, differed from the gas-exchange and labeled-release experiments in two respects. First, it attempted to measure the synthesis of organic matter from atmospheric gases rather than its decomposition. Second, it was designed to operate under the conditions of pressure, temperature and atmospheric composition that actually obtain on Mars, since those are the conditions under which any form of Martian life must exist. In practice the conditions in the chamber were a reasonably good approximation of Martian conditions except for the temperature, which stayed warmer than the outside temperature because of heat sources within the spacecraft.

A sample of Martian soil was sealed in a chamber along with some Martian atmosphere. A quartz window in the chamber admitted simulated Martian sunlight from a xenon arc lamp. Into this Martian microcosm small amounts of radioactive carbon dioxide and radioactive carbon monoxide were introduced. Both gases are present in the Martian atmosphere but not in radioactive form. After five days the lamp was turned off, the atmosphere was removed from the chamber and the soil was analyzed for the presence of radioactive organic matter.

First the soil was heated in the pyrolysis furnace to a temperature high enough to crack any organic compounds into small volatile fragments. The fragments were swept out of the chamber by a stream of helium and passed through column that was designed to trap organic molecules but allow carbon dioxide and carbon monoxide to pass through. The radioactive organic molecules were thus transferred from the soil to the column and at the same time were separated from any remaining gases of the incubation atmosphere. The organic molecules were released from the column by raising the column's temperature. Simultaneously the radioactive organic molecules were decomposed into radioactive carbon dioxide by copper oxide in the column. The carbon dioxide was then carried by the stream of helium into a radiation counter. If organic compounds had been synthesized in the soil, they would be detected as radioactive carbon dioxide; if no organic compounds had been synthesized, no radioactive carbon dioxide would have been formed.

Surprisingly, seven of the nine pyrolytic-release tests executed on Mars gave positive results. The two negative results were obtained at the Utopia site, but a third sample tested at Utopia was positive. This third sample was actually incubated in the dark, implying that light may not be required for the reaction. The amount of carbon fixed in the soil by the experiment was small: enough to furnish organic matter for between 100 and 1,000 bacterial cells. The quantity is so small, in fact, that it could not have been detected by the organic analysis experiment. The quantity is nonetheless significant; it was surprising that in such a strongly oxidizing environment even a small amount of organic material could be fixed in the soil.

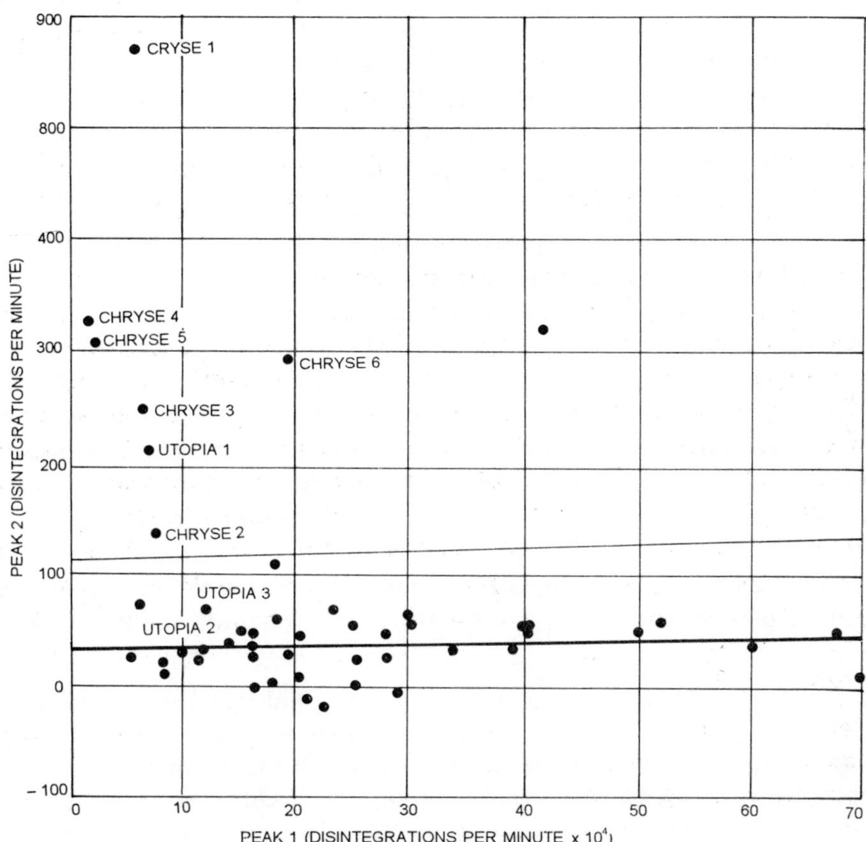

Fig. 13.8: Results of Pyrolytic-release Experiment are shown for all the samples tested on Mars (*dots in colour*). The axis labeled Peak 1 shows how much radioactivity in the form of carbon dioxide and carbon monoxide passed through the column during the pyrolysis of each sample. The axis labeled Peak 2 shows how much radioactivity, representing newly synthesized organic matter, remained attached to the column in each case. Each dot in colour is labeled and numbered according to the site at which it was tested and which experiment it represents. (For example, "Chryse 1" means the result of the first experiment at Chryse, "Utopia 1" means the result of the first experiment at Utopia, and so on). The dots in black are the data obtained from tests of sterilized soil samples in a duplicate of the Viking pyrolytic-release instrument on the earth. The black line drawn through those points represents the best fit to the points. The coloured line above the black line is a statistically significant dividing line; any point lying above the coloured line is a positive result. The single black point above the coloured line is believed to be due to a technical error in performing that particular test. Seven of the nine pyrolytic-release experiments performed on Mars, however, yielded firmly positive results.

Even more significant, the pyrolytic-release instrument had been rigorously designed to eliminate nonbiological sources of organic compounds. During the development of the experiment it had been found that in the presence of short-wavelength ultraviolet radiation, carbon monoxide spontaneously combined with water vapour to form organic molecules on glass, quartz and soil surfaces in the experimental chamber. In order to avoid those reactions and the confusion they would have caused, the short-wavelength ultraviolet was filtered out of the radiation allowed to enter the incubation chamber. To receive positive results from the soil on Mars in spite of that precaution was startling.

Nevertheless, it appears that the findings of the pyrolytic-release experiment must also be interpreted nonbiologically. The reason is that the reaction detected was less sensitive to heat than one would expect of a biological process. In two of the nine pyrolytic-release experiments performed on Mars the soil sample was heated before the radioactive gases were injected and the incubation was begun. In one case the sample was held at 175 degrees C. for three hours and in the other it was held at 90 degrees for nearly two hours. The effect of the higher temperature was to reduce the reaction by almost 90 per cent but not to abolish it. The effect of the lower temperature was nil. When it is recalled that the temperature at the surface of Mars at the two landing sites does not rise above zero degrees C. at any time, and that the temperature below the surface is even lower, it becomes difficult to reconcile the results with a biological source. Any organisms living in the Martian soil should have been killed by those temperatures.

On the other hand, it is not easy to point to a nonbiological explanation for the positive results. Investigations into the problem are now under way in terrestrial laboratories with synthetic Martian soils formulated on the basis of the data from the inorganic analysis carried out by the Viking landers. The solution to the puzzle will probably also explain why the organic-analysis experiment detected no organic material in the Martian surface. Until the mystery of the results from the pyrolytic-release experiment is solved, a biological explanation will continue to be a remote possibility.

Even though some ambiguities remain, there is little doubt about the meaning of the observations of the Viking landers: At least those areas on Mars examined by the two spacecraft are not habitats of life. Possibly the same conclusion applied to the entire planet, but that is an intricate problem that cannot yet be addressed. The most surprising finding of the life-seeking experiments is the extraordinary chemical reactivity of the Martian soil: its oxidizing capacity, its lack of organic matter down to the level of several parts per billion and its capacity to fix atmospheric carbon (presumably into organic molecules) at a still lower level. It seems Mars has a photochemically activated surface that, due to the low temperature and the absence of water, is maintained in a state far from chemical equilibrium.

These conclusions drawn from the results of the life-seeking experiments on the Viking landers are undeniably disappointing. The discovery of life would have been much more interesting, to say the least. There are doubtless some who, unwilling to accept the notion of a lifeless Mars, will maintain that the interpretation I have given is unproved. They are right. It is impossible to prove that any of the reactions detected by the Viking instruments were not biological in origin. It is equally impossible to prove from any result of the Viking experiments that the rocks seen at the landing sites are not living organisms that happen to look like rocks Once one abandons Occam's razor the field is open to every fantasy. Centuries of human experience warn us, however, that such an approach is not the way to discover the truth.

Chapter—14

Life Outside the Solar System

The evolution of stars and the evolution of living organisms appear to be completely dissimilar processes. The differences between the two can be explained in terms of how the particles of matter are bound together and how energy is exchanged among them. Indeed, the evolution of living organisms represents one outcome of stellar evolution. It was the steady flow of energy from the sun for four or five billion years that brought about the biological developments on earth, culminating in the emergence of intelligent organisms able to contemplate the whole remarkable story. If the sun had a different history, life would not have appeared in its immediate vicinity.

Astronomical evidence acquired in recent years indicates that what has happened here is probably not unique. Two decades ago it was thought that the solar system might have originated in a near-collision of the sun and another star, which event supposedly pulled away enough matter from the sun to from the planets. Because such encounters must be rare events, they would give rise to few stars with a company of planets. Today most astronomers believe that a star is formed by the condensation of a cloud of dust and gas. This hypothesis much more readily explains the origin of the solar system and is supported by the observation that more than half of the stars in our galaxy are double or multiple systems. In fact, it now appears that most stars are accompanied by other stars or by planets, though the latter must be so small as to escape sure detection by the present instruments of astronomy. Thus the appearance of life—even the appearance of mind—may be far from unusual events in the universe.

On the other hand, certain critical conditions must be satisfied if life processes are to be initiated and maintained. In the first place the star must shine long enough and steadily enough to permit life to evolve. The star must also be hot enough to warm up a habitable zone deep enough to offer a reasonable chance that a planetary orbit will fall within it. And the planet must ply a stable orbit within this zone. The number of stars that have given rise to life must therefore be considerably smaller than the immediately suggested by the dust-cloud hypothesis.

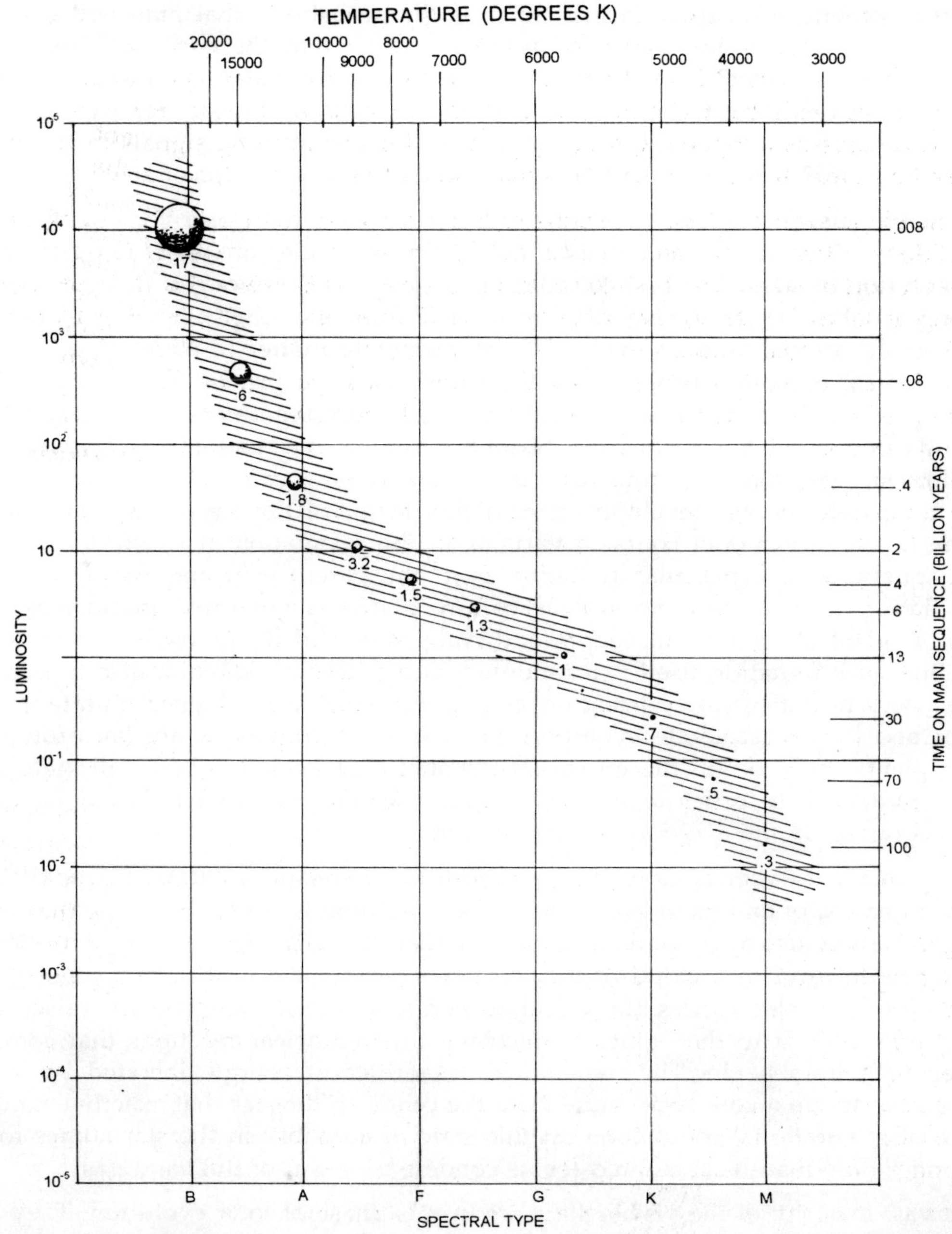

TEMPERATURE (DEGREES K)

20000 15000 10000 9000 8000 7000 6000 5000 4000 3000

LUMINOSITY

10^5, 10^4, 10^3, 10^2, 10, 10^{-1}, 10^{-2}, 10^{-3}, 10^{-4}, 10^{-5}

17, 6, 1.8, 3.2, 1.5, 1.3, 1, .7, .5, .3

TIME ON MAIN SEQUENCE (BILLION YEARS)

.008, .08, .4, 2, 4, 6, 13, 30, 70, 100

B A F G K M

SPECTRAL TYPE

Fig. 14.1: Typical Stars on the Main Sequence are represented on this diagram. Vertical scale at left indicates luminosity (the sun is unity). Scale at right shows time on the main sequence. The numbers at the top denote temperature in degrees Kelvin; the letters at the bottom, spectral type. Small figures beneath each star give its mass with respect to the mass of the sun.

Taking account of all these factors, what is the probability that man will ever be able to visit the life-bearing planet of another star, or that the earth will receive a visitor from such a planet? Does the possibility justify taking measures now-in advance of a visit—to put existing technology to the task of scanning the sky for signals from intelligent organisms outside the solar system, or for transmitting signals in the hope they may be heard? If so, what sort of signals should be listened for or sent?

A reliable answer to these questions calls for a somewhat closer consideration of the conditions critical for life and an estimate of how often they are likely to be satisfied in the evolution of stars. The first condition is a steady and prolonged flow of energy. How long it takes life to evolve may be judged from the single instance available: Here on earth rational animals evolved from inanimate matter in about three billion years. Biological evolution proceeds by the purely random process of mutation, that is, the unpredictable occurrence of novel chemical processes among the fantastically numerous reactions that constitute life. Since the process is a random one, the laws of probability suggest that the time-scale of evolution on earth should resemble the average time-scale for the development of higher forms of life anywhere. The rate at which mutations occur is of course a variable in the calculation. It is affected by the electromagnetic and corpuscular radiation from the parent star, and the amount of such radiation that reaches the surface of any planet is governed in turn by the magnetic field of the planet, the depth and composition of its atmosphere and other factors. But since mutation itself is a random process, the introduction of a few more variables does not affect the calculation greatly. Furthermore, a higher mutation rate does not necessarily accelerate evolution, because most mutations are harmful. The more frequently they occur, the greater the chance that an individual will suffer an injurious mutation. In order to favour natural selection, mutations should be rare, perhaps as rare as in the evolution of life on earth.

It is somewhat easier to estimate the time-scale of stellar evolution. In contrast to the random nature of biological evolution, stellar evolution is governed by the universal law of gravitation and by a relatively small number of thermonuclear reactions. When a star begins to form in a cloud of dust and gas, gravitational attraction among the gas and dust particles causes the cloud to condense until the pressure raises the temperature within it to the point at which the thermonuclear reactions that convert hydrogen to helium begin. The tremendous quantities of energy liberated by these reactions now set up a counterpressure from the center of the star that exactly balances the force of gravitational contraction. In this state of equilibrium the star shines for a much longer time than that required for its condensation out of dust and gas.

The vast majority of the visible stars are in this phase of their evolution. They are called "main-sequence" stars because their luminosity (energy output per unit of time) plotted on graph against their surface temperature places them in sequence in a narrow band. What the chart shows is that the hottest stars are also the most luminous. Luminosity depends upon mass, and so the point at which a star appears on the main sequence depends primarily upon the mass of material incorporated in it during

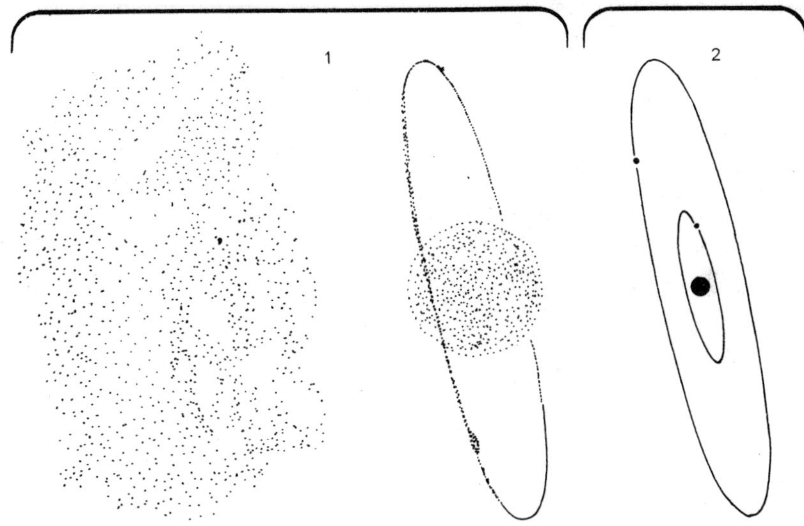

Fig. 14.2: Possible Evolution of a Star 1.2 times the mass of the sun begins with a dust cloud condensing into a star and proto-planets (1), a process taking perhaps 10 million years. Star enters "main sequence" (2) and remains there for approximately eight billion years. Then it expands (3) into red-giant stage (4), first destroying life on its inner planet, then

condensation. The hottest stars are designated by the letter O, followed in descending order by stars classified B, A, F, G, K and M; these classifications are usually called spectral type. The adjectives "early" and "late," which have nothing to do with the age of the star, are often used before the spectral-type designation to denote further relative temperature differences. An early F-type star is hotter than a late one, which in turn has a higher temperature than an early G-type star such as our sun. The classification is further refined by a number from 0 to 9 written after the letter. Thus the class of early B-type stars includes those from B0 to B4; and the late B-type stars, those from B5 to B9.

The length of time a star remains in the equilibrium state on the main sequence can be calculated from the total mass of hydrogen in its core and the rate at which the hydrogen is consumed. Both factors in the calculation are determined by the position of the star on the main sequence. Luminosity, which is an index of the rate of fuel consumption, increases as the fourth power of the star's mass. The most massive stars thus use up their substance most rapidly and so have the shortest lifetimes in the equilibrium state. In general a star will evolve away from the main sequence when the core in which the hydrogen has been consumed has a mass of about 12 per cent of that of the entire star. With the exhaustion of fuel in the central furnace, gravitational contraction takes over again, heating up the interior until the thermonuclear reaction spreads to outer layers. The star now leaves the main sequence and in a comparatively brief period of time evolves into a red giant or supergiant. Its evolution thereafter cannot presently be predicted in detail. But somehow, perhaps through the rapid loss

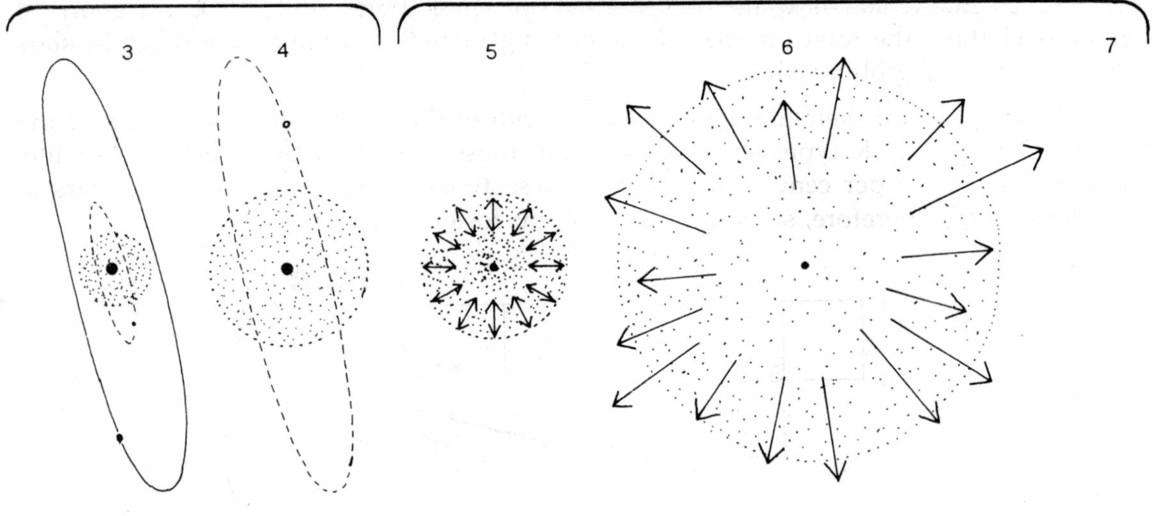

burning up the planets in turn. This period may last about 100 million years. Then star may pulsate in luminosity every few hours (5) for thousands of years, finally exploding into a nova (6) and eventually collapsing into white dwarf (7). Time period for final stages (5, 6 and 7) is not known.

of mass by ejection, it ends up as a hot, faint, dense object known as a white dwarf. A majority of the intrinsic variable stars, of novae and of nova-like objects are apparently in the stage between red giants and white dwarfs.

In leaving the main sequence a star releases so much energy that it would destroy life on any of its planets. Thus the main-sequence stage of the star's evolution is the only important one so far as life is concerned. The O and early B stars are the most massive, but since they burn much more rapidly than the smaller, less luminous stars, their life on the main sequence lasts only a million to 10 million years. The small M stars, in contrast, remain on the main sequence for more than 100 billion years. None of the early stars in the O, B and A groups has a stable lifetime longer than three billion years; they cannot therefore sustain biological evolution long enough for intelligent organisms to appear on any of their planets. Closer calculation shows that only the stars farther down the sequence than F4 maintain their equilibrium for a sufficient length of time to bring biological evolution to its culmination.

If time were the only critical condition, then the late-K and M stars would stand the greatest chance of having life-bearing planets. But these stars have low luminosity, and the habitable zones in which planets might travel about them must be quite narrow. A more luminous star can obviously warm up a larger space than a less luminous one. It is like a fire in a field on a cold night: the bigger the fire, the wider the zone around it in which the temperature will be neither too hot nor too cold. Thus the chance of finding a planet and intelligent life within the habitable zone of any particular M or even late-K star is quite small. However, since there are 10 times as many M stars as G stars, the total number of life-bearing planets traveling around the M stars may not be negligibly small.

On the basis of their lifetimes and the depth of their habitable zones, stars of the late-F, G and early-K types seem to offer the most favourable environments for life. Approximately 10 per cent of stars fall in these types. Out of the 200 billion stars in the Milky Way, therefore, some 20 billion might foster intelligent life.

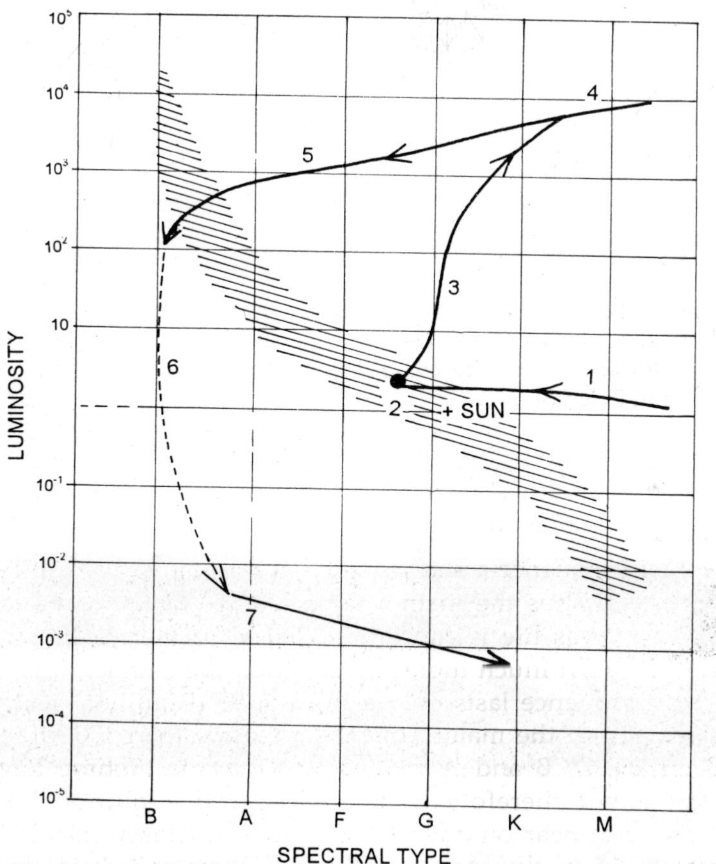

Fig. 14.3: Evolutionary Path of a Star with a mass 1.2 times that of the sun is traced. The vertical scale is luminosity (the sun is unity); the horizontal scale, spectral type. The main sequence is outlined by the gray hatching. The numbers on the path correspond to the stages of stellar evolution depicted in the illustration at the top of the preceding two pages. At 1 the star is in the stage of gravitational contraction; at 2 it is on the main sequence; at 3 it is expanding; at 4 it is a red giant; at 5 it pulsates. The broken line (6) indicates uncertainty as to the path the star follows in reaching the white-dwarf stage (7).

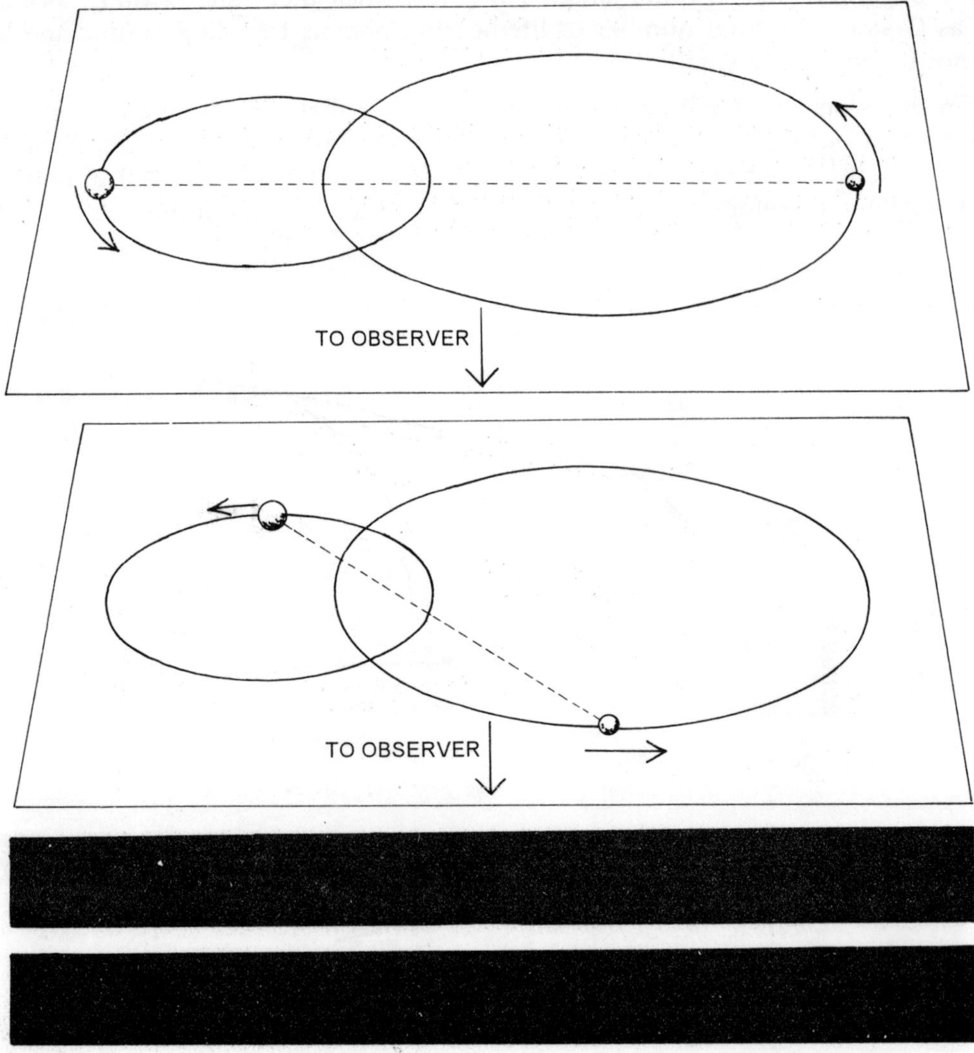

Fig. 14.4: Spectroscopic Detection of Close Binaries is depicted in these diagrams and spectra. At top the stars are orbiting around center of mass, the one at left moving toward the earth, the one at right away from the earth. Spectral lines made by star moving toward earth shift toward violet end of the spectrum, while lines from star moving away shift toward red end, splitting spectral lines as seen in upper spectrum below diagrams. In lower diagram the binaries are moving across line of vision as seen from earth. This gives rise to normal single spectral lines seen in the lower spectrum. Spectroscopic binaries are so close together that no telescope can resolve them into separate bodies, and only their spectra enable us to detect them and to calculate their orbital movements.

This number is reduced considerably upon consideration of the third critical condition; the maintenance of stable planetary orbits. The dust-cloud mechanism that increases the likelihood of planets has also brought most stars into existence as double or multiple systems. It is obvious that the presence of two or more stars in a system will profoundly perturb the orbits of planets in the system. A few double-star systems have members that are so far apart that if they are sufficiently near to us, the stars can be distinguished by a telescope or even by the naked eye. Such systems are called visual binaries. Much more common are the "spectroscopic binaries." In these systems the stars are so close together that even if they are relatively near to us, they cannot be separated by the largest telescope. We can tell that they are double stars only by regular changes in their spectra. These changes also enable us to calculate their orbits.

The orbits of planets in such systems are so complicated that astronomers have been able to work them out for only a few idealized cases. A life-bearing planet would have to travel on an obrbit close to one of the stars in a widely separated system, or on an orbit at a large distance from the stars in a close system. In the case of a hypothetical binary composed of two stars with the same luminosity as our sun and revolving around a common center in a nearly circular orbit, a planet would find the thermally habitable zone dynamically stable only if the stars were more than 10 astronomical units or less than .05 astronomical unit apart (an astronomical unit is the distance between the earth and the sun). With a separation between the stars of .5 to 2 astronomical units there is no overlapping of the habitable zone and the dynamically stable zone. Therefore no inhabitable planets can exist in such systems. Taking everything into consideration, only 1 to 2 per cent of all double and multiple stars may possess inhabitable planets, and perhaps 3 to 5 per cent of all the stars in our galaxy have such planets.

For the present there is no hope of detecting on a photographic plate the existence of a planet of another star. Such planets as may attend even the nearest star are completely lost to view in the brilliance of the star's light. Some-day there may be a telescope on a platform in space, free of the interfering effects of the earth's atmosphere. As has been suggested by Nancy G. Roman of the National Aeronautics and Space Administration, the instrument would produce a sharp star image that could be blocked out so that a planet near the star could be detected by means of a long photographic exposure. Even the subtle methods used to detect spectroscopic binaries are not refined enough to find a planet, because the effect of a planet upon the motion of its parent star is so slight.

There is nonetheless other well-established evidence to support the contention that planets are common. In the first place, no sharp distinction can be drawn between binary or multiple stars and stars with planetary systems. According to Gerard P. Kuiper of the Yerkes Observatory, the mean distance of separation between the components of all binaries so far investigated is about 20 astronomical units. This is of the same order of magnitude as the distance between the sun and its major planets (Jupiter, Saturn, Uranus and Neptune). Kaj Aa. Strand at the U.S. Naval Observatory

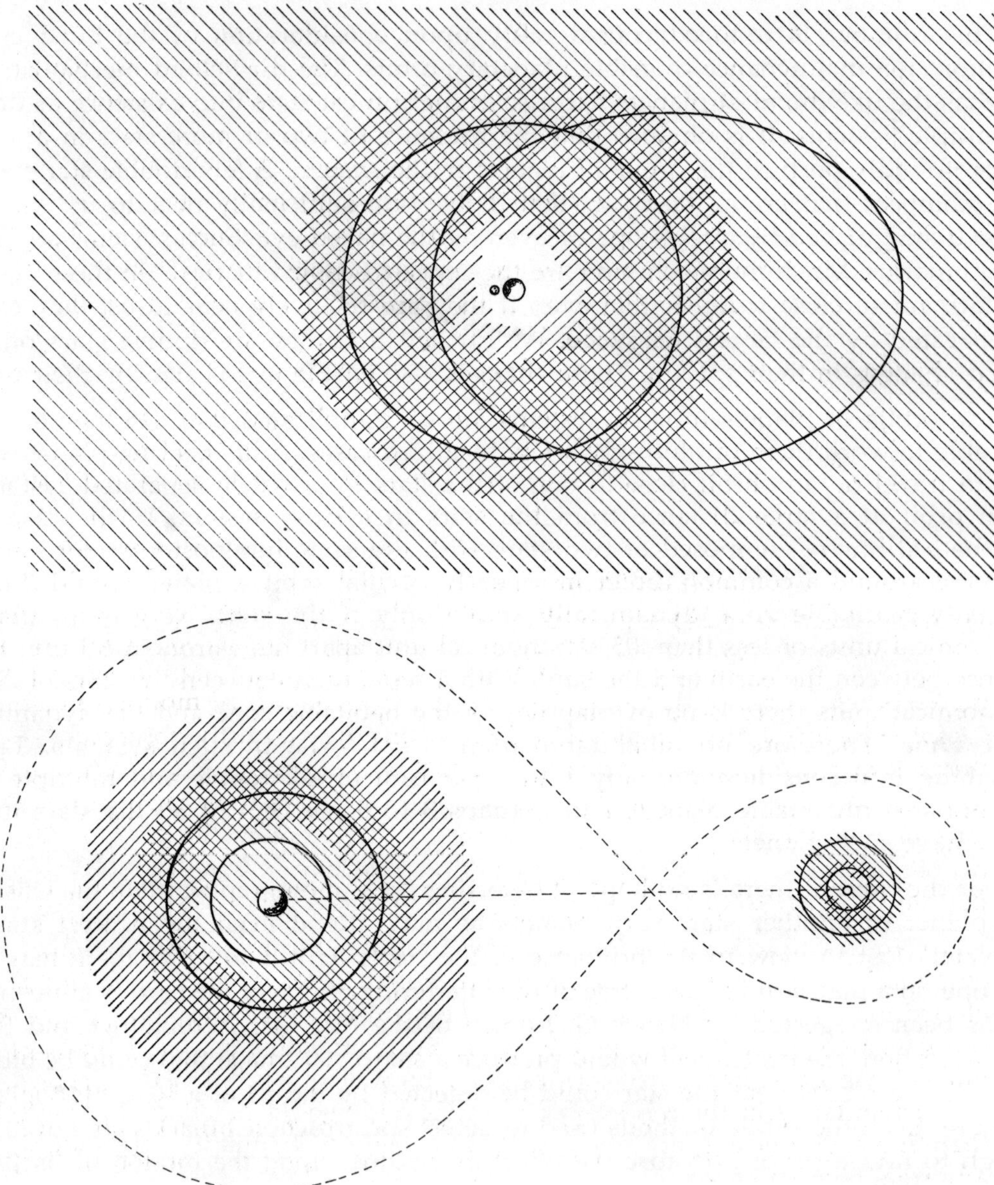

Fig. 14.5: Planetary Orbits Around Binary Stars are depicted here in an idealized manner. The figure eight in the top diagram is simply a mathematical projection from which the dynamically stable planetary orbit is calculated for widely separated binaries. At bottom are close binaries. For a planet to bear life its orbit would have to fall completely within both the dynamically stable zone (*gray hatching*) and the thermally habitable zone (*coloured hatching*). At the bottom the highly elliptical planetary orbit is dynamically stable, but it does not fall completely within the thermally habitable zone; the planet therefore could not harbour any life.

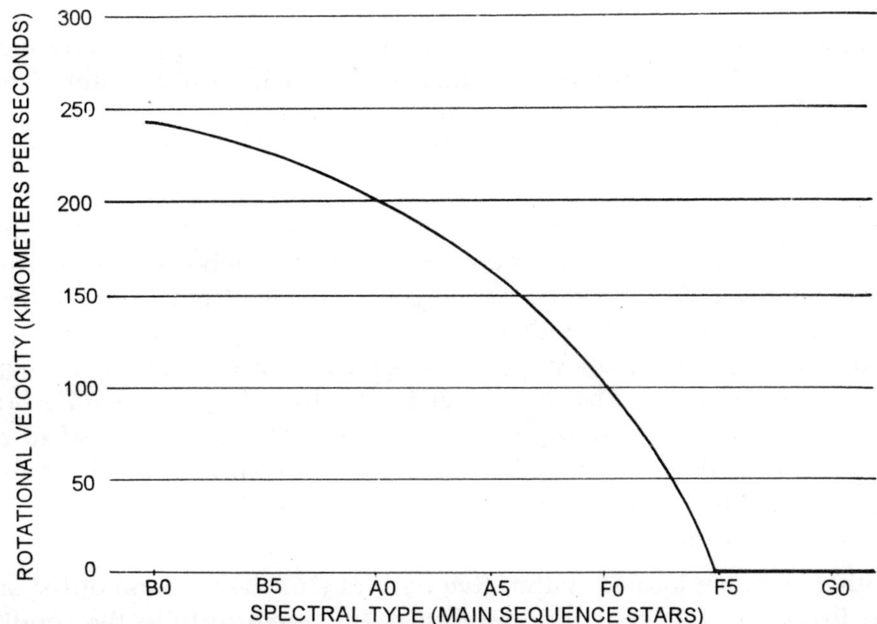

Fig. 14.6: Rotational Velocity of Stars on the main sequence is diagrammed here according to spectral type. Slow spin after Type F5 may indicate that planets are present.

in Washington has studied small perturbations in the orbital motion of a star called 61 Cygni, which has a companion that is too faint to be directly observed. He has found that the mass of this unseen object is about a hundredth of that of the sun. This mass lies between that of stars and of Jupiter. It is therefore reasonable to believe that the masses of small stars in binary systems grade continuously down to the masses of planets. Since binaries are so common in our galaxy, it would seem that many stars that now appear to be alone actually possess planets.

Harold C. Urey of the Scripps Institution of Oceanography has found additional evidence for planets in a certain kind of meteorite. According to Urey, diamonds embedded in these objects show that they must at one time have been under high pressure in a body the size of the moon. Such moonlike objects, known as "pestellar nuclei," would enhance the formation of both stars and planets from a dust cloud. Any irregularity in the motions of a dust cloud should be expected to produce more than one such nucleus, and the formation of planets or multiple stars would follow as a normal consequence.

Finally, measurements of the angular momenta of many stars give every indication that planets exist outside the solar system. Otto Struve of the National Radio Astronomy Observatory has pointed out that main-sequence stars more massive than Type F5 usually rotate rapidly, but starting with this type the rotation of stars slows down abruptly. In other words, the average angular momentum per units mass of the main-sequence stars exhibits a conspicuous break at Type F5. The most reasonable explanation of this strange phenomenon is that unobservable planets have absorbed the angular momentum, just as Jupiter and other planets of the sun carry 98 per cent of the angular momentum of the solar system, leaving the sun with only 2 per cent and a comparatively long period (27 days) of rotation. If planets do indeed account for the slow spin of these otherwise sunlike stars, then planets appear just where life is most likely to flourish.

Thus it seems that intelligent life may be scattered throughout the Milky Way and the universe as a whole. In our immediate neighbourhood, however, we may be its only representatives. The sun's nearest neighbour, Alpha Centauri, is only 4.3 light-years away. It is a triple system with two massive components (a G4 star and a K1) revolving around each other about 20 astronomical units apart; at a considerable distance is a small third star. The two larger bodies have highly eccentric orbits, and if there is a stable zone for a planet in this system, it is extremely hard to compute. Moreover, recent investigations indicate that the system may be much younger than the sun, so that higher forms of life might not have had time to evolve even if a habitable planet does exist in it.

Forty other stars are located within five parsecs (16.7 light-years) of the sun. Only two-Epsilon Eridani (a K2 type) and Tau Ceti (G4) —seem to fulfil the conditions for the existence of advanced forms of life, and Epsilon Eridani may not be exactly on the main sequence. Tau Ceti is 10.8 light-years distant, has an apparent visual magnitude of 3.6, is located on the celestial sphere about 16 degrees south of its equator and appears above the horizon in the northern sky only in the winter.

Since intelligent life is probably not a rare phenomenon, and since at least one star in our vicinity meets the specifications of a life-fostering star, it may seem odd that we have had no visitors from other worlds. The idea would have drawn ridicule 20 years ago, but today it deserves consideration. There are, however, several reasons for believing that we have had no visitors from outer space. For one thing, the 10.8 light-years that separate us from Tau Ceti—astronomically a short distance—is an extremely long distance in terms of human experience. Even traveling at the speed of the artificial satellites that man has launched, space voyagers would need hundreds of thousands of years to traverse it. It is possible that organisms from Tau Ceti might have a far longer life-span than man, but this supposition invokes radical assumptions that cannot be supported by present knowledge. Furthermore, if a meeting is to occur, our high technological civilization would have to be contemporary with that created by intelligent organisms on other planets. The cultural evolution that brought mankind to its present technical competence began only a few centuries ago. And even if human civilization endures for hundreds of thousands of years, it would be a brief episode in the time-scale of biological evolution. Therefore the chance that advanced civilizations might be flourishing at the present time on the planets of one or two nearby stars is excessively small.

Some workers have concluded, however, that the chance is good enough and certainly intriguing enough—to institute radio surveillance of signals originating outside

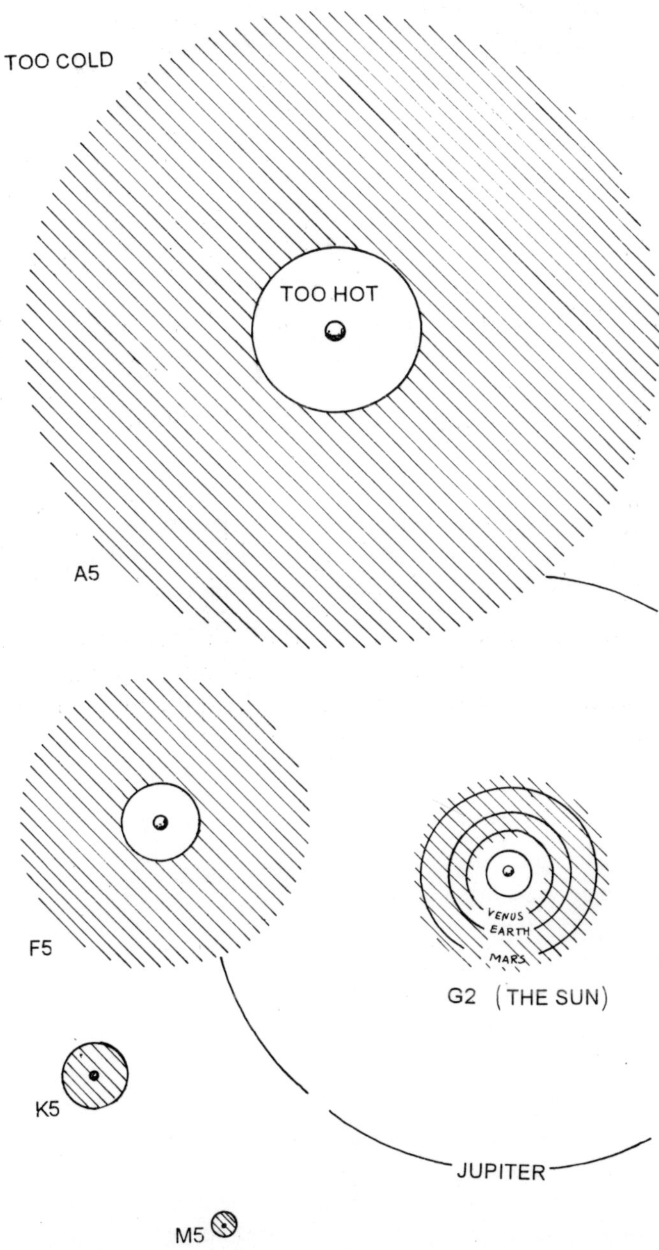

Fig. 14.7: Thermally Habitable Zone of various types of star is here represented by hatched area around star. Here the A5 spectral type has largest zone, but star does not remain stable long enough for evolution to take place on a planet near it. Habitable zone of our sun (a G2 type) extends from orbit of Venus to orbit of Mars. Tiny M5 star has the smallest zone.

the solar system. Guiseppe Cocconi and Philip Morrison of Cornell University have pointed out that the most favourable wavelength would be one close to but outside of the 21-centimeter line emitted by hydrogen in space, because in this region of the spectrum the galactic noise and the noise produced in the earth's atmosphere are at a minimum. Moreover, this important wavelength would be of as much interest to astronomers on another planet as to those on earth. Cocconi and Morrison have urged that radiation picked up by the 600-foot radio telescope, now being built by the Navy in West Virginia, be analyzed for the presence of signals. According to them, that telescope will be capable of detecting signals generated 10 light-years away by a technology no more advanced than our own. Meanwhile Frank D. Drake of the National Radio Astronomy Observatory, who makes a more generous estimate of the distance at which signals sent by intelligent organisms could be detected, is in charge of an actual project that is employing a smaller telescope to detect any radio signals transmitted by living beings in other "solar systems."

What kind of signals may we expect to receive or should we send out? Probably the most abstract and the most universal conception that any intelligent organisms anywhere would have devised is the sequence of cardinal numbers: 1, 2, 3, 4 and so on. The most likely signal would be a series of pulses indicating this sequence repeated at regular intervals. Such a signal may upon first consideration appear to be too simple for the sophisticated task of communicating with other beings far away among the stars. It would sound like baby talk. But after-all interstellar communication is surely still in the baby-talk stage.

Chapter—15
The Anatomy and Relationships of Dinosaurs

For 150 million years, from the Late Triassic to the end of the Cretaceous, dinosaurs dominated the terrestrial environment. They were unquestionably the most successful of all reptiles and rivaled the Cenozoic mammals in their structural diversity and wide distribution. Dinosaurs included the largest terrestrial animals that have ever lived, as well as species that may have equaled many mammals in their high metabolic rate and relative brain size.

The Posture and Ancestry of Dinosaurs

All thecodont groups above the level of proterosuchians tend to bring the limbs under the body and to move them in a more strictly fore-and-aft direction, without the large degree of lateral extension that is evident in primitive tetrapods. However, most retain vestiges of a sprawling posture in the structure of the hip and ankle joints. In the earliest members of all dinosaur groups, these joints show specialization that resulted in a nearly vertical posture.

The head of the femur is angled anteromedially so that it articulates with the dorsal margin of the acetabulum. This margin is strongly buttressed to accommodate the entire weight of the body. The central portion of the acetabulum is not ossified but appears fenestrated in the fossils. In contrast with thecodonts, it no longer needs to resist the medially directed force generated by the femur, which was angled toward the midline.

The shaft of the femur is straight or bowed slightly anteriorly, in contrast with the gently sigmoidal configuration of thecodonts and crocodiles. Just distal to the head are two tuberosities, the greater and lesser trochanters, that serve for insertion of the iliofemoralis internus and the iliofemoralis externus muscles. Further distally, on the posterior surface, is the fourth trochanter, which serves as the area of insertion for the major retractor of the femur, the caudifemoralis. The head of the tibia is twisted to produce a nearly vertical orientation of the shaft with respect to both its proximal and distal articulating surfaces.

A mesotarsal ankle joint is established, with the line of flexion passing transversely between the proximal and distal tarsals. The astragalus and calcaneum are reduced to relatively simple elements that are closely integrated with the ends of the tibia and fibula. The metatarsals are elongate and function as an additional unit of the lower limb. The digits form the main surface that contacts with the ground. The posture in dinosaurs may hence be considered to be digitigrade.

These changes result in a nearly vertical limb, as viewed anteriorly, with both the ankle and knee functioining primarily as hinge joints. Structurally and functioinally, the closest modern homologue is provided by birds.

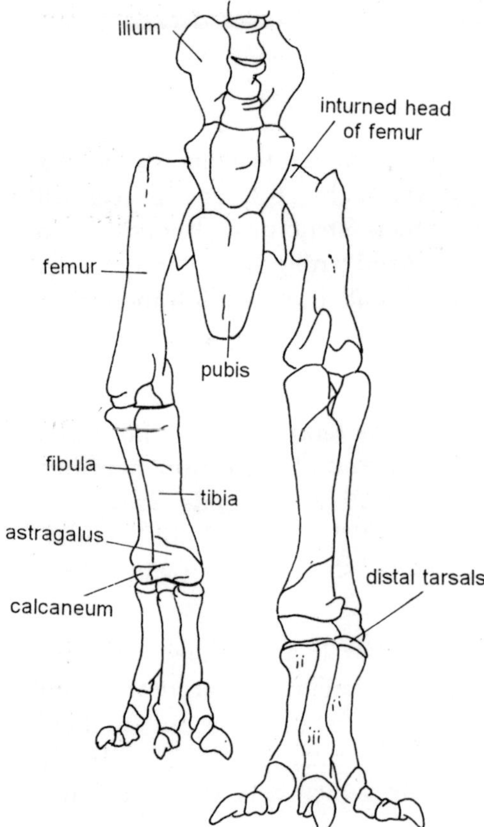

Fig. 15.1: Anterior View of the Rear Limbs of the Dinosour *Tyrannosaurus*. Note the vertical posture and the transverse knee and ankle joints that are characteristic of both saurischians and ornithischians.

Among the thecodonts, the greatest similarity can be found among the Lagosuchidae, in which the rear limb achieved many features that are common to the earliest dinosaurs. *Lagosuchus* is *more* specialized than early dinosaurs in the greater relative length of the metatarsals, but this is a character that might have been reversed in descendants with a larger body size. An animal that was very similar to *Lagosuchus* may have given rise to all the dinosaur groups. However, *Lagosuchus* is more primitive than any dinosaurs in the small degree of fenestration of the acetabulum. We do not know the skull well enough to make a detailed comparison with the major dinosaur groups. The exact phylogenetic position of this genus cannot yet be established.

The upright posture of the limbs in dinosaurs has long been confused with the question of their bipedality. A clear distinction between these features was made by Charig (1972). Among the thecodonts, the ornithosuchids and rauisuchids brought the rear limbs close to a vertical orientation. It was once thought that they were habitually bipedal, but information from footprints as well as skeletal anatomy indicates that they were primarily quadrupedal. *Lagosuchus* may also have been facultatively, rather than habitually, bipedal.

Many Triassic and early Jurassic dinosaurs show strong skeletal evidence for bipedality in the very great disparity in the length and structure of the forelimbs and hind limbs. On the other hand, some early dinosaurs that approached the Jurassic sauropods in their great body size were almost certainly quadrupedal, although their rear limbs were also longer than the front limbs. Romer (1956, 1966) argued that this limb disparity reflects a reversioin to a quadrupedal stance from a primitive bipedal condition. If this were the case, it would provide further evidence that all dinosaurs shared a close common ancestry above the level of the thecodonts. Unfortunately, the fossil record does not demonstrate unequivocally that the early quadrupedal dinosaurs descended from fully bipedal ancestors. The early dinosaurs may have diverged from a primitive stock that was facultatively, but not habitually, bipedal. The clear distinction between these patterns of locomotion may have evolved after the occurrence of the changes in the structure of the rear limb and girdle by which we recognize dinosaurs and after the initial divergence of the major dinosaur lineages.

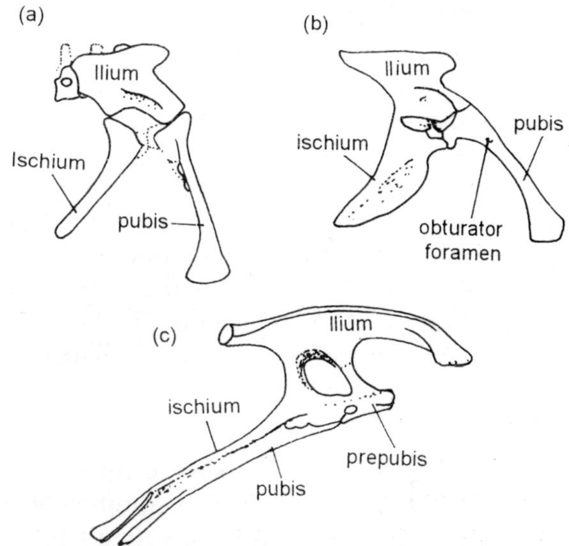

Fig. 15.2: The Configuration of the Pelvis, the Primary Criterion for Distinguishing the Orders Saurischia and Ornithischia. The saurischian pelvis, exemplified by *(a) Staurikosaurus* is similar to that of thecodonts such as *(b) Lagosuchus* in which the elongate pubis and ischium together with the ilium form a triradiate structure. The pelvis of ornithischians, as exemplified by the early genus *(c) Heterodontosaurus* is advanced in having the pubis oriented posteriorly, alongside the ischium. In some advanced ornithischians, a new structure, the prepubis, has developed anterior to the pubis. This results in a tetraradiate configuration.

Dinosaurs have long been divided into two major groups, the ornithischians and the saurischians, primarily on the basis of the structure of the pelvis. The configuration of the pelvis and most other features that distinguish early saurischians are primitive, asdemonstrated by their similarity with those of thecodonts. In contrast, the ornithischian pattern of the pelvis and many features of the skull are clearly derived relative to those of the saurischians. Ornithischians may have evolved from primitive saurischians, but there is no fossil evidence to support the long-held assumption (which Charig reiterated in 1976) that saurischians and ornithischians had separate origins from distinct groups of thecodonts. No thecodonts other than the lagosuchids share significant derived features with either of the two major groups of early dinosaurs.

Saurischians

The ornithischians are a diverse assemblage and their interrelationships are not fully known, but they unquestionably have a common ancestry, established on the basis of a host of shared derived characters. We cannot state with equal assurance that the several groups of saurischians share a unique common ancestry since the features they share are primarily primitive. Two major groups have long been recognized, the carnivorous and obligatorily bipedal theropods and the typically quadrupedal and herbivorous sauropods. The theropods have been further separated into carnosaurs and coelurosaurs, primarily on the basis of size. Typical sauropods are only known from the Jurassic and Cretaceous. In the late Triassic and early Jurassic, a distinctive group of more primitive herbivorous saurischians are common. These are termed the plateosaurs or prosauropods; their relationships with other dinosaurs are subject to continuing speculation.

Unfortunately, the late Triassic record of dinosaurs remains very incomplete; the few fossils that we know share a number of primitive features, but the nature of their relationships with later forms is difficult to establish.

THEROPODS

Primitive Triassic Carnivores

The only dinosaur that we recognize from the Middle Triassic is *Staurikosaurus* from southern Brazil. It is missing the skull, the forelimbs, and the tarsus and foot, but the lower jaw, vertebrae, pelvis, and rear limb are comparable to those of later saurischian dinosaurs. The lower jaw is as long as the femur, with sharp piercing teeth. The trunk vertebrae are short and the rear limb is fully erect. Galton (1977) suggests that the posture may have been fully bipedal. The tibia is longer than the femur, as is common in cursorial animals, and the limb bones are hollow. Together with the size of the jaw, these characters suggest that *Staurikosaurus* was an active predator. As restored, the skeleton is slightly more than 2 meters long.

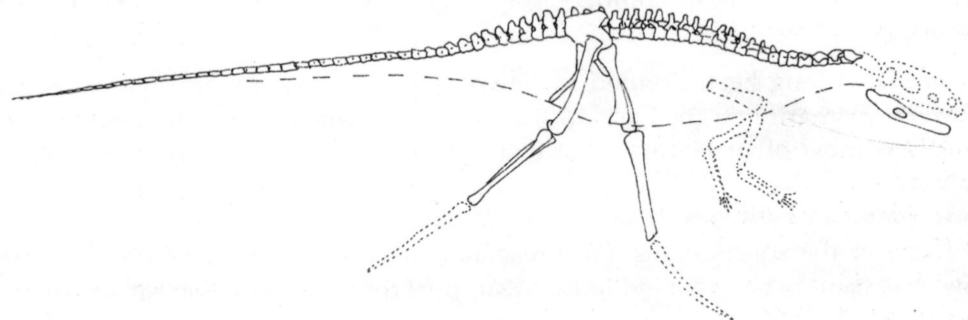

Fig. 15.3: Restoration of *Staurikosaurs*. This is the earliest-known dinosaur from the late Middle Triassic of South America. Total length is 2.1 meters.

A possibly related genus from the Upper Triassic is *Herrerasaurus,* which was about 4 meters in length. The ankle and foot are typical of primitive dinosaurs in the integration of the astragalus and calcaneum with the tibia and fibula, the reduction of the distal tarsals, and the elongation of the metatarsals. The fifth digit is reduced and the weight is supported evenly on the other four digits.

The large size of the skull and the expansion of the distal end of the pubes into a vertical plate are features that are shared with the large carnivorous dinosaurs of the Jurassic and Cretaceous. However, these features also evolved independently in large carnivorous thecodonts.

Because these Triassic genera are so incompletely known and primitive, we cannot ally them with certainty with later dinosaur groups. What we know of their anatomy and general habitus suggests that they may be related to the large carnivorous dinosaurs of the Jurassic and Cretaceous.

"Coelurosaurs" and "Carnosaurs"

Late Triassic, Jurassic and Cretaceous carnivorous dinosaurs have customarily been divided into two infraorders, the Carnosauria and the Coelurosauria, that are based primarily on differences in size and proportions. The term carnosaur has been applied to the largest carnivorous dinosaurs, which are further characterized by the disproportionately large skull and relatively short forelimbs. Most genera that have been recognized as coelurosaurs are smaller, with a skull that is typically smaller relative to the length of the trunk and relatively larger forelimbs.

Genera that have been described within the past 15 years show a combination of features that were previously though to distinguish these two groups. *Deinocheirus* and *Therizinosaurus* from the Cretaceous of Mongolia are still incompletely known but demonstrate the existence of very large theropods with extermely long

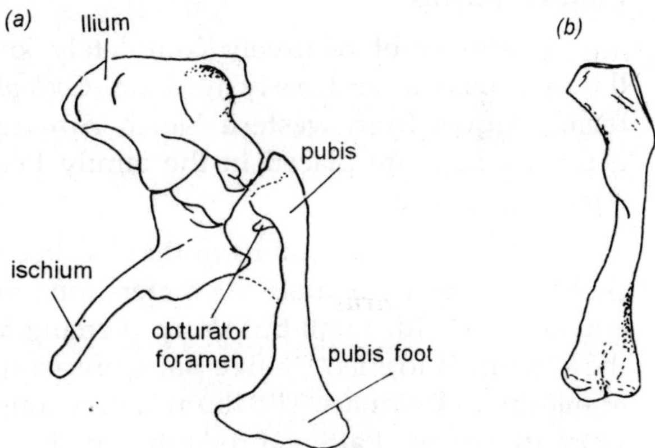

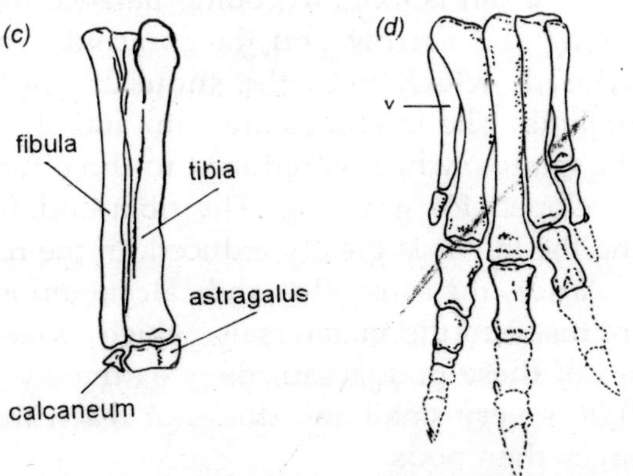

Fig. 15.4: (*a*) Pelvis, (*b* and *c*) rear limb, and (*d*) foot of *Herrerasaurus*, a carnivorous dinosaur from the Upper Triassic of South America.

forelimbs. Nearly complete skeletons of *Dilophosaurus* and *Deinonychus* combine a nearly equal number of "carnosaur" and "coelurosaur" traits. Use of these two infraorders to classify the theropods has become difficult to justify. Unfortunately, we have not been able to establish an alternative classification that encompasses all the theropod families in a consistent phylogenetic scheme. The better-known theropod families are each readily distinguished by unique features, but shared, derived characters that unite them with one another are more difficult to recognize. The "carnosaur" families may each have evolved separately from different groups that have been classified as coelurosaurs. Or perhaps several separate lineages of both large and small theropods evolved independently from among primitive forms such as *Staurikosaurus* and *Herrerasaurus*.

Podokesaurids

A number of relatively completely known theropods have been described from the late Triassic and early Jurassic. *Coelophysis* and *Syntarsus* are small, lightly built forms known from western North America and eastern Africa, respectively. Similar genera, which are placed in the family Podokesauridae, are present in Europe, South America, and Asia.

Coelophysis is a small form that has been considered typical of the coelurosaurs. The skeleton is approximately 2½ meters long and extremely lightly buuilt. The skull is low and slender with small but sharp piercing teeth. The neck is longer than the trunk, and the sacrum is formed by five pairs of sacral ribs that are fused to the vertebrae and ilia in mature individuals. The ilium is very long and low, and the pubes extend far forward. They are narrow flattened rods that are fused at the midline, but, in contrast with larger theropods, they are not expanded into a vertical plate distally. The ischium, as in the other early dinosaurs, is a long, posteroventrally directed rod.

The tail is long, to counterbalance the weight of the presacral region. The scapula is long and narrow, and the coracoid is a small oval plate. The clavicle is present but reduced, which frees the shoulder girdle to allow greater maneuverability of the forelimb. The forelimbs are slim and show no evidence that they supported the body. The manus, which is reduced to three functional digits and a trace of a fourth, is well developed for grasping. The tibia and fibula are 20 per cent longer than the femur, and the tarsus is greatly reduced. In the related genus *Syntarsus* from the Lower Jurassic of Africa, the astragalus and calcaneum are fused to one another and the distal tarsals are fused to the metatarsals, which strictly limits the joint to a fore-and-aft hinge. The pes of these podokesaurids is extremely birdlike, with three principal digits. The first digit is very small and does not reach the ground; it is not reversed in the manner of larger theropods.

Coelurids and Compsognathus

We have not found any fossils of small theropods in the Middle Jurassic. In the late Jurassic and early Cretaceous several distinct lineages may be recognized.

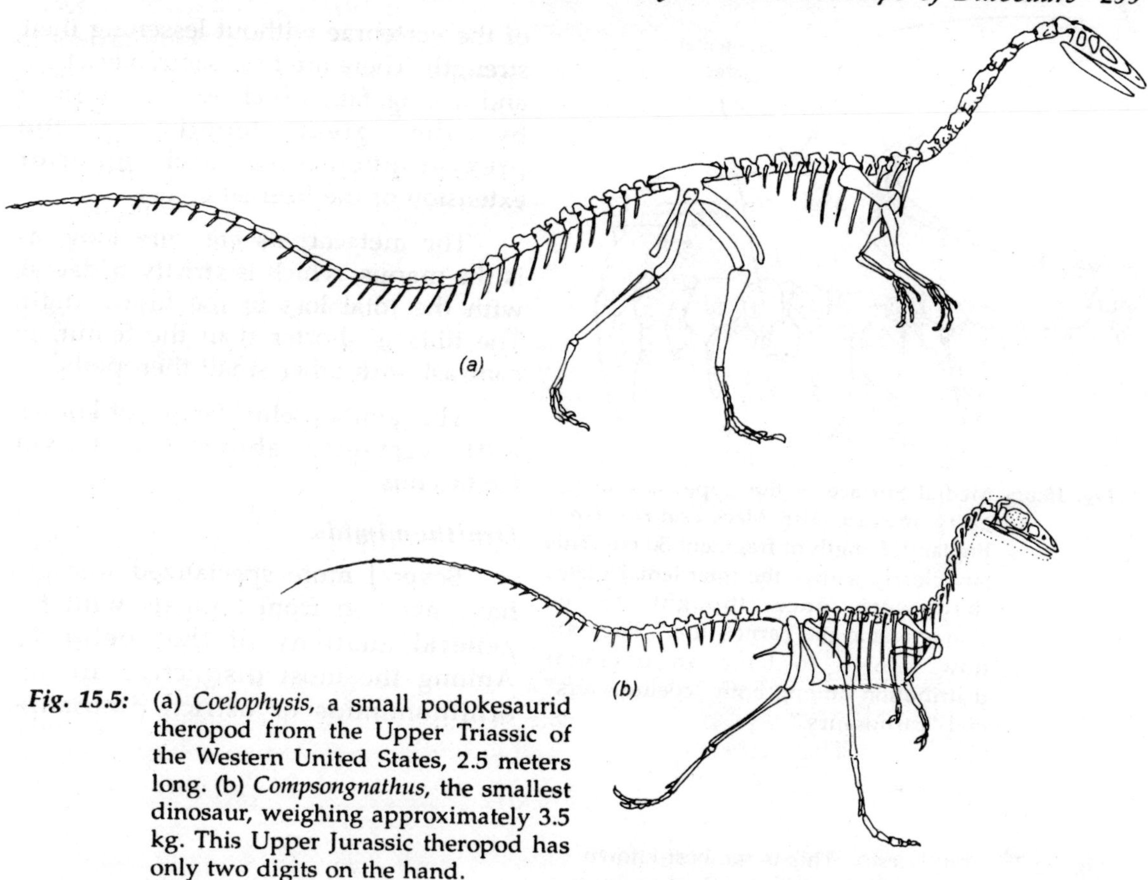

Fig. 15.5: (a) *Coelophysis*, a small podokesaurid theropod from the Upper Triassic of the Western United States, 2.5 meters long. (b) *Compsongnathus*, the smallest dinosaur, weighing approximately 3.5 kg. This Upper Jurassic theropod has only two digits on the hand.

Compsognathus, which is represented by two specimens from the Upper Jurassic of Europe, is the smallest known dinosaur. Ostrom (1978) estimates that the body weight was 3 to 3½ kilograms. In contrast with all other small theropods, the manus is reduced to two digits, with a phalangeal count of 2, 2, 0. The tibia is considerably longer than the femur, and the metatarsus is elongated as well, which suggests that *Compsognathus* was an effective cursorial predator. The type specimen has the bones of an agile lizard in its abdominal cavity. In contrast with other small theropods, interdental plates, which once were considered diagnostic of "carnosaurs," are present in both the upper and lower jaws.

The family Coeluridae represents a continuation of the pattern that was established by small podokesaurids, *Ornitholestes* is the best-known member of the group. The restored skeleton is about 2 meters long. The skull is relatively high; the teeth are small and limited to the front of the jaws. The cervical vertebrae are relatively short and are recessed posteriorly to form a more effective articulating surface, a pattern termed opisthocoelous. They are further characterized by the presence of large depressions on the lateral surface called ***pleurocoels,*** which would reduce the weight

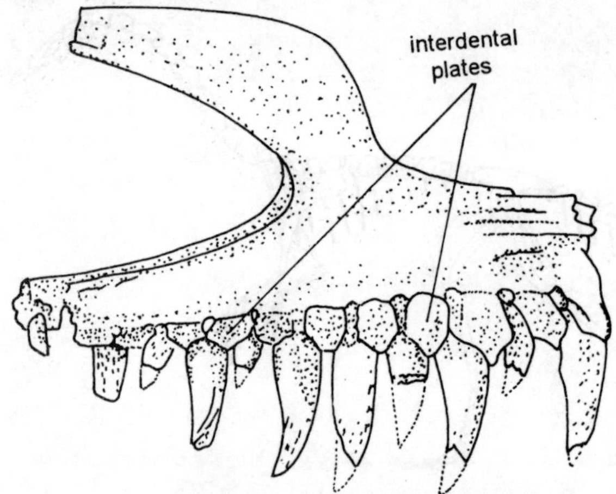

Fig. 15.6: Medial Surface of the upper jaw of the early megalosaur *Megalosaurus* from England. Length of fragment 30 cm. This jaw clearly shows the interdental plates that were once thought to be characteristic of "carnosaurs." They are now known to have an irregular distribution among both "coelurosaurs" and "carnosaurs."

of the vertebrae without lessening their strength. There are four sacral vertebrae and a long tail, which is characterized by the great length of the prezygraphophyses and anterior extension of the haemal arches.

The metacarpals are very long, as is the manus, which is strictly tridactyl, with the total loss of the fourth digit. The tibia is shorter than the femur, in contrast with other small theropods.

The family coeluridae is not known with certainty above the Lower Cretaceous.

Ornithomimids

Several more specialized lineages have evolved from animals with the general anatomy of the coelurids. Among the most distinctive are the ornithomimids or ostrich dinosaurs,

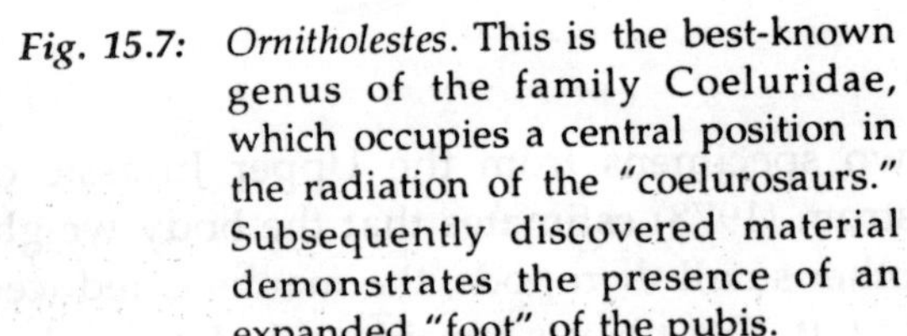
Fig. 15.7: *Ornitholestes*. This is the best-known genus of the family Coeluridae, which occupies a central position in the radiation of the "coelurosaurs." Subsequently discovered material demonstrates the presence of an expanded "foot" of the pubis.

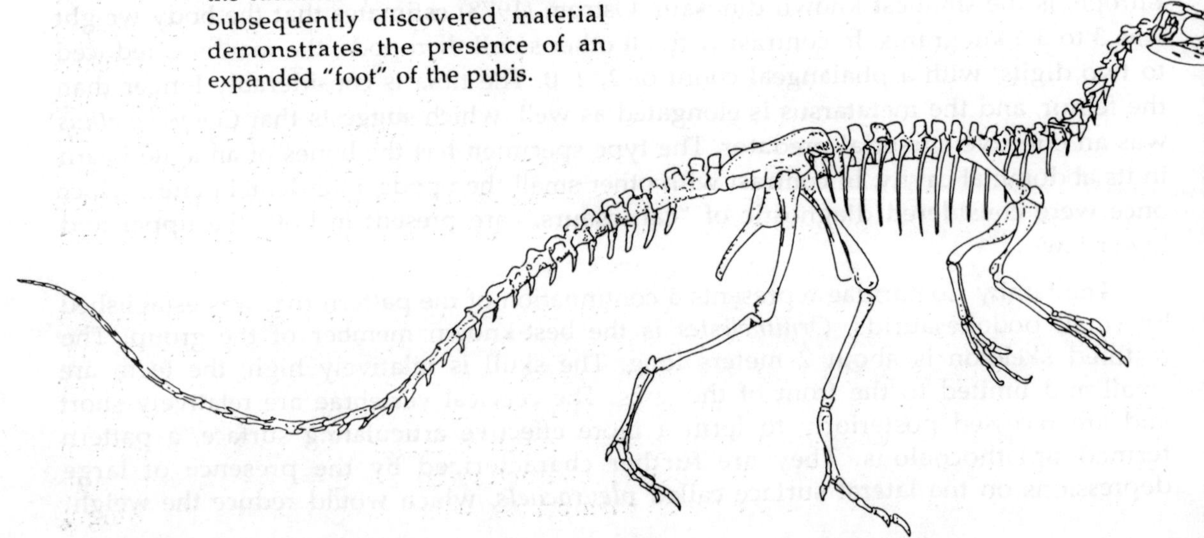

which are found primarily in the Cretaceous of North America and East Asia. Well-known genera such as *Ornithomimus* and *Struthiomimus* have the size and proportions of an ostrich. They are distinguished from all other dinosaurs by the total absence of teeth. The skull is smaller relative to the vertebrae and trunk than in other coelurosaurs. The bones are extremely thin, and the skull is thought to be kinetic in the manner of birds (Bock, 1964). The orbits and the braincase are both very large, with the estimated ratio of brain size to body weight approaching that of some birds. The cervical centra are elongated and fused to the cervical ribs. There are six coossified sacral vertebrae, which include elements that are part of the trunk and tail in more primitive dinosaurs. The forelimbs are very long, but the carpals are reduced and allow little mediolateral movement of the digits. In contrast with most other small theropods, there is a large public boot. The tibia is 20 per cent longer than the femur and the metatarsals are very long. The first digit of the pes is lost, which results in a strictly tridactyl foot.

Specimens of the Upper Jurassic genus *Elaphrosaurus* from East Africa lack a skull, but the postcranial skeleton suggests a transitional position between early coelurids and ornithomimids. Ornithomimids were poorly represented in the early Cretaceous but became widespread later in that period; only a few fragmentary specimens are found in the uppermost Mesozoic.

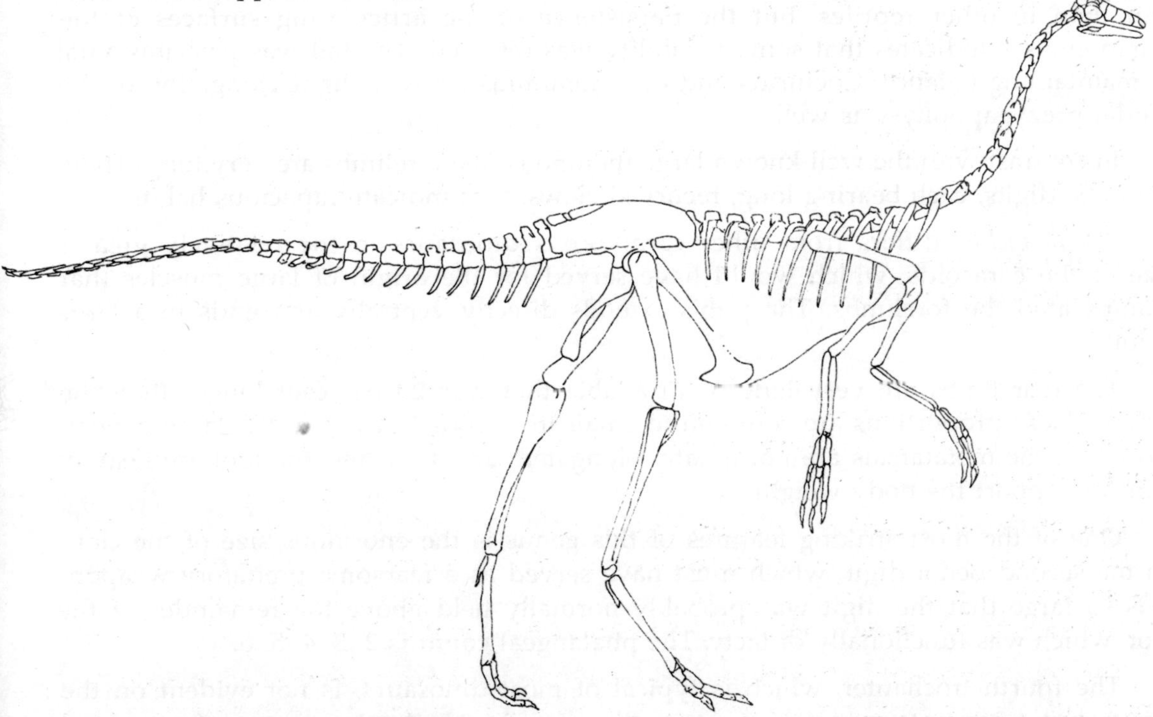

Fig. 15.8: *Struthiomimus*. This is a member of the ostrich dinosaur family Ornithomimidae from the Upper Cretaceous.

Dromaeosaurs and the Saurornithoididae

Small podokesaurids, coelurids, *Compsognathus,* and ornithomimids fit the old concept of coelurosaurs. In contrast, dromaeosaurs and saurornithoidids appear to be structurally intermediate between the primitive members of that assemblage and the large "carnosaurs." One of the most thoroughly described of all dinosaurs is *Deinonychus* from the Lower Cretaceous of Montana. It is a lightly built animal, about 3 meters long, with an estimated weight of 60 to 75 kilograms. It appears to combine almost equally characters of carnosaurs and coelurosaurs. The skull is relatively large, with a deep maxilla and lower jaw to accommodate large, bladelike teeth. However, the interdental plates that are characteristic of most large theropods are missing.

The centra of the cervical vertebrae are not opisthocoelous, as in almost all other advanced dinosaurs, but flattened on both ends and beveled to produce a fixed curvature that maintains the large head above the level of the trunk.

The trunk was held nearly horizontal and supported by five sacral vertebrae. The posterior caudal vertebrae are highly distinctive in having the prezygapophyses and anterior processes of the haemal arches elongated as narrow rods that run the length of 8 to 10 vertebrae and form supporting bundles around the centra for much of the length of the tail. They would have stiffened the tail so that it would move more as a unit than in other reptiles, but the persistence of the articulating surfaces of the zygapophyses indicates that some flexibility was retained. The tail was probably vital in maintaining balance. Coelurids and ornithomimids show a slight elongation of the caudal prezygapophyss as well.

In contrast with the well-known large theropods, the forelimbs are very long. There are three digits, each bearing long, recurved claws, that indicate rapacious habits.

Deinonychus differs from other carnivorous dinosaurs in the relatively greater size of the coracoids, which would have served for the origin of large muscles that manipulated the forelimbs. The pubis extends directly ventrally and ends in a large "foot".

The rear limbs are very long, with a tibia that was 20 per cent longer than the femur. These proportions are common to small theropods but not to the large genera. However, the metatarsals are not greatly elongated, and they and the foot are heavily built to support the body weight.

One of the most striking features of this genus is the enormous size of the claw on the second pedal digit, which must have served as a fearsome predatory weapon. It is so large that the digit was probably normally held above the remainder of the foot, which was functionally didacty. The phalangeal count is 2, 3, 4, 5, 0.

The fourth trochanter, which is typical of most dinosaurs, is not evident on the femur, but a separate proximal process, the posterior trochanter, is conspicuous and may have served for the attachment of the ischiotrochantericus muscle. Ostrom (1969) suggests that this muscle may have served to retract the leg when the great claw dug into the prey. The use of this muscle, which originated on the ischium, would be less likely to upset the balance of the animal as it stood on one foot than would the use of the principal retractor, the caudifemoralis, which moved the tail laterally as it was contracted.

Ostrom (1969) tabulated an almost equal number of "carnosaur" and "coelurosaur" characters in *Deinonychus* and argued that these groups could not be clearly distinguished. On the other hand, he accepted that the ancestry of *Deinonychus* probably lay among the Coeluridae and hence that this genus was phylogenetically allied with the "coelurosaurs" despite the presence of many "carnosaur" features. The hands and feet in particular point to an ancestry that was shared with *Ornitholestes*. *Deinonychus* belongs to the family Dromaeosauridae, other members of which are known primarily from the Upper Cretaceous of North America and central Asia.

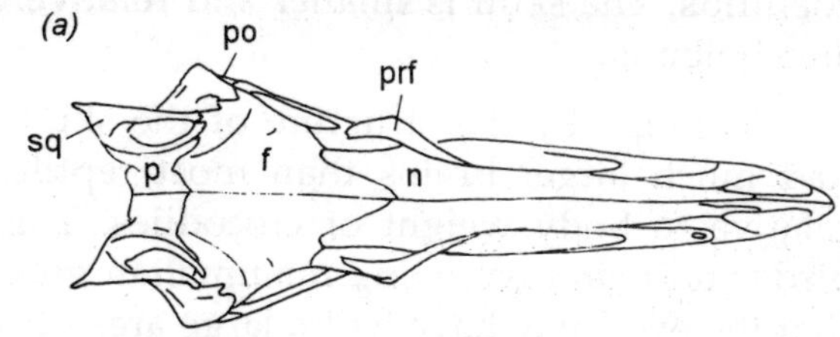

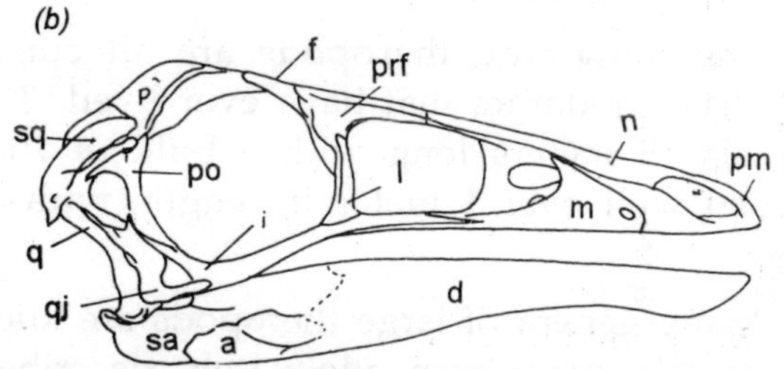

Fig. 15.9: Skull of the Ostrich Dinosaur *Dromiceiomimus*. *(a)* Dorsal and *(b)* lateral view.

The Saurornithoididae constitute a related family of rapacious coelurosaurs from the Upper Cretaceous. They are somewhat smaller than dromaeosaurs, with lighter limbs but a similar enlargement of the ungual phalanx of the second pedal digit. They lack pleurocoels in the cervical vertebrae, which are present in dromaeosaurs and

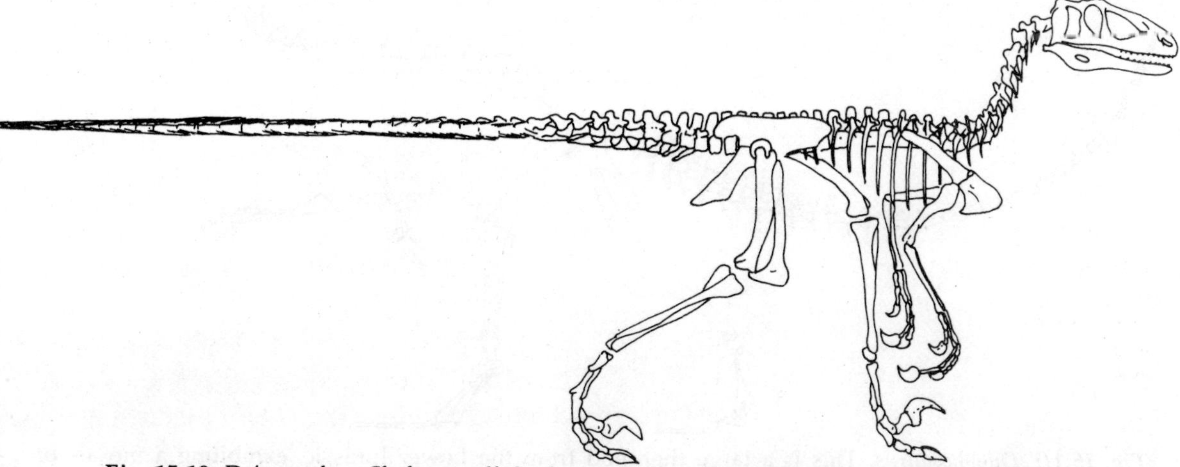

Fig. 15.10: Deinonychus. Skeleton off the large rapacious "coelurosaur," 3 meters long.

coelurids. The skull is smaller and relatively lower, but the braincase is relatively large and bulbous.

To judge by the structure of the skull, both dromaeosuaurs and saurornithoidids had much larger brains than most reptiles, reaching about seven times the volume relative to body weight of crocodiles. This ratio is the same as that found in many birds and primitive living mammalian groups. The configuration of the orbits suggests that the eyes may have had a large area of overlap for binocular vision. In their sensory acuity, degree of cerebral integration, and agility, the saurornithoidids represent a high point of dinosaur evolution.

The remaining theropods are all considerably larger and include the largest terrestrial predators that have ever lived. They culminate in the Upper Cretaceous in animals 15 meters long with a bulk of 7000 kilograms. The skull of *Tyrannosaurus* reached well over 1 meter in length, with individual teeth that were 15 centimeters long.

Many genera of large theropods are known, from as early as the late Triassic, but only a few have been adequately described and their relationships are not firmly established. As few as two or as many as five families may be recognized. The tyrannosaurids are a clearly defined group from the Upper Cretaceous. Most other large theropods may be placed in the most primitive large theropods, with separate families recognized for a series of more derived lineages: the Ceratosauridae, the Spinosauridae, and the Allosauridae.

The genus *Dilophosaurus* from the Lower Jurassic of North America (Welles, 1984) occupies a particularly isolated position among early theropods. With a length of over

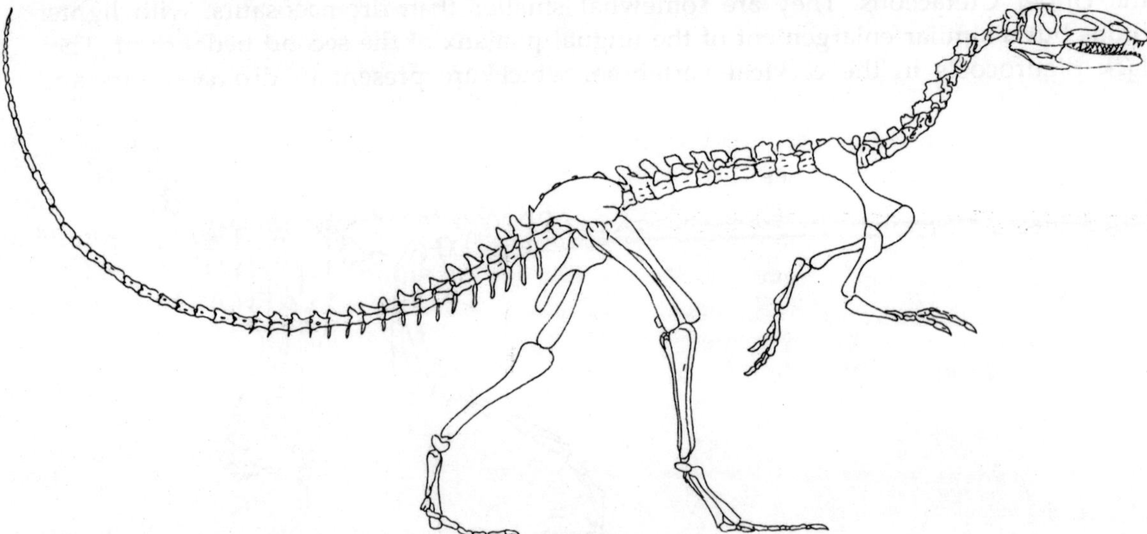

Fig. 15.11: Dilophosaurus. This is a large theropod from the Lower Jurassic, exhibiting a mosaic of "coelurosaur" and "carnosaur" traits. Approximately 6 meters long.

6 meters, its bulk rivals animals that were previously classified as carnosaurs. The skull is large relative to the length of the cervical vertebrae but is lightly constructed. The forelimbs are relatively large.

The skull is distinctive in having a pair of narrow, parasagittal crests above the nasals and frontals, and the premaxilla is only weakly attached. The cervical vertebrae are short and planoconcave rather than amphicoelous, as they are in small podokesaurids. They are lightened by pleurocoels. There are four sacral vertebrae that remain unfused. The manus consists of three large and one reduced digit. In contrast with other large theropods, the pubis lacks a boot. The femur is slightly longer than the tibia, and the metatarsals are not greatly elongated. Three large toes face forward and a small one turns to the rear. *Dilophosaurus* shows no clear affinities with any later theropods.

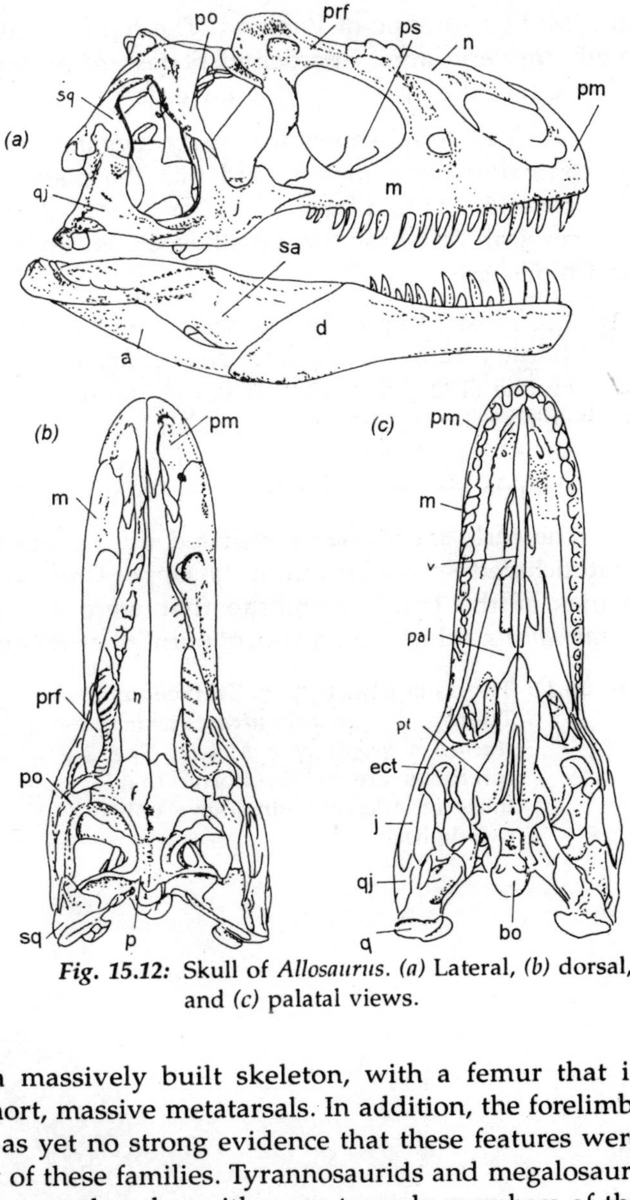

Fig. 15.12: Skull of *Allosaurus*. *(a)* Lateral, *(b)* dorsal, and *(c)* palatal views.

In relationship to their large size, all tyrannosaurids and megalosaurs (if one uses that term in the more general sense) share a massively built skeleton, with a femur that is typically longer than the tibia and short, massive metatarsals. In addition, the forelimbs are always relatively short. There is as yet no strong evidence that these features were uniquely derived within the ancestry of these families. Tyrannosaurids and megalosaurs may be no more closely related to one another than either are to early members of the "coelurosaur" assemblage.

Megalosaurs

We find megalosaur remains from as early as the base of the Jurassic and possibly even in the late Triassic, but few animals earlier than the late Jurassic are represented by relatively complete remains and fewer yet have been adequately described. Jaw and limb elements that are placed in the genus *Megalosaurus* are known from the Lower and Middle Jurassic of England. The femur is almost 1 meter long, and the jaws and teeth are very large. Interdental plates are present at the base of the inside surface of the teeth of the upper and the lower jaw.

Other material shows that early megalosaurs were fully bipedal, reached a length of approximately 6 meters, and that the distal end of the pubis, as in *Herrerasaurus*, was expanded into a large "foot" or "boot." As in other large theropods, the centra of the cervical vertebrae are opisthocoelous. Four digits are retained in both the front and hind limb.

The most completely known of the early megalosaurs is *Eustreptospondylus* from the Middle Jurassic of England, a specimen of which is on display at Oxform University. Von Huene (1926) described it under the name *Streptospondylus*. It consists of a nearly complete skeleton, although the skull is fragmentary. In general, it resembles the Upper Jurassic *Allosaurus*, although it is more primitive in having only three sacral vertebrae and a single antorbital opening. The femur is slightly longer than the tibia.

The early megalosaurs radiated during the Jurassic and led to specialized forms that included *Spinosaurus* from the Upper Cretaceous and its relatives, which had neural spines of the trunk vertebrae that were 2 meters long, and the Upper Jurassic *Ceratosaurus*, which had a short "horn" that was borne on the nasal bones.

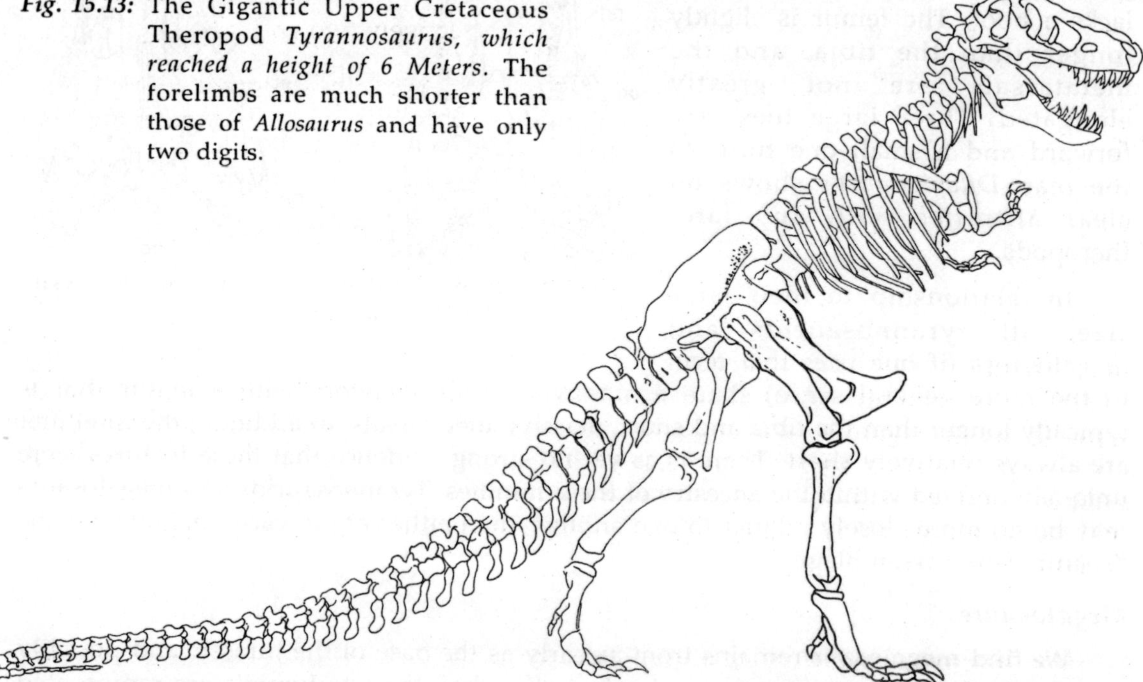

Fig. 15.13: The Gigantic Upper Cretaceous Theropod *Tyrannosaurus, which reached a height of 6 Meters*. The forelimbs are much shorter than those of *Allosaurus* and have only two digits.

The upper Jurassic genus *Allosaurus* is the best-known representative of a conservative lineage that extends into the Upper Cretaceous. *Allosaurus* reached 12 meters in length and was among the most powerful carnivores of its time. The body was strengthened to support the extra weight, and the sacrum had five fused vertebrae.

The skull is high and laterally compressed; the orbital opening is triangular and smaller than the principal antorbital fenestra. The teeth are long, laterally compressed, and recurved. In contrast with more primitive megalosaurs, the cervical vertebrae have well-ossified anterior condyles that contribute to well-defined ball-and-socket joints between the vertebrae. The more posterior trunk vertebrae retain a shallowly amphicoelous configuration. The tail is long, and the posterior pre-zygapophyses are considerably elongated.

The forelimb is considerably shorter than the rear limb and could not possibly have supported the body. Only three digits of the manus are retained. Each has a long recurved claw. Metatarsals II, III, and IV are closely integrated elements. The first digit, like that of birds, was oriented posteriorly.

Tyrannosaurids

The gigantic tyrannosaurids of the Upper Cretaceous were the largest of all theropods. Specializations of this family include the further reduction in the length of the forelimbs (which could not have reached the mouth) and the retention of only two functional digits, the first and second, which bear respectively two and three phalanges.

The skull is distinguished by the shape of the parietals, which form a sharp sagittal crest between the upper temporal openings. In megalosaurs, the parietals are flat in this area. Little, if any, movement was possible between any of the elements of the

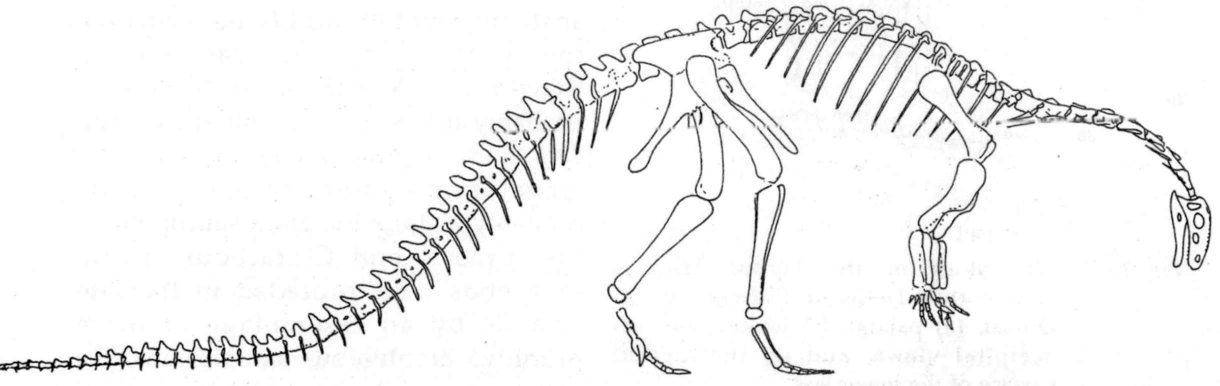

Fig. 15.14: *Plateosaurus*. Skeleton of the large plateosaur from the Upper Triassic of Germany. Length is about 7 meters. Plateosaurs are among the most primitive dinosaurs. Their dentition indicates that they were herbivores. They may have been facultatively, but not habitually, bipedal.

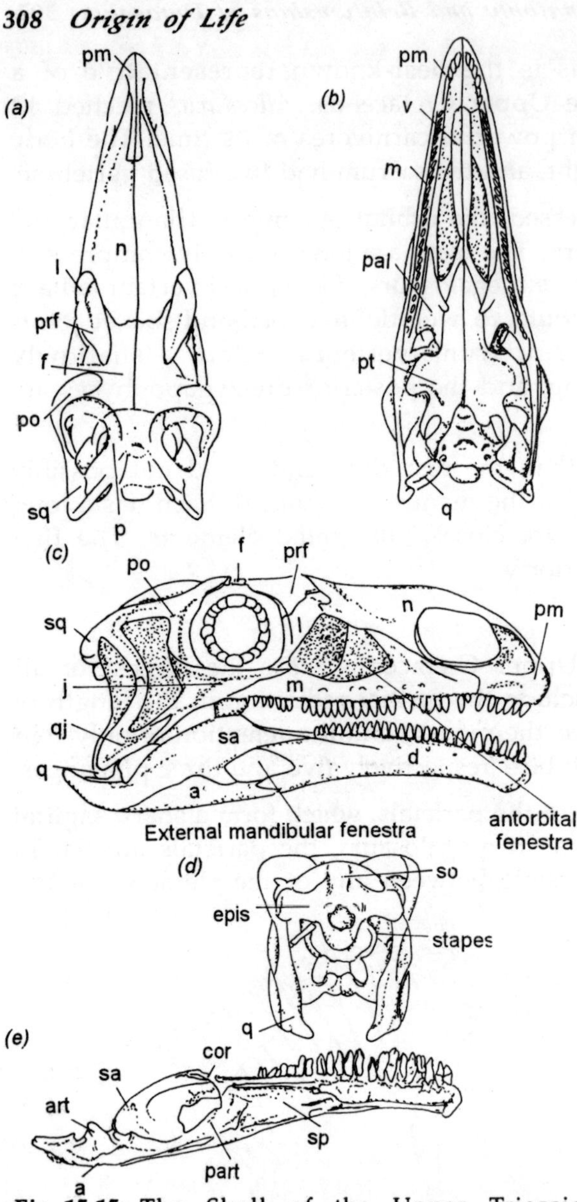

Fig. 15.15: The Skull of the Upper Triassic Herbivorous Dinosaur *Plateosaurus*. (*a*) Dorsal, (*b*) palatal, (*c*) lateral, and (*d*) occipital views, and (*e*) the medial surface of the lower jaw.

skull in tyrannosaurids. In contrast, the lower jaw had some mobility between the dentary and more posterior elements, which allowed the gape to be increased so that the teeth could bite into extremely large prey.

Russell (1970a) recognizes three genera in North America: *Albertosaurus, Daspletosaurus*, and *Tyrannosaurus*. *Tyrannosaurus* also occurs in central Asia, where it was originally described under the name *Tarbosaurus*, together with *Alioramus* and *Alectrosaurus*. The genus *Indosuchus* is recognized in India.

We assume that tyrannosaurids evolved from among the megalosaurs, but no specific ancestor has been recognized. Chatterjee (1985) proposes that they may have evolved directly from Triassic thecodonts.

SAUROPODOMORPHS

Plateosaurs

The theropods were the first dinosaurs to be described because their anatomy could be readily derived from the pattern of the carnivorous thecodonts. These large predators probably fed on comparable-sized prey. The major food sources for the megalosaurs and tyrannosaurids probably include the giant sauropods of the Jurassic and Cretaceous. These sauropods were preceded in the late Triassic by an assemblage of more primitive herbivorous saurischians, the plateosaurs or prosauropods.

The best known plateosaur is *Plateosaurus* from the Upper Triassic of Germany, but we find other genera in eastern and western North America, South America, southern Africa, and China. Plateosaurs were 1 to 7 meters long and of moderately heavy build. They were initially restored in a bipedal pose because of the greater length of their hind limbs, but footprints demonstrate that they were usually quadrupedal.

The skull is small relative to the trunk, and the neck is long. The teeth have laterally compressed crowns with crenulated edges that resemble those of herbivorous lizards. In advanced genera, the jaw articulation is well below the level of the tooth row, a feature that frequently developed in herbivorous groups to provide stronger leverage for the jaw muscles.

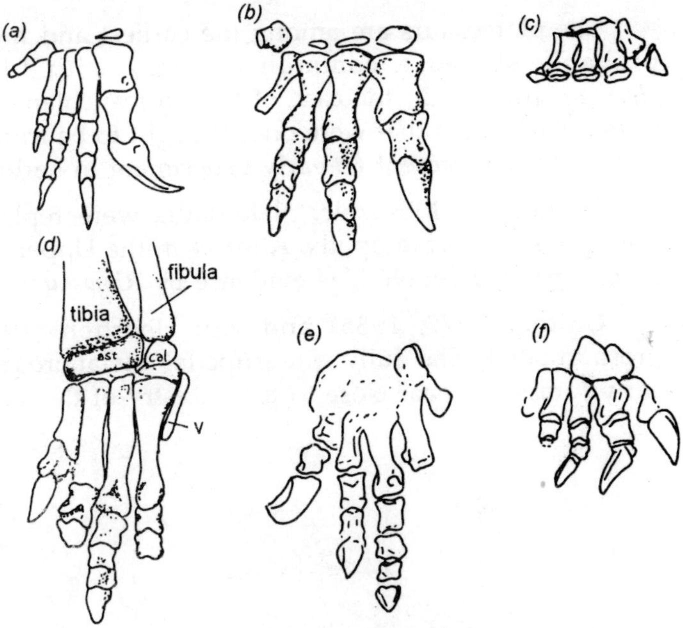

Plateosaurs were once thought to include carnivorous genera because of the discovery of large, bladelike teeth in association with their skeletons, but these teeth were probably left by other dinosaurs or large thecodonts that preyed upon them or fed on their carcasses.

Fig. 15.16: Hands and Feet of Sauropodomorphs. *(a)* Manus of the plateosaur *Ancbisaurus*, Upper Triassic, ×¼. *From Galton, 1976. (b)* Manus of *Riojasaurus*, ×¼, a form that is structurally transitional between plateosaurs and sauropods, from the Upper Triassic. *(c)* The Upper jurassic sauropod *Apatosaurus*. *(d)* The foot of *Riojasaurus*, ×1/12. *From Bonaparte, 1971. (e)* Foot of *Vulcanodon*, a form that is structurally intermediate between plateosaurs and sauropods, from the Lower Jurassic of East Africa. *(f)* Foot of the Upper Jurassic sauropod *Apatosaurus*. ast, astragalus; cal, calcaneum; v, fifth metatarsal.

Plateosaurs had three sacral vertebrae. Like *Staurikosaurus* and *Herrerusaurus*, the iliac blade is very short; the pubes are long narrow bones that fused at the midline but were not expanded vertically at the distal end. Proximally, each surrounds a large obturator foramen.

The forelimb is approximately two-thirds the length of the hind limb. The humerus, ulna, and radius are stout, and the manus is short and retains all five digits. It probably served for support, although the first digit is larger than the remaining ones and carries a huge claw.

The femur broadly resembles that of other early dinosaurs but is relatively thicker. The tarsus shows a well-developed mesotarsal hinge. The foot has four strongly developed, but short, digits. The fifth is reduced to a splint. According to van Heerden (1979), the posture was not fully improved, with the femur angled at about 25 degrees from the vertical.

The plateosaurs are among the earliest and most primitive of all dinosaurs. They probably shared a common ancestry with forms such as *Staurikosaurus* and *Herrerasaurus.* As in the case of the protosuchian crocodiles, we now recognize many of the plateosaurs that were once thought to be from the Triassic as being early Jurassic in age. They represent an early experiment in herbivory among the dinosaurs.

In the later Mesozoic, plateosaurs were replaced as effective herbivores by the sauropods and ornithopods. Already in the Upper Triassic, more advanced saurischian herbivores had evolved, as evidence by *Riojasaurus.*

Galton (1977, 1985) and van Heerden (1979) recognized four families of prosauropods. The fully quadrupedal Melanorosauridae from the Late Triassic and early Jurassic appear close to the ancestry of the sauropods.

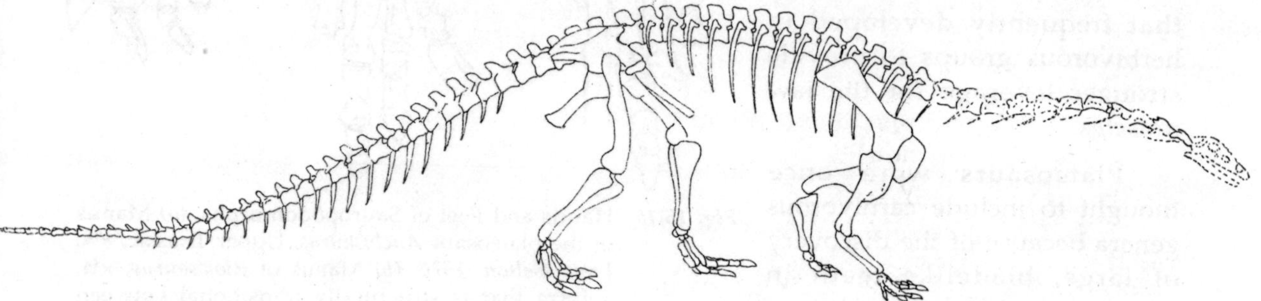

Fig. 15.17: Skeleton of *Riojasaurus.* This primitive sauropodomorph, from the Upper Triassic of South America, is 10 meters long.

Sauropods

Sauropods, which are exemplified by the Upper Jurassic genera *Diplodocus, Camarasaurus* and *Brachiosaurus,* are characterized by their enormous size, with weights as great as 80,000 kilograms and body lengths approaching 30 meters. They were obligatorily quadrupedal, and the limb bones are greatly thickened, solidly ossified, and held in a nearly vertical position. Like elephants, they did not bend much at the elbow and knee, and the stride would have been short. The feet were graviportal, with five short digits on both the manus and pes.

Surprisingly, the joint surfaces of the limbs are poorly defined, and there must have been a good deal of cartilage in the carpus and tarsus as well. The capacity for the cartilage to yield under pressure and conform to a shape that would most effectively distribute the force produced by the weight of the body was apparently more important than the greater per unit strength of bone.

The skull is strikingly small, and the teeth are simple, incisorlike structures that form a single functional row. The teeth show wear that could be associated with cropping vegetation, but they lack the grinding or shearing surfaces that are provided by the cheek teeth of ornithischians. The teeth are set at the margins of the jaw, which precludes the presence of a fleshy cheek that would help to retain the food within the mouth cavity. Processing of food in sauropods may have occurred primarily within the digestive tract, either with the help of gizzard stones, which are reported in a few sauropods, or through a symbiotic relationship with intestinal microorganisms.

All adequately known sauropods have long necks, with 12 to 19 cervical vertebrae, each of which are elongated relative to those of the trunk. There are usually about 12 trunk vertebrae and from four to six sacrals.

We recognize nearly 50 genera of sauropods that are grouped in 6 subfamilies or families, but few are well known. Because of their great size, sauropod remains are easily found but difficult to collect and prepare. Nevertheless, we have an excellent record in the Upper Jurassic from geographically diverse localities: the Tendagaru in East Africa, the Szechwan basin in China, and the Morrison Formation in Utah, Colorado, and Wyoming. In contrast, their history in the early Jurassic and for most of the Cretaceous is poorly known.

Because of their gigantic size, the habitat of sauropods has long been cause for speculation. Some authors have suggested that their enormous bulk could only have been supported if they lived in the water. An aquatic way of life also seems to be suggested by the dorsal position of the nostrils in several genera. On the other hand, there is nothing in the skeleton that can be associated with aquatic locomotion, and the limbs and girdles have many attributes that are found in large terrestrial mammals such as elephants. Some strictly terrestrial mammals have dorsal narial openings, including genera with a trunk or complex nasal structures.

Coombs (1975) concludes that sauropods may have been quite varied in their habits, both as a group and seasonally within individual species. The presence of five or six well-defined genera within a single fossil locality suggests that they must have differed considerably in their specific ways of life to avoid competition for particular food sources. Most sauropods may have lived near streams, swamps, and lakes to judge by footprints and depositional evidence, but it is unlikely that they habitually lived in deep water, as has been suggested in some restorations. No direct evidence is available regarding their diet. With such small heads, they must have spent most of their lives eating.

Sauropods have long been thought to have evolved from the plateosaurs. In addition to their common herbivorous habits, they are united by similar features of the postcranial skeleton. Some of these similarities may be attributed to common specializations for a massive body. The presence of a very large first digit of the manus with a strong claw may have been a unique specialization of the common ancestor of both groups. In most sauropods, especially the most primitive forms, the forelimbs are significantly shorter than the rear limbs, which suggests an ancestry among facultatively bipedal animals.

Most of the well-known plateosaurs are too late to be the actual ancestors of the sauropods, which appear to have diverged from the ancestral sauropodomorph lineage before the end of the Triassic.

Primitive Sauropods

The oldest known genus that can be closely associated with the later Mesozoic sauropods is *Riojasaurus* from the Upper Triassic of South America. The skeleton is only about 6 meters long, but the girdles and limbs are much more massive than those of the plateosaurs, which indicate an obligatorily quadrupedal stance. The articulating surfaces of the humerus are enormously expanded. The ulna is massive, with little development of the olecranon. These features contrast with the quadrupedal ankylosaurs and stegosaurs that are discussed later in this chapter, both of which have a well-developed olecranon that may have been associated with a more sharply angled elbow joint. As in the plateosaurs, all five digits of the forelimb are retained. The first digit bears a very large claw, but the other phalanges of all the digits are relatively short.

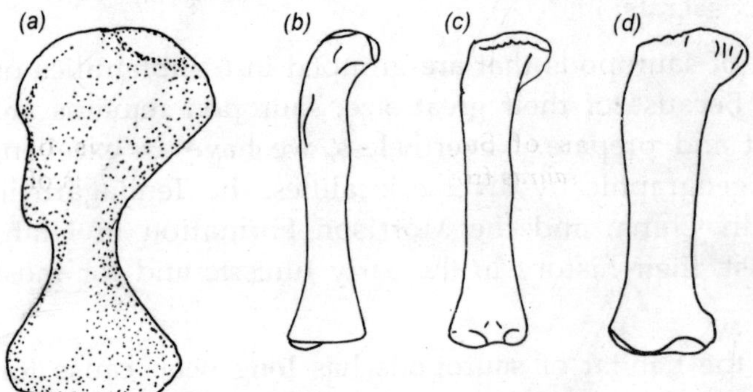

Fig. 15.18: Limb Bones of Sauropodomorphs. (*a*) Humerus of *Riojasaurus* from the Upper Triassic, ×1/10. *From Bonaparte, 1971.* (*b, c,* and *d*) Femora of *Plateosaurus* (Upper Triassic), *Barapasaurus* (Lower Jurassic) and *Apatosaurus* (Upper Jurassic).

The ilium is massive but short from back to front and is associated with only three sacral vertebrae. The pubis and ischium are short relative to those of theropods and generally resemble those of the plateosaurs. The puboischiac plate is extensive, as in primitive archosaurs. The femur is massive and considerably longer than the tibia. The tarsus and pes closely resemble those of the plateosaurs, with wide but flat astragalus and calcaneum and two small distal tarsals. The metatarsals of digits I through IV are long and fairly slender, and the fifth metatarsal is much shorter. The digits are all short but retain claws.

The skull is not known. There are estimated to be 10 or 11 cervical and 15 trunk vertebrae. In contrast with later sauropods, the vertebrae remain primitive in their general form but, like plateosaurs and advanced sauropods, they have extra articulating surfaces—the hyposphene and hypantrum, which are medial to the zygapophyses—that would contribute to the rigidity of the column.

Vulcanodon from the early Jurassic of Zimbabwe is even more similar to later sauropods (Cooper, 1984). Much of the girdles, limbs, and posterior axial skeleton is known but, unfortunately, not the head or neck. The trunk and tail would have been approximately 6 meters long. Most skeletal features are close to, but slightly more primitive than, typical sauropods and similar to, but advanced over the condition in robust prosauropods. The pelvis and rear limbs exhibit a clear mosaic of prosauropod and sauropod features. The sacrum is advanced over prosauropods in incorporating four fused vertebrae. However, the shape of the pubes, which form an anteriorly facing "apron," is typical of prosauropods. The femur is straight rather than sigmoidal, and the distal tarsals are not ossified. In contrast, the large size of the ungual of the hallux resembles that of prosauropods.

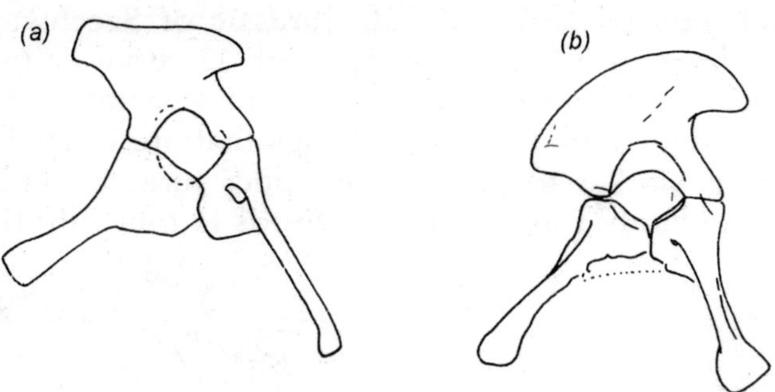

Fig. 15.19: Pelves of Sauropodomorphs. *(a)* pelvis of *Plateosaurus* from the Upper Triassic. *(b)* pelvis of *Barapasaurus* from the Lower Jurrasic.

The structure of both the forelimbs and hind limbs indicates that *Vulcanodon* was strictly quardrupedal and shifting from digitigrade to semiplantigrade posture. Cooper emphasized particularly close similarities in primitive features to *Plateosaurus* and the melanorosaurid *Euskelosaurus* combined with advanced characters that presage the pattern in diplodocid and camarasaurid sauropods.

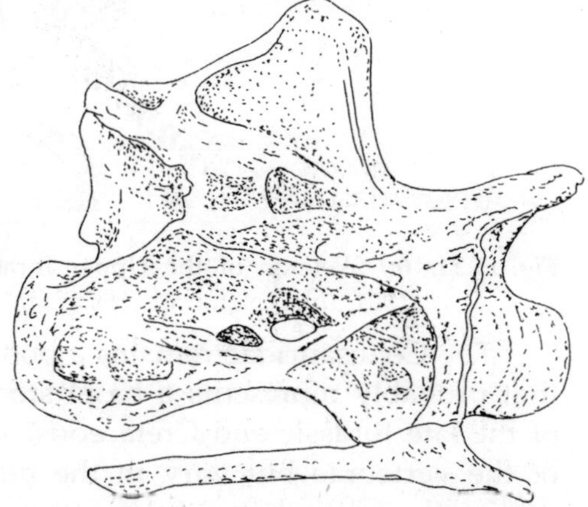

Fig. 15.20: Cervical Vertebra of the Upper Jurassic Sauropod *Diplodocus*, ×1/10. The very open structure conserves weight.

Cetiosauridae

Barapasaurus from the Liassic of India is contemporary with *Vulcanodon*. This genus is represented by a great deal of disarticulated material but no reconstruction has yet been attempted. It has already reached sauropod dimensions, approaching the size of *Diplodocus*. The individual bones are generally similar to those of the later sauropods but are primitive in some respects. Unlike those of *Riojasaurus* and *Vulcanodon*, the vertebrae resemble those of advanced sauropods in the development of large cavities in the centra and neural arches that serve to lighten these structures. The cervical vertebrae are opisthocoelous and elongate, but most of the trunk vertebrae remain shallowly amphicoelous. Four fused sacral vertebrae are attached to a long and high iliac blade of typical sauropod proportions.

Barapasaurus is placed in the family Cetiosauridae, which spans the length of the Jurassic. Another member of this family was briefly described on the basis of numerous skeletons from the Middle Jurassic of Szechwan Province, China. *Shunosaurus* is approximately 7 meters long, with 13 cervicals (which are one and one-half times as long as the dorsal), 12 dorsals, and 4 sacral vertebrae. The skull has spatulate teeth and large, paired narial openings that are located high on the snout—a configuration that is similar to that of the Upper Jurassic genus *Camarasaurus*. Within this family, the forelimbs range from two-thirds to four-fifth the length of the hind limbs.

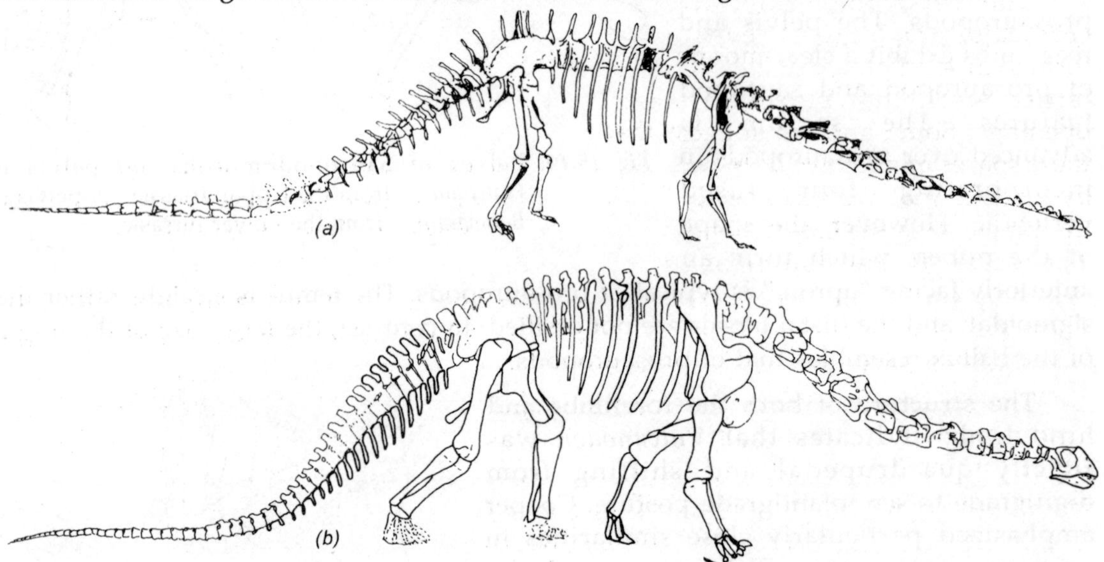

Fig. 15.21: (a) Skeleton of the Upper Jurassic sauropod *Diplodocus*, 30 meters long. *From Hatcher, 1901. (b)* Skeleton of *Camarosaurus*, from the Upper Jurassic.

The Cetiosauridae remains a poorly known and incompletely described assemblage that probably represents a basal stock that is ancestral to the more specialized forms of the late Jurassic and Cretaceous. The later forms are all advanced in the excavation of the vertebrae but vary in the number of sacral vertebrae; the proportions of the neck, tail, and limbs; and the nature of the skull and its dentition.

Diplodocidae

The Diplodocidae are found throughout the world in the Upper Jurassic, but this family is limited to eastern Asia in the Upper Cretaceous. Both *Diplodocus* and *Apatosaurus [Brontosaurus]* are known from complete, thoroughly described skeletons. Berman and McIntosh (1978) most recently reviewed this group. The body reaches approximately 30 meters in length, with the skull about 55 centimeters long at the end of a very long neck. Within the family, the number of cervical vertebrae varies from 12 to 19, with the individual vertebrae two and one-half times the length of the trunk vertebrae. There are five coossifed sacrals, and the tail has a maximum of over 80 vertebrae; the last are long, narrow, and form a thin whiplash.

The skull has large orbits with a median narial opening between them at the top of the skull. The teeth are long slender structures that are limited to the front of the mouth. They would be effective for cropping food and are rapidly replaced, if we judge by the number of replacement teeth that are present in the jaws.

The vertebrae, especially the cervicals, are huge. They are not formed of solid bone but by a complex lattice work that surrounds large openings. Some authors suggest that these spaces might have been occupied by air sacs that were connected with the lungs, as is the case among birds. The cervical and anterior trunk vertebrae are strongly opisthocoelous. There are only 10 trunk vertebrae that have tall neural spines.

The rib cage is high and narrow. Compared with other sauropods, the limbs of *Diplodocus* are relatively slender. The front limbs are two-thirds to three-fourths the length of the rear limbs. In *Diplodocus,* there are only two carpals, and the rest of the carpus is formed entirely of cartilage. In *Apatosaurus,* there is only a single carpal. In both genera, the only bone of the tarsus to ossify is the astragalus. The weight is borne by five metacarpals and five metatarsals of the hand and foot. The phalanges are greatly shortened. The first toe in the hand bears a large claw, like that of the plateosaurs. Two or three of the toes on the foot end with claws.

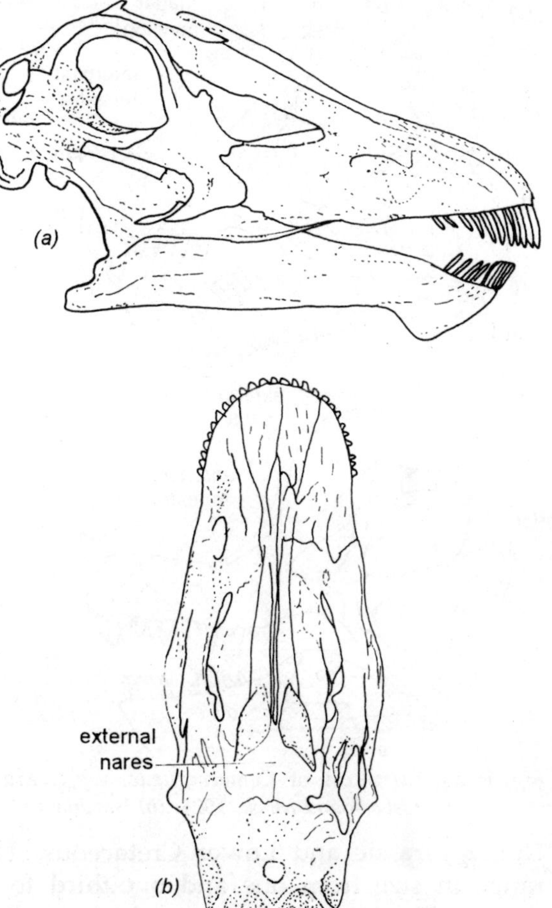

Fig. 15.22: Skull of *Diplodocus.* *(a)* Lateral and *(b)* dorsal views, 55 centimeters long.

Camarasauridae

The Camarasauridae is represented by *Camarasaurus* from the Upper Jurassic of North America and *Opisthocoelocaudia* from the Upper Cretaceous of Mongolia. *Camarasaurus* has a short neck with only about 12 cervical vertebrae that are only a little longer than the dorsals (which also number 12). The tail is short. The skull is high, and the teeth are spoon shaped and extend for much of the length of the jaw margin, in contrast with those of the Diplodocidae. The nasal openings are high but paired. The front limbs are two-thirds to three-fourth the length of the rear. The caudal centra of *Camarasaurus* are procoelous, as in most sauropods, but they are opisthocoelous in the later, Asian genus.

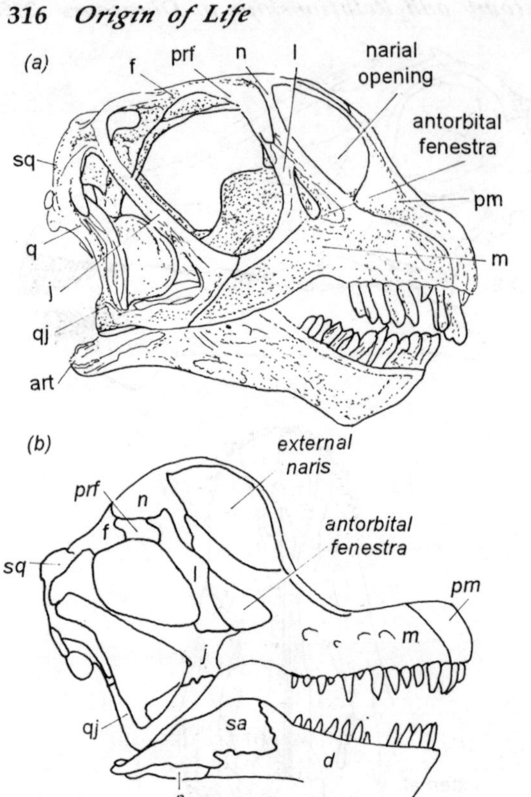

Fig. 15.23: *(a)* Skull of *Camarasaurus*, × 1/10. *From Osborn and Mook, 1921. (b) Brachiosaurus.*

In camarasaurids, as in all other Upper Jurassic sauropods, the sacrum is composed of five vertebrae: one dorsosacral, three true sacrals, and one caudosacral.

Brachiodauridae

The Brachiosauridae are among the largest of all sauropods, with estimated weights of 80,000 kilograms. The skull is high, with the nasal openings bulging up in front of the orbits. The neck is extremely long, and the individual vertebra are three times as long as the dorsals. The front limbs are as long or longer than the rear limbs. We find brachiosaurs from the Middle Jurassic into the Lower Cretaceous in North America, Africa (including Madagascar), Europe, and perhaps in Australia but not in Asia.

Euhelopodidae

In contrast, the Euhelopodidae are known only in eastern Asia from the Upper Jurassic and Lower Cretaceous. These genera exhibit 17 to 19 cervicals that range in size from one and one-third to two and one-third times the length of the dorsals. There are 14 dorsals, which results in a very long presacral column. The teeth are spatulate and the skull of *Euhelopus* is similar to that of *Camarasaurus*.

Titanosauridae

The Titanosauridae are a primarily Cretaceous assemblage that had a possible forerunner in the Upper Jurassic of East Africa. Although occurring in North America and Europe, they are more common in the southern continents. The most important defining character is the strongly procoelous nature of the anterior caudal vertebrae, which allowed the tail to move more freely from side to side. There are six sacral vertebrae, and the neck is comparatively short. The skull of the South American genus *Antarctosaurus* is decidedly like that of *Diplodocus* in character.

Ornithischians

We have difficulty establishing the specific interrelationships of the major groups of saurischian dinosaurs because of the incomplete fossil record in the late Triassic. The early history of the ornithischians is also incompletely known, but this assemblage shares a number of derived features that clearly demonstrate a common origin.

The most significant feature is the presence of a unique bone at the symphysis of the lower jaws, the predentary. This element does not bear teeth but may have had a horny covering like the beak of turtles. The predentary is not firmly attached to the more posterior bones of the jaws in primitive genera and may have allowed them to rotate slightly on their long axis to help manipulate food.

The teeth of all ornithischians have laterally compressed crowns that are crenulated along the edges. As with the plateosaurs, this pattern is associated with feeding on plant material. In contrast with plateosaurs and sauropods, the tooth rows of most ornithischians are medial to the margins of the dentary and maxilla, which indicates the presence of fleshy cheeks that would have assisted in retaining the food in the mouth as it was chewed. In most genera, a specific occlusal relationship is achieved between the upper and lower teeth. According to Galton (1973), these advances in the feeding apparatus were probably the main reason for the success of the ornithischians throughout the Jurassic and Cretaceous.

Crocodiles and some other tetrapods have bony plates in the eyelid or along the dorsal margin of the orbit, while in primitive ornithisachians, a narrow bone, the supraorbital, attaches to the anterior margin of the orbit. A comparable bone is not present in saurischians. Two supraorbitals are present in several ornithischian groups, including the iguanodontids and pachycephalosaurs.

The character that is most commonly used to distinguish ornithischians from saurischians, and that Seeley (1888) used as the basis for recognizing the two dinosaur orders, is the configuration of the pelvis. Saurischians retain the primitive, 'Lizardlike' pelvic pattern that is modified beyond that of early archosaurs by the extension of the pubis and ischium anteriorly and posteriorly to form, with the ilium, a triradiate structure. Ornithischians are specialized in having the pubis lie alongside the ischium in a posterior orientation. It is commonly said that the pubis has rotated posteriorly, but without any fossils that show an intermediate condition leading to that of the early ornithischians, we cannot say with assurance why the change occurred or how it was achieved.

Romer (1956), Galton (1969), Charig (1972), Walker (1977), and Santa Luca (1980) have all tried to explain why the pubis is in a posterior position. But their explanations are not fully satisfactory since it is very difficult to understand why the shift would have occurred in ornithischians but not among the obligatorily bipedal saurischians, such as *Coelophysis,* which resemble them in many other skeletal features. This change is particularly difficult to understand since many later ornithischians, including the small bipedal hypsilophodontids, redevelop an anterior pubic process that compares topographically with the primitive saurischian pubis.

Charig (1972) suggests that all dinosaurs had to modify the configuration of the pelvis from the pattern of advanced thecodonts because of the greater stride developed as a result of a fully upright rear limb. In primitive archosaurs, the swing of the femur was limited by the anterior extent of the pubis because the major muscles that protract the limb originated on that bone. If the femur moved anteriorly to the level of the pubis or even beyond, the efficiency of these muscles would be progressively reduced.

Several solutions to this problem are possible. If the body were to assume a habitually bipedal stance, the pubis would be tipped upward and out of the way of the femur. This solution was achieved in the theropods. The sauropods, which are graviportal animals, would not have moved the limbs in a wide arc, so that the configuration of the primitive pelvis would not have been a problem.

Ornithischians may have solved this problem by shifting the site of origin of the femoral protractors to other bones. The ilium of primitive ornithischians extends far forward of the acetabulum, which would provide space for the origin of the protractors of the femur, including the sartorius (iliotibialis) and the puboischiofemoralis. As in modern crocodiles, the latter muscle could also have originated from the posterior trunk vertebrae.

If the pubis were not functionally necessary for the origin of these muscles, it would not need to have remained in its primitive position. Selection may have occurred to reverse its position, possibly to allow the abdominal cavity to extend posteriorly and increase the volume of the gut in the herbivorous ornithischians (but not in the carnivorous theropods) or to shift the center of gravity posteriorly to facilitate a bipedal stance.

Two major groups of ornithischians have a well-developed pattern of dermal armour that might have been inherited from the thecodonts. In this one feature ornithischians may be more primitive than saurischians. Ornithischians, but not saurischians, have ossified tendons lateral to the vertebral column in the posterior trunk and the sacral and caudal regions.

We have not found any fossils that link early ornithischians with any of the saurischian groups. Bakker and Galton (1974) pointed out features of the dentition in which ornithischians resemble plateosaurs and aspects of the rear limb and feet that they share with early coelurosaurs, but none of the known saurischians combine the skull and postecranial features that we would expect to find in the immediate ancestors of ornithischians. Similar derived features of the rear limb and tarsus suggest that ornithischians and saurischians share a common ancestry among animals that are similar to *Lagosuchus*.

The earliest known ornithischian is *Pisanosaurus* from the Upper Triassic of South America. It is very incompletely known but shows the specializations of the lower jaw and dentition that characterize the much-better-known early Jurassic genera.

Fabrosaurids

We find several distinct ornithischian lineages in the lowest Jurassic from beds that were previously thought to be late Triassic in age. Although they are distinct from one another in important features of the dentition, their basic skeletal anatomy is very similar. The fabrosaurids show a pattern that could be close to that from which most other ornithischians evolved.

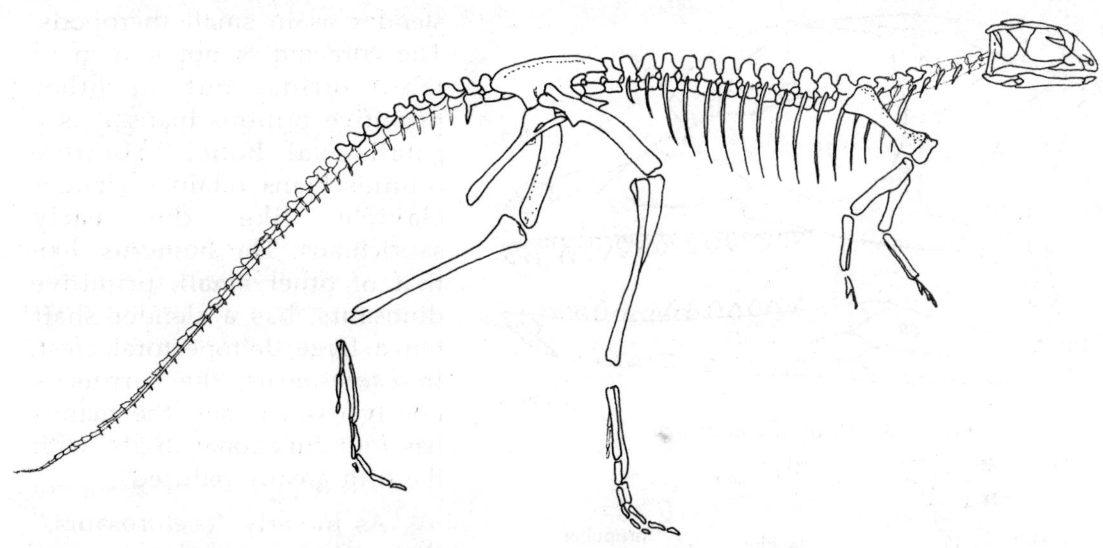

Fig. 15.24: Skeleton of *Fabrosaurus*. This is one of the most primitive ornithischians, from the Lower Jurassic of southern Africa, 50 centimeters tall.

The best-known genus is *Fabrosaurus [Lesothosaurus]* from southern Africa. Like the prodokesaurid *Coelophysis*, the skeleton is light and has exceedingly slender limbs. The forelimbs are even shorter, indicating obligatory bipedality. The total length is under 1 meter.

The skull is small and short with a large orbit that is crossed by a slim supraorbital bone. The antorbital opening, as in most ornithischians, is small and partially occluded by the maxilla. The premaxilla bears a continuous row of small incisiform teeth; the teeth in the maxilla and dentary extend along the margin of the bones. The tooth shape is common to that of all ornithischians. Like those of plateosaurs, the crowns are laterally compressed and leaf shaped. The upper and lower teeth occlude alternately between one another and have enamel on both the medial and lateral surfaces.

The cervical centra are short, which results in a shorter neck than is found in the early theropod dinosaurs. The trunk centra remain shallowly amphicoelous. There are approximately 22 presacral vertebrae and 5 sacrals. Alongside the posterior trunk and sacral vertebrae are numerous rod-shaped, ossified tendons that stiffened the column. They may have evolved in relationship to a habitually bipedal stance in primitive ornithischians but are retained in later quadrupedal forms as well.

There is a long, slender tail; the zygopophyses at its base are nearly vertical, which would have limited movement of the tail to the vertical plane. This limitation may have facilitated balance on the rear limbs. The high angle of the zygopophyses would also have reduced the tendency for the tail to be bent laterally during contraction of the caudifemoralis musculature, which served as a major retractor of the rear limbs.

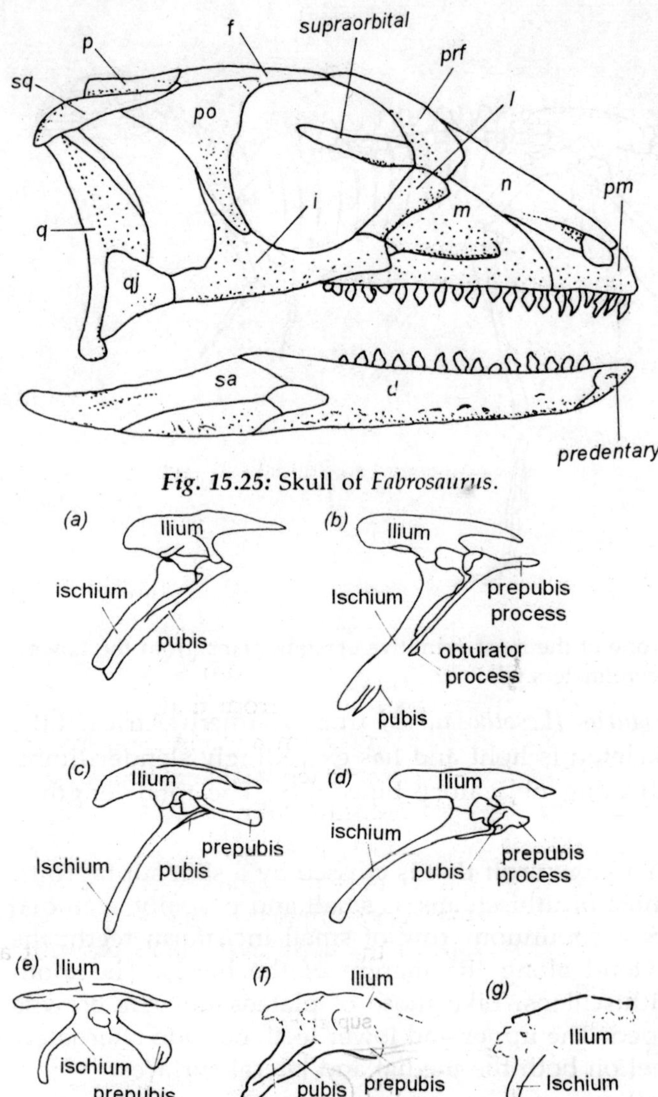

Fig. 15.25: Skull of *Fabrosaurus.*

Fig. 15.26: Pelves of Ornithischians. (*a*) *Scelidosaurus,* a primitive ornithischian without a prepubic process. (*b*) *Hypsilophodon,* a typical ornithopod with a prominent obturator process. (*c*) *Homoalcephale,* a pachycephalosaurid. (*d*) *Protoceratops.* (*e*) *Triceratops.* (*f*) *Stegosaurus.* (*g*) *E.**locephalus,* and ankylosaur.

The scapula is thin and slender as in small theropods. The coracoid is not known in fabrosaurids, but in other primitive ornithischians it is a small oval bone. Primitive ornithischians retain a slender clavicle like the early saurichians. The humerus, like that of other small, primitive dinosaurs, has a slender shaft but a large deltopectoral crest. In *Fabrosaurus,* the carpus is poorly ossified and the manus has four functional digits, with the fifth greatly reduced.

As in early "coelurosaurs," the ilium is long and accommodates five pairs of sacral ribs. The ischium, like that of small theropods, is slender and extends posteroventrally. Proximally it bears a ventral projection, the obturator process, that is also present in a major assemblage of later ornithischians, the ornithopods. The very slender pubis lies just below the ischium. In contrast with most later ornithischians, there is only a short anterior pubic process.

The head of the femur is *Fabrosaurus* is not strongly inturned. The shaft is slightly bowed anteriorly, and there is a long pendant fourth trochanter for insertion of the strong retractor muscles that originated at the base of the tail. The tibia is considerably longer than the femur and bears most of the weight of the lower limb. The fibula is reduced to a narrow splint. As in other ornithischians, the astragalus and calcaneum are so strongly integrated with the ends of the tibia and fibula that they are functionally an extension of the crus. The distal tarsals are separate but closely associated with the heads of the metatarsals. As in early "coelurosaurs," the pes is functionally tridactyl, with the fifth digit lost and the first reduced and turned slightly to the rear.

Fabrosaurus shows no trace of dermal armour. In contrast, *Scutellosaurus,* a possibly related from North America, has an extensive covering along the back and onto the flanks.

Well-known fabrosaurids are restricted to the Lower Jurassic, but Galton (1978) described jaws and teeth from the Jurassic-Cretaceous boundary that appear to represent a continuation of this group and similar teeth are known from both the Upper Triassic and Upper Cretacesous. The teeth in later genera are slightly inset, which demonstrates the initial development of cheeks within this group.

Heterodontosauridae

Another group of primitive ornithischians, the heterodontosaurids, accompany the fabrosaurids in the early Jurassic. This family probably did not give rise to any later ornitheschians, but they are important in establishing the primitive anatomy of this assemblage since we know their skeleton in greater detail than that of the fabrosaurids.

Heterodontosaurus is known from two complete skeletons from southern Africa. The body is slightly more than 1 meter long. The skull differs from that of *Fabrosaurus* in several significant features. There are conspicuous caniniform teeth in both the upper and lower jaw. The upper "canine" originates from the back of the premaxilla (and not the front of the maxilla, as do the canine teeth of primitive amniotes and mammals). The lower canine originates at the front of the dentary, just behind the predentary bone, and fits into a notch in the upper jaw that is part of a diastema between the premaxillary teeth and the cheek teeth. The other very important feature is the fact that the cheek teeth are not located at the margin of the skull and lower jaw but are inset, leaving a space lateral to the tooth row. The bone flares outward and is ridged above and below the tooth row in the maxilla and dentary, as if to support a lateral sheet of tissue that would have functioned like the mammalian cheek to retain food within the oral cavity.

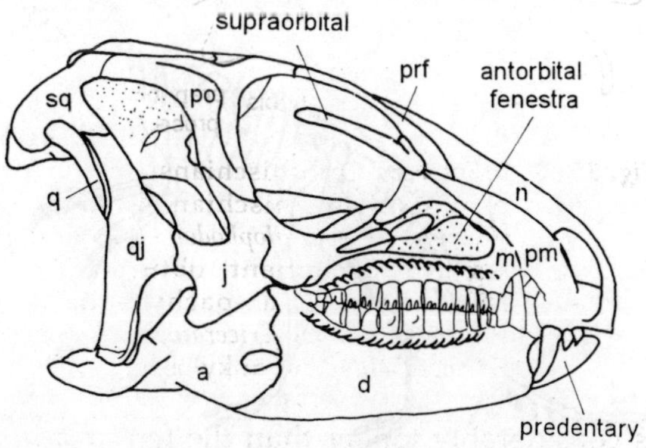

Fig. 15.27: Skull of *Heterodontosaurus*. From the Lower Jurassic of Southrn Africa.

Heterodontosaurus shows tooth wear that results from a specific occlusal pattern between the chisel-shaped upper and lower teeth. The lateral surface of the lower teeth wear against the medial surface of the maxillary teeth. Enamel is present only on the lateral surface of the upper teeth and the medial surface of the dentary teeth to provide a resistant cutting edge. The rate and pattern of tooth replacement is modified relative to more primitive dinosaurs to maintain an even cutting edge. The late Triassic genus *Pisanosaurus* has a similar pattern of tooth wear and may belong to this family, although it does not have a lower caniniform tooth (the anterior part of the upper dentition is not known).

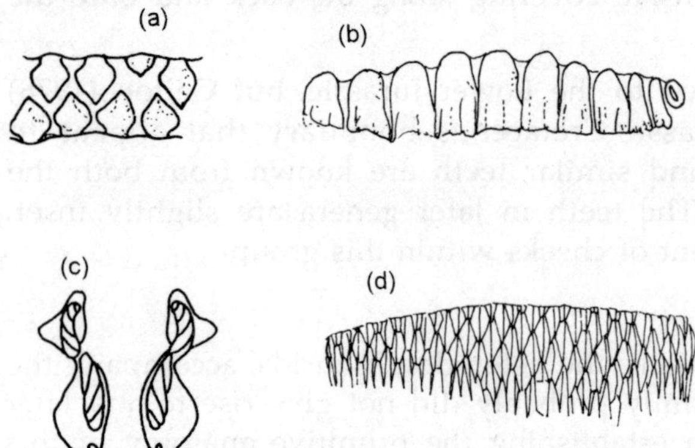

Fig. 15.28: Tooth Pattern and Occlusion in Ornithischian Dinosaurs. *(a) Fabrosaurus*, in which the upper and lower teeth alternate with one another. *(b) Heterodontosaurus. (c)* Diagrammatic cross-section to show occlusal pattern in hadrosaurs. The same basic pattern also applies to most other ornithopods. *From Ostrom, 1961. (d)* Hadrosaur tooth battery in medial view showing many ranks of replacement teeth.

The forelimbs of *Heterodontosaurus* are considerably longer and more powerfully built than those of *Fabrosaurus* and may have been used for support in quadrupedal locomotion. The nature of the elbow joint, with a prominent olecranon, indicates that the manus was capable of powerful grasping. The carpus is more fully ossified than in other ornithischians and may provide a model for the primitive pattern in the group. Interestingly, there is a small pisiform, that is missing or unossified in most other archosauromorphs. Digits four and five are much smaller than the first three.

The pelvis resembles that of fabrosaurids but lacks an obturator process. The joint surfaces between the femur and the fused tibiofibula indicate that the limb was not held absolutely vertically but, as in modern birds, the femur was slightly abducted. The knee is angled well forward. The astragulus and calcaneum and almost indistinguishably fused to the ends of the crus, and the distal tarsals are fused to the ends of the metatarsals that are, in turn, fused to one another proximally in a nearly avian pattern. The fifth digit is lost. The first is shorter than the next three, and the phalanges are somewhat divergent.

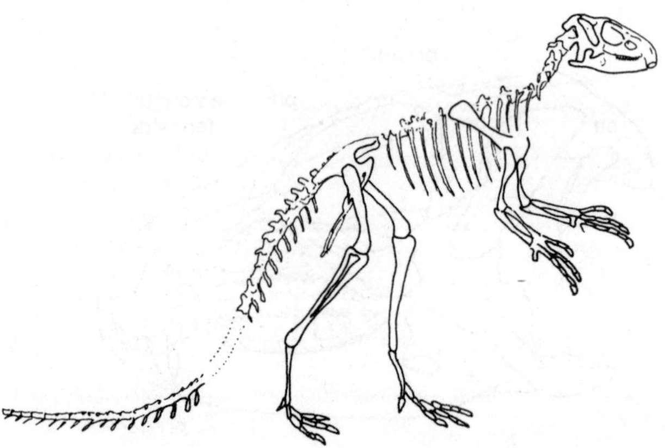

Fig. 15.29: Restoration of the Skeleton of *Heterodontosaurus*. 50 centimeters tall.

Hypsilophodontids

The fabrosaurids may have given rise directly to the late Jurassic and Cretaceous hypsilophodontids, which in turn gave rise to the iguanodontids and hadrosaurs. The hypsilophodontids form a conservative group that is widespread but neither common nor diverse. They appear as a continuation of the fabrosaurid habitus as modest-sized, light-bodied, cursorial bipeds.

All are advanced over the early fabrosaurids in the medial position and regular occlusion between the upper and lower teeth. In *Hypsilophodon*, the teeth were replaced in groups of three and maintained a functional shearing surface more effectively than did the alternate replacement of primitive reptiles, although the resulting wear pattern appears less regular than that of *Heterodontosaurus*.

The upper temporal openings of hypsilophodontids are larger than in fabrosaurids. The fused parietals form a medial crest. The lower jaw exhibits a high coronoid process. Teeth are retained in the premaxilla.

The best-known genus is *Hypsilophodon* from the Lower Cretaceous of England, which is less than 5 meters long. In contrast with the fabrosaurids, an anterior prepubic process extends upward toward the tip of the ilium. The ischium has a strongly developed obturator process near the midpoint of its length. Ossified tendons sheath the end of the tail.

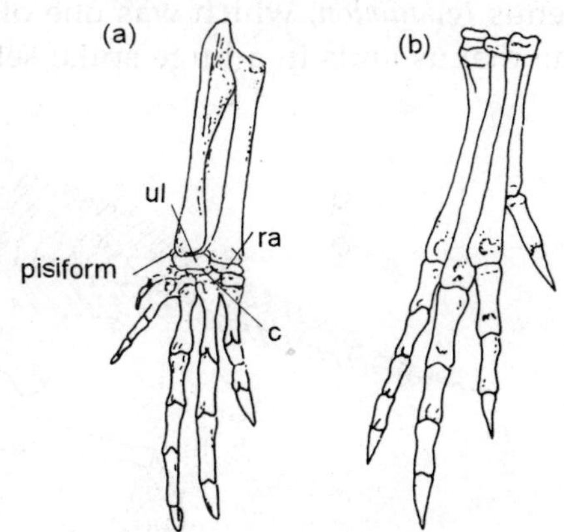

Fig. 15.30: *Heterodontosaurus.* *(a)* Forelimb and *(b)* hind-limb. Abbreviations as follow: c, centrale; ra, radiale; ul, ulnare.

Iguanodontids

Two other ornithopod groups that are characterized by larger body size and a graviportal posture evolved from animals that resembled the hypsiolophodontids. The more primitive group, the iguanodontids, appear in the Middle Jurassic, reach their greatest diversity at the end of the Lower Cretaceous, and continue to the end of the Mesozoic.

Camptosaurus is a well-known early iguanodontid. Its body length of approximately 6 meters exceeds that of most hypsilophodontids. The preorbital region of the skull is

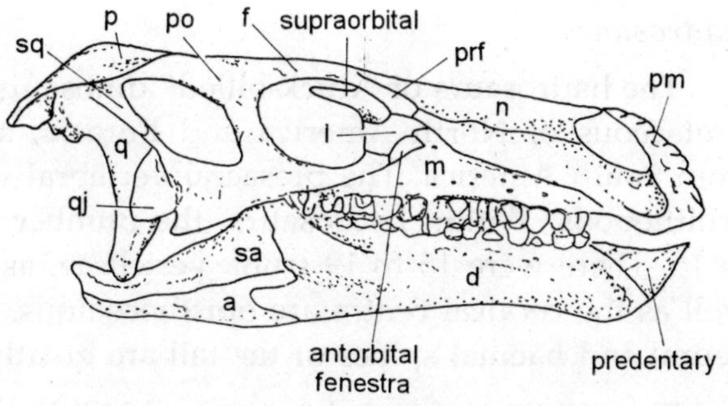

Fig. 15.31: *Camptosaurus.* The skull of the primitive Upper Jurassic iguanodontid.

considerably elongated, the size of the antorbital fenestra is greatly reduced, and the premaxillary teeth are lost. The single row of marginal teeth resemble those of hypsilophodontids, except for the complete loss of the enamel on the wear surface.

The neck is long and the centra of the cervical vertebrae are opisthocoelous, as in advanced surischians. There are 26 to 28 presacral vertebrae, an increase of 4 to 6 over primitive ornithischians.

The limbs are more massive, the tibia is shorter than the femur, and the metatarsals are short and stout. The manus ends in blunt claws that were probably capable of supporting the body in awkward quadrupedal locomotion. In the Lower Cretaceous genus *Iguanodon*, which was one of the first dinosaurs to be described, the first digit of the manus ends in a huge spike set at right angles to the other digits (Norman, 1980).

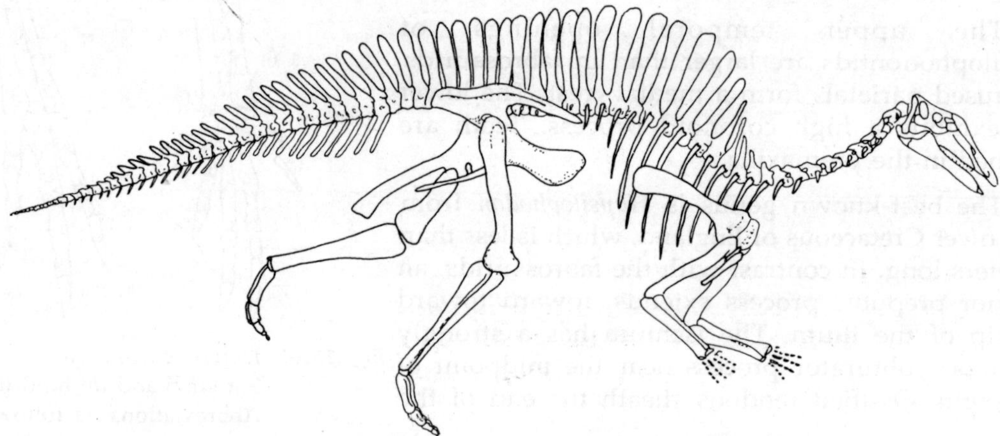

Fig. 15.32: Skeleton of the Iguanodontid Ourano-saurus from the Lower Cretaceous. The genus represents the pattern from which hadrosaurs may have been evolved.

The late Lower Cretaceous genera *Ouranosaurus* and *Probactrosaurus* provide appropriate ancestors for the next advance in ornithopod evolution, the family Hadrosauridae.

Hadrosaurs

The hadrosaurs or "duck-billed" dinosaurs were common and varied in the Upper Cretaceous of North America and Eurasia, and a single species has been reported from South America. The presacral vertebral was elongated relative to more primitive ornithopods. Within hadrosaurs, the number of cervical vertebrae increased from 12 to 15. There were 15 to 19 trunk vertebrae, as well as 8 sacrals. The anterior trunk as well as the cervical centra are opisthocoelous. The tail is laterally compressed, and the neural and haemal spines of the tail are greatly elongated.

The appendicular skeleton resembled that of the iguanodontids but was somewhat heavier and had longer forelimbs that bore small hoofs on the digits rather than claws. Nevertheless, we think that they were habitually bipedal.

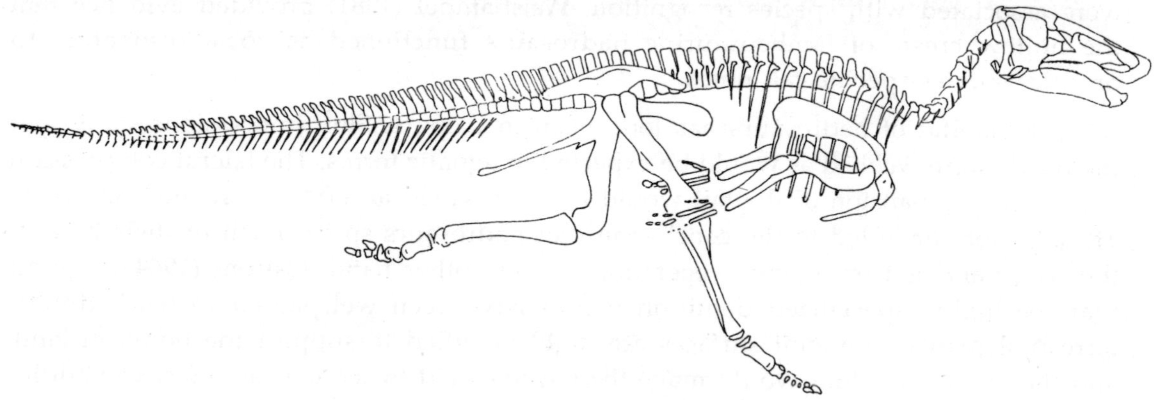

Fig. 15.33: *Anatosaurus*. The primitive, noncrested hadrosaur is shown in running pose.

Much more important changes were evident in the skull. Instead of a single row of functional teeth in each jaw ramus, a series of replacement teeth were exposed simultaneously. There were 45 to 60 tooth positions in each jaw, each with several replacement teeth in sequence, giving a total of as many as 700 teeth that were exposed at once. Ostrom (1961) argued that the upper and lower tooth batteries sheared past one another, with the pterygoideus acting as a protractor and the posterior adductor acting as an antagonist to retract the jaws. In contrast, Weishampel (1983) provided evidence that the jaws were closed primarily vertically but that chewing also involved transverse movements of the dentition that were produced by rotating the maxillae.

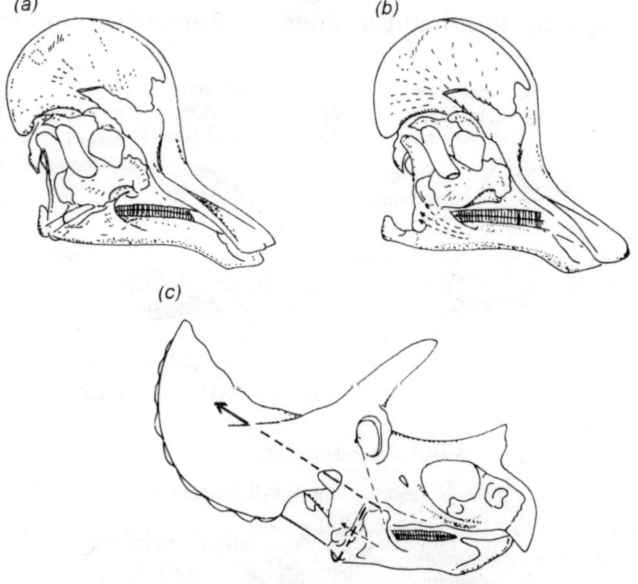

Fig. 15.34: Occlusal Pattern of the Hadrosaur *Corythosaurus*. Shown is propalineal movement of the lower jaw *(a)* protracted and *(b)* retracted. *(a, b)* Weishampel (1983) presents evidence that jaw closure was primarily vertical in this group. *(c)* Jaw mechanics in the ceratopsian *Triceratops*. Ceratopsians had a scissorslike jaw closer with no propalineal movement. Arrows show direction of force of major jaw muscles.

In many, but not all, hadrosaur genera, the skull was greatly elaborated with crests of various sizes and shapes. Hopson (1975) and Dodson (1975) argued that the crests and other specializations of the nasal region were associated with species recognition. Weishampel (1981) provided evidence that the hollow crests of lambeosaurine hadrosaurs functioned as vocal resonators to produce species-specific call notes.

The habitat of hadrosaurs has long been in dispute. Skin impressions show that the hands were webbed as would be expected in aquatic forms. The lateral compression and vertical expansion of the tail would have provided an effective swimming organ. These factors have led to the conclusion that hadrosaurs spent much of their time in the water and fed on aquatic vegetation. On the other hand, Ostrom (1964a) argued that the highly specialized dentition would have been well suited to hard, fibrous terrestrial plants. The joint surfaces are highly ossified to support the body on land, and the ossified tendons would make the tail too rigid to serve as an effective paddle.

Most of the remains of hadrosaurs are from lowlying deposits of coastal planes and the margins of rivers. The absence of immature individuals led to the hypothesis that hadrosaurs migrated to higher land to reproduce. This assumption has been strikingly confirmed by the discovery of nesting sites in the foothills of the ancestral Rocky Mountains. The great success of hadrosaurs may have been the result of their capacity to inhabit a range of different environments. We may associate the change in dental patterns and subsequent success of hadrosaurs with the proliferation of the angiosperms at the beginning of the Upper Cretaceous.

There were at least 26 genera of hadrosaurs in the Upper Cretaceous. Of two subfamilies, only the Hadrosaurinae, which lacked a crest, continued to the very end of the period.

We can include fabrosaurids, hypsilophodontids, iguanodontids, and hadrosaurs in the single suborder Ornithopoda. They are united by a unique derived character, the presence of an obturator process on the ischium, and are also linked by a series of morphological intermediates. The remaining ornithischian groups lack an obturator process and may have evolved separately from the base of this assemblage.

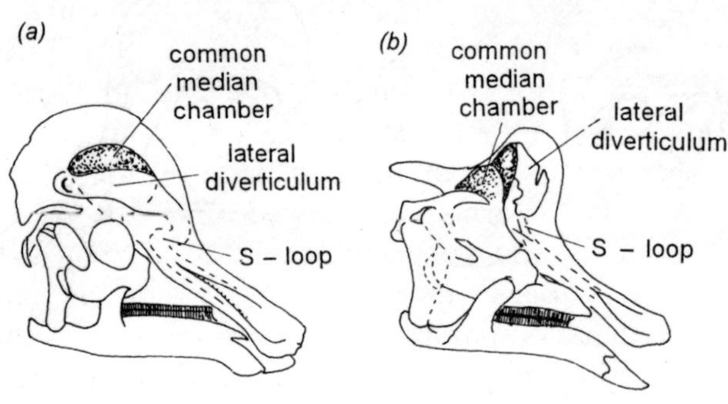

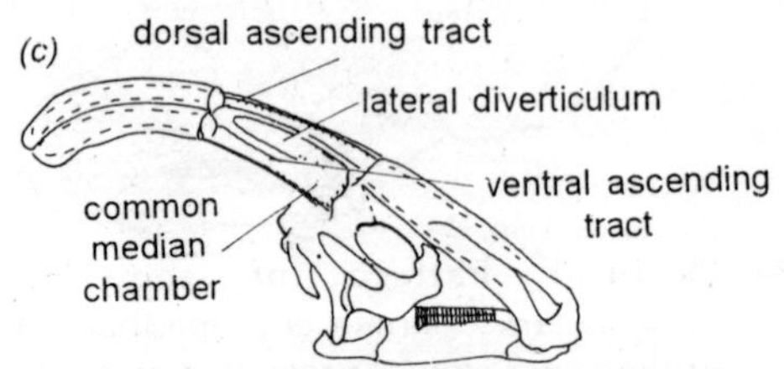

Fig. 15.35: Internal Anatomy of the Crest of Lambeosaurine Hadrosaurs. Weishampel described the acoustic properties which would provide for species recognition. *(a) Corythosaurus. (b) Lambeosaurus. (c) Parasaurolophus.*

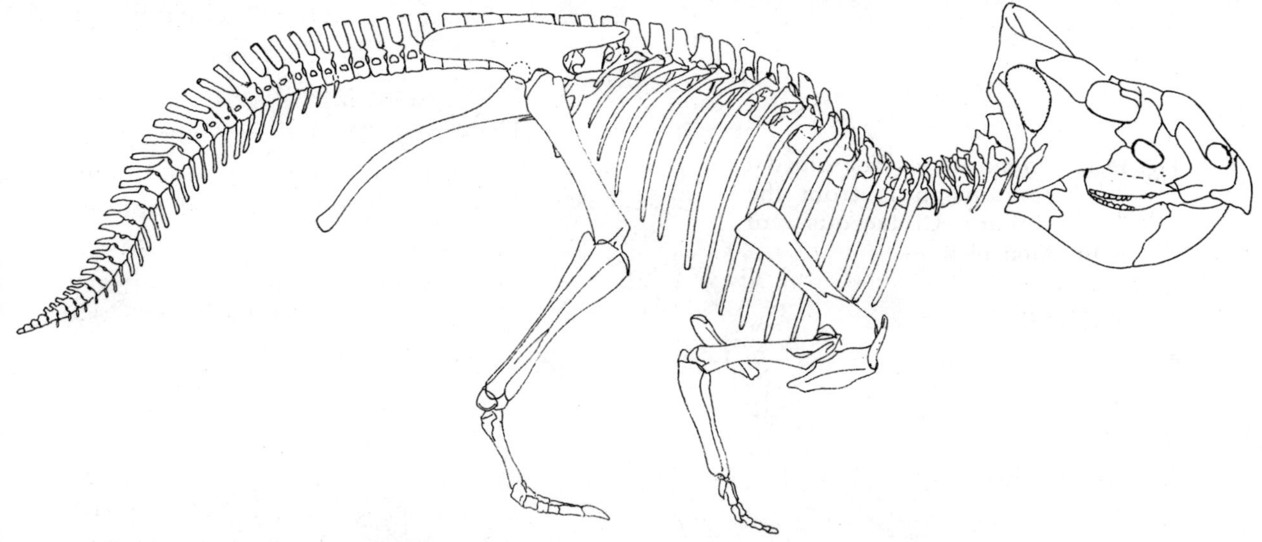

Fig. 15.36: *Leptoceratops,* the North American Upper Cretaceous ceratopsian.

Ceratopsians

Our concept of the ceratopsians is based primarily on the well-known Upper Cretacous genera from North America, as epitomized by *Triceratops.* These animals are stocky, quadrupedal herbivores. The posterior portion of the skull is extended as a huge frill that spreads over the neck region. These forms are further distinguished by a variable pattern of "horns" in the nasal region and above the eyes. The end of the snout is in the form of a laterally compressed "beak" that overhangs the lower jaw and terminates in a unique rostral bone that is comparable to the predentary in the lower jaw. We find the most advanced of the ceratopsians, the family Ceratopsidae, only in western North America.

Ceratopsians with somewhat less-specialized cranial anatomy, placed in the family Protoceratopsidae, are known in both North America and eastern Asia in the Upper Cretaceous. A further Asian genus, *Psittacosaurus,* from the Lower Cretaceous is thought to have been bipedal rather than quadrupedal and may provide a link with the base of the ornithischian stock. In both *Psittacosaurus* and the protoceratopsids, the tibia is much longer than the femur, which suggests cursorial locomotion. The skull of *Psittacosaurus* is comparable with that of later ceratopsians in having a beaklike nasal area, including a rostral bone, and one species has a nasal horn, but there is no evidence of the frill that characterizes the Upper Cretaceous families. This genus cannot be directly ancestral to the later families since it is more advanced in the closure of the antorbital fenestrae that are present in protoceratopsids and the reduction of the fourth and fifth digits of the manus. These three groups appear to represent distinct radiations from an earlier ceratopsian stock.

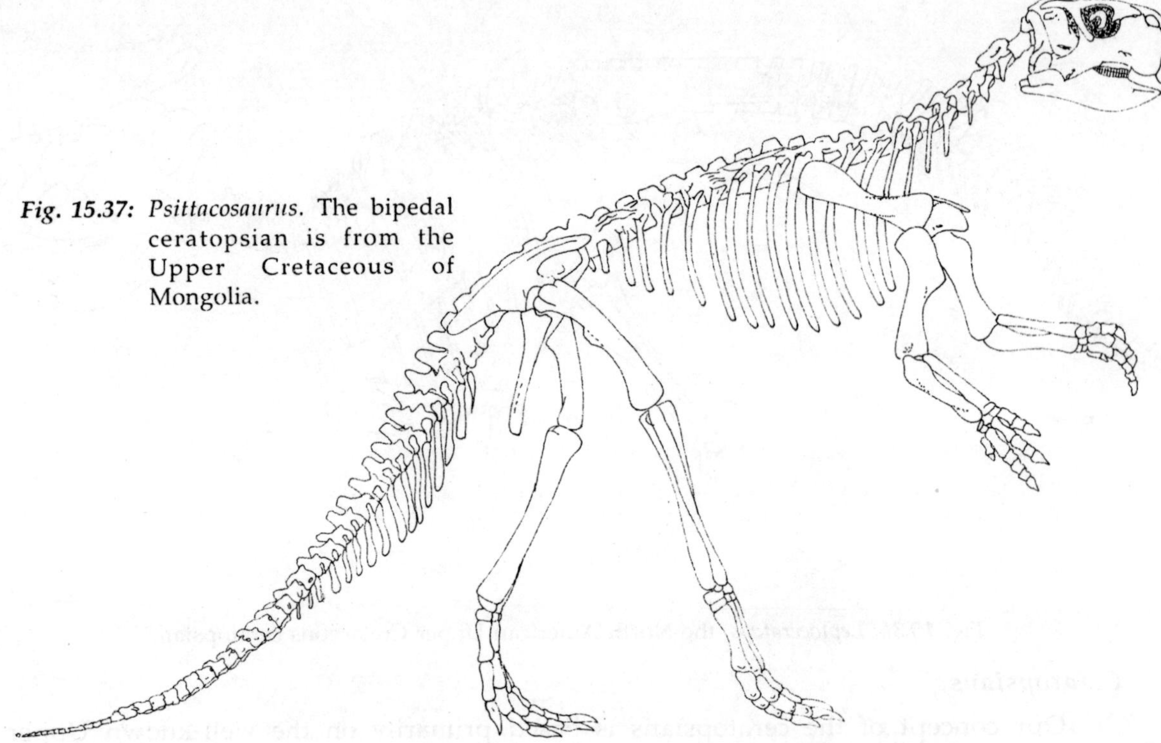

Fig. 15.37: *Psittacosaurus.* The bipedal ceratopsian is from the Upper Cretaceous of Mongolia.

The oldest putative ceratopsian is a single specimen of *Stenopelix* from the base of the Cretaceous in Germany. Its specific affinities with later ceratopsians are difficult to establish because the skull is missing, but the structure of the pelvis and rear limbs suggests that it is a member of this suborder. Despite its early appearance, it differs from psittacosaurids in having a femur that is longer than its tibia. The pelvis closely resembles that of the other members of the suborder in having a short prepubic process and a short posterior pubic process but no obturator process. As in other ceratopsians, the feet remains primitive. The metatarsals are slender but not greatly elongated, and the phalangeal formula is 2, 3, 4, 5, 0.

The pelvis of these ornithischians does not possess the one key feature that unites the ornithopods, an obturator process, but this absence might be expected in forms that have a short posterior portion of the pubis. The short prepubis together with the retention of premaxillary teeth in primitive ceratopsians suggest an early divergence from the ornithischian assemblage. We have not found any possible ceratopsian ancestors in the Jurassic; no fossils of this suborder are known in the southern hemisphere.

The success of ceratopsians in the Upper Cretaceous may be attributed to a highly specialized feeding mechanism. This apparatus involves the dentition, the configuration of the lower jaw, and the long frill formed by the squamosal and parietal. The teeth are arranged in a single functional row in each jaw ramus. The individual teeth resemble those of hadrosaurs in having leaf-shaped surface of the lower teeth and the lateral surface of the upper teeth.

Although only a single row of teeth was functional at a given time, the teeth were apparently replaced very rapidly so that the ones at the jaw margin were always sharp. Tooth occlusion was produced by vertical shear that resulted from a scissorslike closure of the jaw. Leverage on the lower jaw was increased by the high coronoid process. Much of the force of jaw closure was provided by the adductor mandibulae posterior that extended posterodorsally behind the jaw, through the upper temporal fenestra, and out over the dorsal surface of the squamosal-parietal frill.

Although the primary function of the frill apparently was for muscle attachment, Rowe, Colbert, and Nations (1983) pointed out that the edge of the skull in some genera extended well beyond the area of muscle attachment. In genera such as *Triceratops,* in which the skull forms a nearly continuous bony surface, it would have served for protection from attack. In other genera such as *Pentaceratops,* there are large openings that would be vulnerable to injury by either predators or members of the same species. The large size of the frill would have made it important for species recognition and sexual selection, as is the case in many large mammalian herbivores.

Pachycephalosaurs

It was long thought that all advanced ornithischian groups might trace their origin to the primitive ornithopods. However, the remaining groups apparently evolved from an even more primitive stage at the very base of ornithischian evolution, because they share no derived features with even the most primitive ornithopods that are not present in all ornithischians.

All retain a simple dentition that is composed of a single row of teeth with leaf-shaped crowns that have enamel on both surfaces. The predentary is small, and the upper teeth continue onto the premaxilla

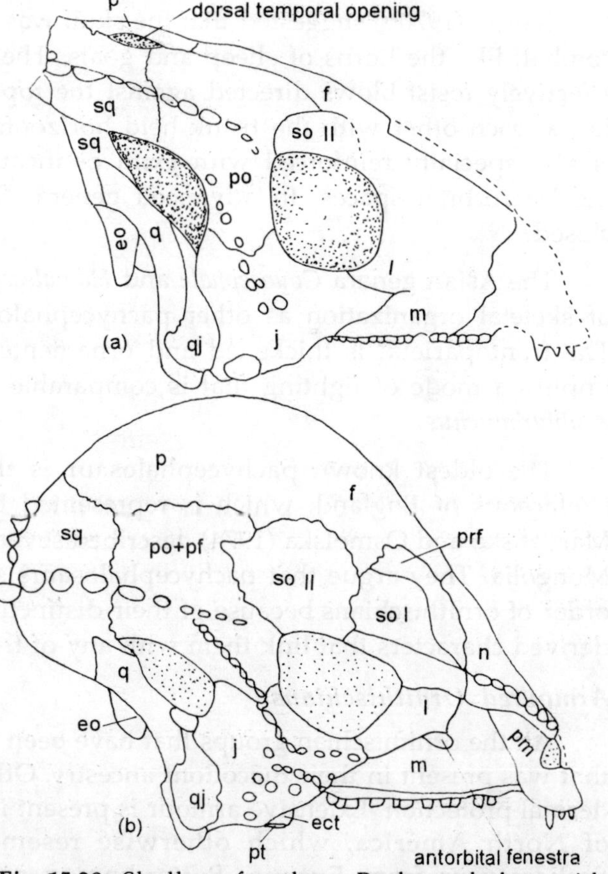

Fig. 15.38: Skull of the Pachycephalosaurids *Homoalcephale and Prenocephale*. From the Upper Cretaceous of Mongolia representing *(a)* flat and *(b)* domeheaded genera.

in primitive members of each of the groups. They are all advanced over the early fabrosaurids in having cheeks, but all lack the obturator process that characterizes that group and other ornithopods.

The pachycephalosaurids are relatively small, bipedal forms that are known primarily from the Upper Cretaceous of Asia and North America, with one genus from Madagascar. The posteranial skeleton resembles that of primitive ornithopods except in having a very short posterior process on the pubis and a long prepubis. The acetabulum is formed entirely by the ilium and ischium.

The most striking feature of pachycephalosaurids is the domed appearance of the skull. The very high forehead is not a reflection of a large brain, however, but an enormously thickened frontoparietal. A knobby occipital frill is variably developed and may be ornamented with short spikes.

Galton (1970b) suggested that the skull was used as a battering ram in intraspecific combat, like the horns of sheep and goats. The pattern of the thickening would most effectively resist blows directed against the top of the skull, which suggests that they ran at each other with the trunk held horizontally. The internal structure of the skull is also specially reinforced with extra ossifications around the brain and by closing the interorbital spaces. In advanced genera, the dorsal temporal openings are also closed.

The Asian genera *Goyocephale* and *Homalecephale* have essentially the same pattern of skeletal organization as other pachycephalosaurs but lack a frontoparietal dome. The frontoparietal is thickened and ornamented with either pits or tubercles, which implies a mode of fighting that is comparable to that observed in the marine iguana *Amblyrhynchus*.

The oldest known pachycephalosaur is the genus *Yaverlandia* from the Lower Cretaceous of England, which is represented by a distinctively thickened skull cap. Maryanska and Osmolska (1974) described several genera from the Upper Cretaceous of Mongolia. They argue that pachycephalosaurs should be recognized as a distinct sub-order of ornithischians because of their distinctive skeletal anatomy and the absence of derived characters that link them with any of the other advanced ornithischian groups.

Armoured Ornithischians

All the ornithischian groups that have been discussed thus far have lost the armour that was present in their thecodont ancestry. Other ornithischian retained or elaborated dermal protection. Extensive armour is present in *Scutellosaurus* from the Lower Jurassic of North America, which otherwise resembles the early ornithopods, and in *Scelidosaurus* from Europe. Better-known armoured groups, the stegosaurs and ankylosaurs, are found in the Middle Jurassic through the Cretaceous. Both are strictly quadrupedal, although the rear limbs remain much larger than the front. The toes bear hoofs, rather than claws, and the posture is graviportal.

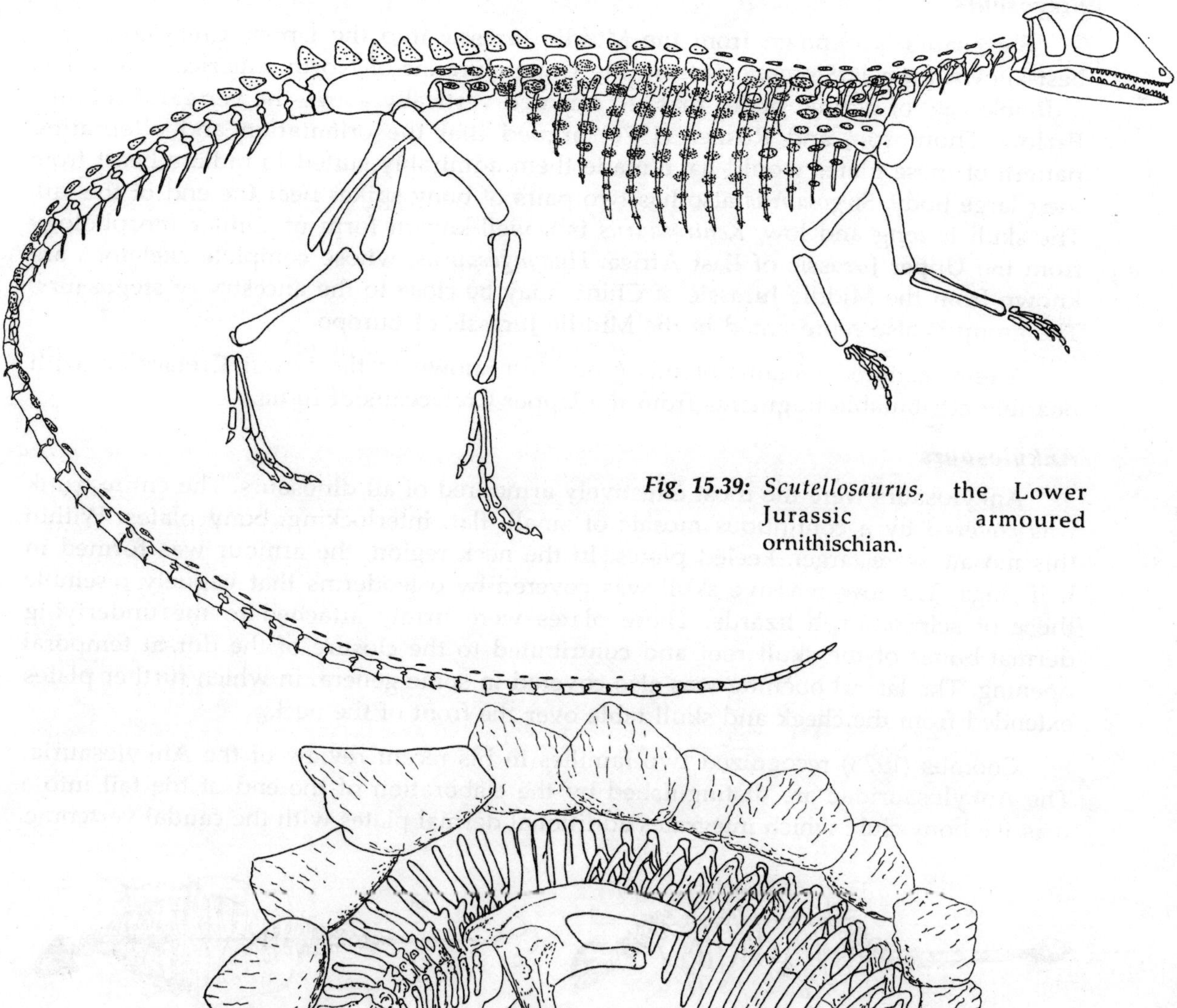

Fig. 15.39: *Scutellosaurus,* the Lower Jurassic armoured ornithischian.

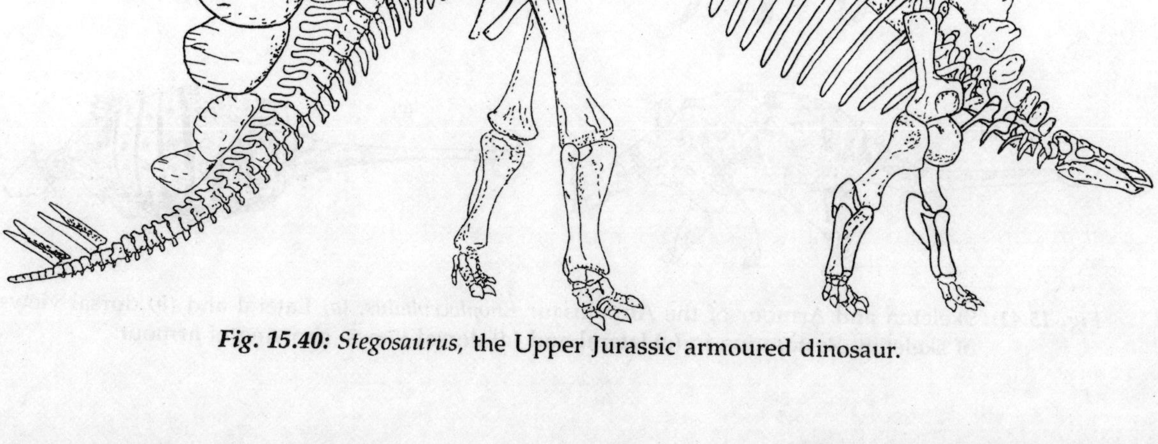

Fig. 15.40: *Stegosaurus,* the Upper Jurassic armoured dinosaur.

Stegosaurs

Stegosaurs are known from the Middle Jurassic into the Upper Cretaceous. The best-known form is *Stegosaurus* from the Upper Jurassic of North America, which has a double row of dermal plates that are oriented vertically above the vertebral column. Farlow, Thompson, and Rosner (1976) argued that the orientation and alternating pattern of these plates would have made them admirably suited to radiate heart from their large body. *Stegosaurus* also has two pairs of bony spikes near the end of the tail. The skull is long and low. *Kentrosaurus* is a well-known form of similar morphology from the Upper Jurassic of East Africa. *Huayngosaurus,* whose complete skeletons are known from the Middle Jurassic of China, may be close to the ancestry of stegosaurs. The group is also represented in the Middle Jurassic of Europe.

A few scattered remains of this group are known in the Lower Cretaceous with possibly attributable fragments from the Upper Cretaceous of India.

Ankylosaurs

Ankylosaurs were the most effectively armoured of all dinosaurs. The entire trunk was covered by a continuous mosaic of small, flat, interlocking, bony plates. Within this mosaic were larger, keeled plates. In the neck region, the armour was formed in half rings. The low, massive skull was covered by osteoderms that vaguely resemble those of scincomorph lizards. These plates were firmly attached to the underlying dermal bones of the skull roof and contributed to the closure of the dorsal temporal opening. The lateral opening was also covered in some genera, in which further plates extended from the cheek and skull table over the front of the neck.

Coombs (1978) recognized two families in his recent review of the Ankylosauria. The Ankylosauridae are distinguished by the elaboration of the end of the tail into a massive bony club, which integrates additional dermal plates with the caudal vertebrae.

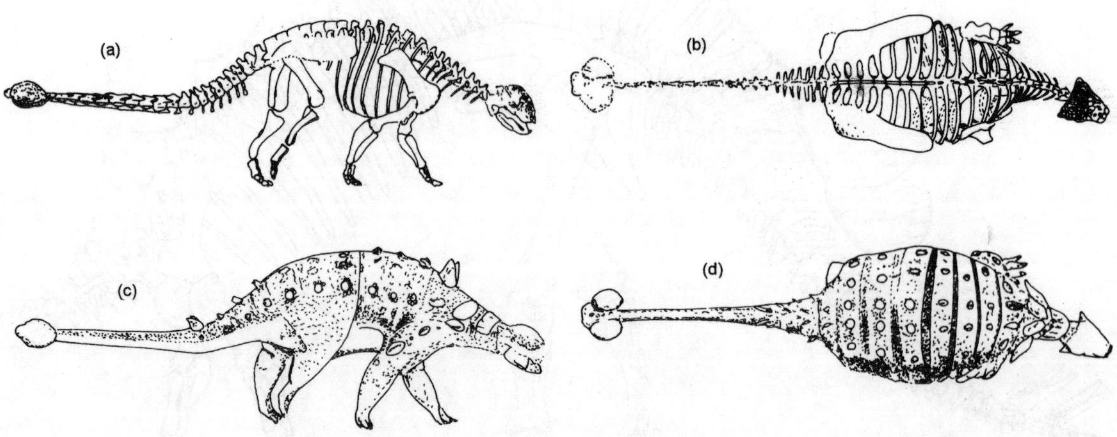

Fig. 15.41: Skeleton and Armour of the Ankylosaur *Euoplocephalus. (a)* Lateral and *(b)* dorsal views of skeleton. Restoration in *(c)* lateral and *(d)* dorsal views, showing of armour.

Members of the Nodosauridae lack the caudal expansion but are distinguished by a solid fusion between the braincase and the palate. Some nodosaurs have long bony spines on the sides of the body that resemble the tail spines of the stegosaurs.

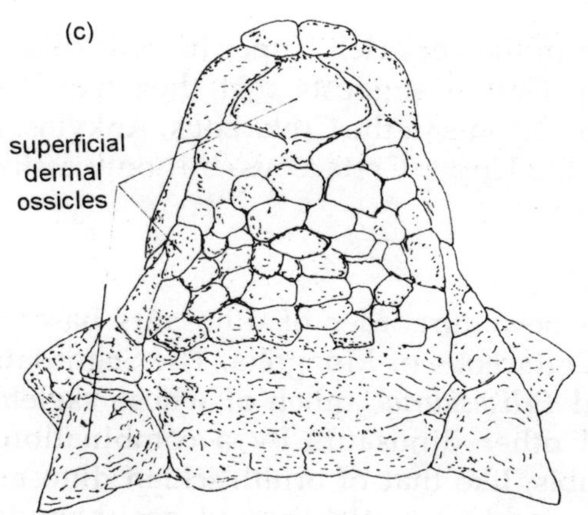

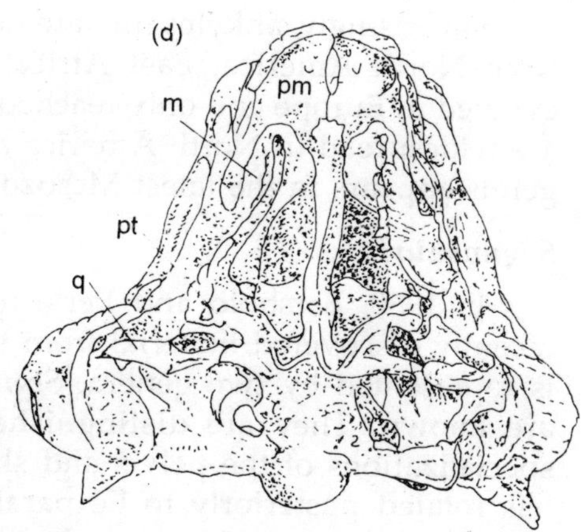

Fig. 15.42: Skull of *Ankylosaurus*. *(a)* Dorsal and *(b)* palatal views.

Both families have short heavy limbs and an obligatorily quadrupedal stance. The ilium is peculiar in that it broadly overhangs the femur. It is greatly expanded anteriorly to support the armour and surround the posterior ribs and greatly enlarged abdominal region. There is no prepubic process. The pubis is small and is virtually excluded from the acetabulum; the acetabulum is imperforate. At least eight vertebrae are incorporated in the sacrum. In advanced ankylosaurs, the head of the femur is terminal, rather than angled medially.

Adequately known ankylosaurs are limited to the Cretaceous of North America, Europe, Asia, and possibly Australia. Galton (1980a,b) recognized several fragmentary specimens from the Middle through Upper Jurassic of Europe as probable ancestors. *Sarcolestes* from the middle Jurassic of England is represented by a nearly complete lower jaw in which the diagnostic ankylosaur

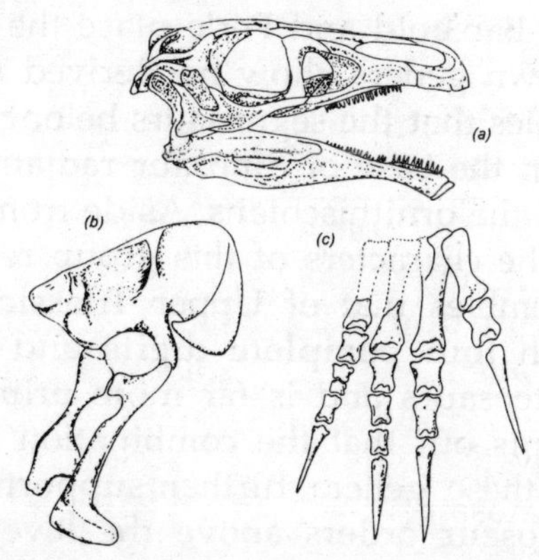

Fig. 15.43: Representatives of the Segnosauria. *(a)* Skull of *Erlikosaurus*. *From Paul, 1984* *(b)* Pelvic girdle of *Segnosaurus*. *(c)* Dorsal view of right pes of *Erlikosaurus*.

dermal plates are fused to the external surface. The dentition extends to the distal end of the bone, with space for only a very small predentary. *Cryptodraco* from the Upper Jurassic is based on a femur that is similar to the North American nodosaurid *Hoplitosaurus*. *Dracopelta* from the Upper Jurassic of Portugal is known from a trunk region that is covered with ankylosaurlike armour. We have not found any forms that might link ankylosaurs with other ornithischians.

Surprisingly, ankylosaurs are not known from the very rich Upper Jurassic deposits from North America, East Africa, and China. Galton suggests that they may have evolved in Europe and only reached the other continents in the Cretaceous. Ankylosaurs are widespread in North America and Asia in the Upper Cretaceous, but only a single genus appears in the latest Mesozoic beds.

Segnosauria

In 1980, Barsbold and Perle recognized a new infraorder of dinosaurs based on recently discovered material from the Upper Cretaceous of Mongolia. The Segnosauria is represented by two genera, *Segnosaurus* and *Erlikosaurus*, much of whose skeletons are known. They are distinguished from all other dinosaurs by a combination of specializations of the pelvis and skull. The pubis, like that of ornithischian dinosaurs, has rotated posteriorly to lie parallel with the ischium, which has an ornithopodlike obturator process. However, the ilium is very short and gives the dorsal portion of the pelvis the appearance of primitive theropod dianosaurs. The skull resembles that of ornithischians in having the tooth row of the maxilla and dentary inset from the skull margin. As in that order, the premaxilla is toothless. Segnosaurs lack the predentary bone that is diagnostic of ornithischians, but the anterior extremity of the dentary is toothless. Barsbold and Perle suggest that the end of the upper and lower jaws was covered by a horny beak.

Barsbold and Perle place the Segnosauria among the theropod dinosaurs, but the known genera show no derived characters that support that assignment. Paul (1984) argues that the segnosaurs belong to an otherwise unrepresented lineage that diverged from the base of dinosaur radiation close to the point of divergence of prosauropods and the ornithischians. Aside from the specializations of the dentition and pelvis, most of the characters of this group reflect a primitive level of evolution that more closely resembles that of Upper Triassic than other Upper Cretaceous dinosaurs. The foot, with four complete digits and an unconsolidated metatarsus, resembles that of plateosaurs and is far more primitive than that of even the earliest theropods. Paul points out that the combination of prosauropod and ornithischian features exhibited by these genera further supports the common ancestry of the two long-recognized dinosaur orders above the level of the thecodonts. The large number of derived characters that are shared by prosauropods, segnosaurs, and ornithischians suggests that all the herbivorous dinosaurs may belong to a single, monophyletic assemblage.